04-HHE-4

DISINFECTION OF WASTEWATER EFFLUENT – COMPARISON OF ALTERNATIVE TECHNOLOGIES

by:

Lawrence Y.C. Leong, Ph.D., QEP
Kennedy/Jenks Consultants

Jeff Kuo, Ph.D., P.E.
California State University, Fullerton

Chi-Chung Tang, Ph.D., P.E., DEE
Sanitation Districts of Los Angeles County

2008

WERF

The Water Environment Research Foundation, a not-for-profit organization, funds and manages water quality research for its subscribers through a diverse public-private partnership between municipal utilities, corporations, academia, industry, and the federal government. WERF subscribers include municipal and regional water and wastewater utilities, industrial corporations, environmental engineering firms, and others that share a commitment to cost-effective water quality solutions. WERF is dedicated to advancing science and technology addressing water quality issues as they impact water resources, the atmosphere, the lands, and quality of life.

For more information, contact:
Water Environment Research Foundation
635 Slaters Lane, Suite 300
Alexandria, VA 22314-1177
Tel: (703) 684-2470
Fax: (703) 299-0742
www.werf.org
werf@werf.org

This report was co-published by the following organizations. For non-subscriber sales information, contact:

IWA Publishing
Alliance House, 12 Caxton Street
London SW1H 0QS, United Kingdom
Tel: +44 (0) 20 7654 5500
Fax: +44 (0) 20 7654 5555
www.iwapublishing.com
publications@iwap.co.uk

Library of Congress Catalog Card Number: 2007941990

IWAP ISBN: 978-1-84339-799-1/1-84339-799-4

Kennedy/Jenks Consultants, Sanitation Districts of Los Angeles County, California State University, Fullerton

ACKNOWLEDGMENTS

The authors thank all of the people and POTWs who participated in the UV and ozone surveys and provided valuable data and insight into these disinfection technologies. The authors would also like to acknowledge the UV and ozone vendors, and Gary Hunter and the WEF Disinfection Committee, for contributing information that was invaluable in developing the survey of the disinfection practice for the major POTWs in the US. The authors also thank Shiaw Huitric, Bruce Mansell, Jeff Weiss, Ramon Gonzalez, and Jason Pivovaroff, all of the Sanitation Districts of Los Angeles County for their help in conducting the surveys and gathering information for this report. Lastly, the authors would like to thank the Sanitation Districts of Los Angeles County and Kennedy/Jenks Consultants who provided significant in-kind contributions.

Report Preparation

Principal Investigator:
Lawrence Y.C. Leong, Ph.D., QEP
Kennedy/Jenks Consultants

Co-Principal Investigator:
Jeff Kuo, Ph.D., P.E.
California State University, Fullerton

Chi-Chung Tang, Ph.D., P.E., DEE,
Sanitation Districts of Los Angeles County

Project Team:
Joseph A. Drago, Ph.D., P.E.
Kennedy/Jenks Consultants

Naoko Munakata, Ph.D.
Sanitation Districts of Los Angeles County

Craig Thompson, P.E.
Kennedy/Jenks Consultants

Project Technical Review Panel:
Ronald Gehr, Ph.D., P.E.
McGill University

William Mitch, Ph.D., P.E.
Yale University

Robert Ryder, P.E.
Kennedy/Jenks Consultants

Richard Sakaji, Ph.D., P.E.
East Bay Municipal Utility District

R. Rhodes Trussell, Ph.D., P.E., DEE
Trussell Technologies, Inc.

Project Subcommittee

Bethany Acquisto, Ph.D.
U.S. EPA

Ricardo DeLeon, Ph.D.
Metropolitan Water District of Southern California

Kimberly Kunihiro
Orange County Utilities, Orlando FL

George Tchobanoglous, Ph.D., P.E.
University of California at Davis

Elliot Whitby, Ph.D.
Calgon Carbon Corporation

Water Environment Research Foundation Staff

Director of Research: Daniel M. Woltering, Ph.D.
Senior Program Manager: Lola Olabode, MPH

ABSTRACT AND BENEFITS

Abstract:

There is a tremendous amount of literature on and experience with wastewater disinfection alternatives. However, it is difficult for wastewater professionals to sift through all of the available information, especially for relatively newer technologies. In addition, there are many factors, some of them site-specific, that influence whether a facility changes disinfection practice, and which alternative it chooses. There are few resources that provide a comprehensive discussion of decision factors with a direct comparison of the disinfection alternatives.

This project developed a singular document that presents the pros and cons and costs of the various technical options for wastewater disinfection. The resulting report is intended to be utilized by wastewater professionals to help evaluate and select the appropriate technology for their application. This report presents a review of the existing literature, a survey of disinfection practice by major POTWs, and surveys of facilities with UV and ozone systems. The known advantages and disadvantages of the mature technologies (chlorine, UV, and ozone), other technologies, and combinations of multiple disinfection alternatives are summarized. The report synthesizes this information and presents a coherent method for selecting a disinfection technology, based on individual priorities and criteria. Finally, the report identifies data gaps that would benefit from additional research.

Benefits:

- Saves staff time in researching the mature technologies (chlorine, UV, and ozone) and the developing technologies (peracetic acid, ferrate, bromine compounds, pasteurization, ultrasonic cavitation, electron beam and gamma irradiation, photocatalysis, and chlorine dioxide) and combinations of those technologies.
- Summarizes in one document the knowledge from the literature and compares the advantages and disadvantages of each of the disinfection alternatives with respect to historical usage, cost, effectiveness for pathogen and indicator organism removal and inactivation, ability to meet disinfection standards, reliability and safety issues, residuals, byproducts, as well as potential impacts to receiving waters.
- Provides a framework for selecting an appropriate disinfection technology, based on the priorities of the POTW and the relative importance of safety, disinfection efficacy, effects on effluent water quality, sensitivity to influent water quality, design and O&M issues, and costs.
- Provides a survey snapshot of the experiences with regard to design, installation, and operation of approximately 10% of the major POTWs with UV systems.
- Identifies gaps in the collective knowledge on each of the disinfection alternatives.

Keywords: Wastewater, disinfection, chlorine, UV, ozone, POTWs.

TABLE OF CONTENTS

LIST OF TABLES

LIST OF FIGURES

LIST OF ACRONYMS

ANOVA	Analysis of variance (statistical analysis)
AOP	Advanced oxidation process
AOX	Adsorbable organohalogens
ATHN	Acetylterahydronaphthaline, or tonalide (a musk fragrance found in wastewaters)
BCDMH	1-bromo, 3-chloro, 5,5-dimethylhydantoin (a disinfectant)
BNR	Biological nitrogen removal
BOD	Biochemical oxygen demand
CCI	Construction cost index
CFD	Computational fluid dynamics
CFU	Colony forming units
COD	Chemical oxygen demand
COEC	Compounds of emerging concern
CPI	Consumer price index
CSO	Combined sewer overflow
CT	Product of concentration and contact time (used to regulate disinfection)
CWNS	Clean water needs survey
DBP	Disinfection byproduct
DEET	Meta-N,N-diethyl toluamide (an insect repellent found in wastewaters)
DMA	Dimethylamine
DMH	Dimethylhydantion
DNA	Deoxyribonucleic acid
DO	Dissolved oxygen
DOC	Dissolved organic carbon
DRM	Dose-regression model
EDC	Endocrine disrupting compound
ENR	*Engineering News Record*
FC	Fecal coliform
HAA	Haloacetic acid
HHCB	Galaxolide (a musk fragrance found in wastewaters)
HTH	High test hypochlorite (calcium hypochlorite)
LACSD	Los Angeles County Sanitation Districts (Sanitation Districts of Los Angeles County)
LOX	Liquid oxygen
LPHO	Low-pressure high output
LPLO	Low-pressure low output
MCL	Maximum contaminant level
MG	Million gallons
MGD	Million gallons per day
MP	Medium pressure
MPN	Most probable number
MTBE	Methyl tertiary-butyl ether
MWRDGC	Metropolitan Water Reclamation District of Greater Chicago
NACWA	National Clean Water Agencies
NDMA	N-nitrosodimethylamine
NIOSH	National Institute of Occupational Safety and Health

NOM	Natural organic matter
NPDES	National pollutant discharge elimination system
NSF	National Sanitation Foundation
NTU	Nephelometric turbidity units
NWRI	National Water Research Institute
O&M	Operation and maintenance
OCC	Off-line chemical cleaning
OMC	On-line mechanical cleaning
ORP	Oxidation-reduction potential
OSHA	Occupational Safety and Health Administration
PAA	Peracetic acid
POTW	Publicly owned treatment works
Ppmv	Parts per million by volume
PSU	Power supply unit
RMP	Risk management plan
RNA	Ribonucleic acid
SCADA	Supervisory control and data acquisition
SMX	Sulfamethoxazole (an antibiotic found in wastewaters)
SS	Suspended solids
TCEP	Tris(2-chloroethyl)phosphate (a flame retardant found in wastewaters)
TDS	Total dissolved solids
TOC	Total organic carbon
TOD	Transferred ozone dose
TOX	Total organic halogen
THM	Trihalomethane
TSS	Total suspended solids
UFC	Uniform Fire Code
U.S.	United States
USD	U.S. dollars
U.S. EPA	U.S. Environmental Protection Agency
UV	Ultraviolet
UVA	UV absorbance
UVT	UV transmittance
WEF	Water Environment Federation
WERF	Water Environment Research Foundation
WTP	Water treatment plant

EXECUTIVE SUMMARY

This report examines a wide range of disinfection technologies, their prevalence of use, and information on their implementation at major POTWs in the U.S. (defined as wastewater treatment plants that are state or municipally owned, with an average dry weather design flow of greater than 0.95 MGD). The report focuses on the mature disinfection methods: chlorine, UV, and ozone. It also discusses a number of emerging alternatives: peracetic acid, ferrate, brominated compounds, pasteurization, ultrasonic cavitation, electron-beam and gamma irradiation, and photocatalysis. In addition, the report includes the relatively new trend of combining individual disinfection technologies ("combined disinfection"). The following sections briefly describe the findings from this project.

ES.1 Disinfection Technologies

The disinfection landscape for major POTWs can be divided into two categories, "mature" and "other". The mature technologies have been installed at a number of full scale facilities for a number of years, and in some cases for decades. The "other" technologies have been investigated at bench to pilot scale or operated at demonstration scale for a short period. In general, the "other" technologies do not have a track record of capital and operation and maintenance (O&M) costs that cover the range of flows (1->100 MGD) of the major POTWs. In addition, there is little or no long term operational experience for these technologies.

Mature Technologies

The survey of practice task for this project identified 4,450 major POTWs. Chlorination remains the most common form of disinfection. Although its use has been declining over the last 20 years, approximately 75% of all major POTWs still use some form of chlorine for disinfection today. Over the past two decades, the use of gaseous chlorine has declined significantly, the use of aqueous hypochlorite has increased significantly, and on-site generation has become a viable alternative for some POTWs.

Meanwhile, UV use has grown tremendously over the last 10 years. The survey of practice conducted for this report indicates that roughly 21% of all major POTWs currently apply UV disinfection, and approximately 40% of all UV systems currently in use were installed between 2001 and 2005 (the last year for which data were available).

Only a small fraction of POTWs currently use ozone, although it is a mature technology that is commonly applied to drinking water treatment. Early ozone technologies were adopted by a number of POTWs, but maintenance and operational problems caused most facilities to abandon ozone. However, ozone technologies have improved since that time, and a few POTWs are planning or designing ozone facilities because it is the only mature disinfection alternative capable of partially or completely removing trace organic compounds (e.g., pharmaceuticals, hormones, and precursors for disinfection byproducts such as THMs) at typical disinfection doses.

Other Technologies

The other disinfection technologies discussed in this report (chlorine dioxide, peracetic acid, ferrate, bromine compounds such as BCDMH (1-bromo, 3-chloro, 5,5-dimethylhydantoin), pasteurization, ultrasonic cavitation, electron beam and gamma irradiation, photocatalysis, and combined technologies) are not yet used for full-scale wastewater disinfection. Based on literature published thus far, chlorine dioxide and cavitation are unlikely to be widely used as stand-alone technologies for wastewater disinfection. Chlorine dioxide does not have significant advantages over chlorine, and retains some of the key disadvantages such as safety issues and the formation of disinfection byproducts. Cavitation is economically unfavorable due to the energy required for sufficient disinfection. However, cavitation does break up particles, which could enhance the effectiveness of the other technologies, and may therefore be useful in combination with those technologies.

Implementation issues remain for photocatalysis and e-beam/gamma irradiation, and must be resolved before these technologies can be widely used at large scales. Photocatalysis has only been tested in batch experiments at the laboratory bench scale, and catalysts that act quickly enough for practical application are still under development. Furthermore, there are not yet data with wastewater effluents, where constituents may deactivate the catalyst through fouling or poisoning. E-beam and gamma irradiation require a water depth of less than 7 mm at the point of use, which may limit its applicability, particularly at large facilities. However, these irradiation methods have been shown to disinfect treated wastewater at the pilot scale, and also have the potential to remove trace organic contaminants in water.

Of the developing alternatives covered in this report, peracetic acid (PAA), ferrate, BCDMH, pasteurization, and the combined technologies have the highest potential to become viable disinfection alternatives in the foreseeable future. All of these alternatives have been tested at the pilot scale, and all provide the benefit of eliminating or significantly reducing the levels of the disinfection byproducts associated with chlorine and ozone. However, each alternative has distinct advantages and disadvantages; as with the mature technologies, their applicability will depend on site-specific constraints and preferences. All would require some equipment and/or process changes: PAA has requirements for safe generation, handling and storage, while BCDMH requires solids handling equipment. Ferrate may require a settling tank following disinfection, if the effluent solids content is regulated. Pasteurization requires the installation of gas turbines or other heat sources.

Finally, the combined disinfection alternatives will pose the same requirements as the individual technologies that comprise them. For all of these developing technologies, use will remain limited until more data are available on their operations and their ability to disinfect effectively across the wide range of conditions found at POTWs.

Technology Selection

It is clear from this report that distribution of disinfection technologies at POTWs across the U.S. is changing. As these POTWs begin to upgrade their disinfection methods, they must consider and prioritize many factors. Each facility has different needs, so the selection process for a disinfection alternative is very site-specific, and the ultimate choices made by facilities reflect this variety. Chapter 8.0 compares the various mature disinfection technologies on a variety of criteria; the developing technologies generally had insufficient information to provide an adequate comparison. Issues covered include safety, disinfection efficacy, effects on effluent water quality, sensitivity to influent water quality, design and O&M issues, and costs. Using these decision factors, Chapter 8.0 provides a framework for choosing a disinfection technology, based on the selection criteria of the facility and the relative importance of those criteria.

ES.2 Survey of UV Use

Database Information

Data obtained in this study indicate that UV use is growing, with roughly 21% of all major POTWs currently applying UV disinfection. The technology is geographically widespread; all states except Delaware had at least one major POTW using UV disinfection. Kansas had the highest rate of adoption of UV technology, with more than 75% of the major POTWs using UV. There are several clusters of high UV use: in the Pacific Northwest, in the central/southwestern U.S. states surrounding Colorado, and near the Ohio River valley. Large plants are less likely to use UV, and those that do use UV are less likely to use the older low-pressure low output (LPLO) lamp technology and more likely to use medium pressure (MP) lamps. These large facilities may be slower to adopt newer technologies and less likely to be using the LPLO lamps associated with older UV systems, for a number of reasons, such as the considerable investment required to upgrade their disinfection systems or less stringent discharge limits for ocean outfalls. Larger plants are also more likely to use medium pressure (MP) lamps, which have higher intensity and a correspondingly smaller footprint. One possible reason for this trend is that these larger facilities tend to be located in urban areas, where space is limited and a smaller footprint is important. The vast majority of UV systems at POTWs use horizontal lamps parallel to flow in an open channel.

Survey Information

The UV survey of 103 POTWs indicated that safety and elimination of gas chlorine and chlorination/dechlorination are key factors driving the change to UV. More than half of the surveyed plants had an average UV Transmittance (UVT) between 60% and 80%, and used a design UVT of 50% to 70%. The median BOD and TSS values in the effluent were each 3 mg/L, and the median turbidity was 1.4 NTU. All surveyed facilities use at least secondary treatment processes, and almost half (47%) use tertiary treatment.

Although more than half (62%) of respondents felt that UV was more complicated than their previous disinfection system, 86% felt that the transition to UV was easier than expected or "not too bad." In evaluating the overall success of a UV project, the consultant's skill and knowledge was most often ranked as the most important factor, followed by the vendor characteristics and visiting and touring other facilities (these two factors were equally important), the skill and knowledge of the POTW's own staff, and finally other resources such as books or articles. For vendor characteristics, system design was most often ranked as the most important factor, followed by technical knowledge, responsiveness, and system features.

Reported training times, maintenance times, and costs varied by several orders of magnitude; median classroom training time was 6 hours (reported values ranged from 1 to 40 hours), hands-on training time was 10 hours (1 to 80 hours), maintenance time was 4 hours/week (0.2 to 120 hours/week), and annual replacement costs for lamps, sleeves, and ballasts were \$21,000/year (\$900 to \$420,000/year). Maintenance times and lamp component costs increased as plant size increased. MP lamps incurred higher expenses for lamp components and had a median lifetime of 5,000 hours, while LPLO and low-pressure high output (LPHO) lamps had median lifetimes of 10,000 hours.

Respondents to the survey also gave advice to facilities considering UV. One of the most common suggestions was to do thorough research by visiting similar facilities to 1) ensure that UV is applicable to the POTW considering UV, 2) get better estimates for the cost and time requirements, and 3) better understand potential problems that might occur. People also recommended covering the UV system with a roof or locating it inside a building to reduce maintenance requirements due to algae growth, to make maintenance easier in bad weather, and to prolong the lifetime of electronics that would otherwise be exposed to the elements. Finally, automatic wipers were strongly recommended to save maintenance time. More complete recommendations can be found in Section 7.3.7.

ES.3 Survey of Ozone Use

Currently there are seven major POTWs using ozone with a median design flow of 10 MGD and a range of 3 to 34 MGD. Two new ozone facilities have been installed in the last 10 years, one to remove color and the other to treat wastewater being discharged to a drinking water source. The remaining five have been operating since 1989 and all have undergone at least one upgrade. All the treatment plants use some form of biological suspended growth (oxidation ditch, air activated sludge, and pure oxygen activated sludge) as the secondary treatment.

By 2010 it is estimated that there will be 10 major POTWs using ozone. Four new ozone systems are under design or in the planning stage and one system will be changed to UV disinfection.

The user community for ozone disinfection is very small in the wastewater industry, in contrast to the drinking water industry. Recent improvements in design and ozone systems have been driven by the growth and needs of the water industry, but many of these improvements should be transferable to the wastewater industry. POTWs that need user experience information should look at the drinking water industry for this type of information.

CHAPTER 1.0

INTRODUCTION

Historically, chlorine gas has been the dominant disinfection technology for the wastewater industry. Recently, wastewater disinfection practices have been shifting away from the use of chlorine, particularly chlorine gas. The drivers for this change include the following:

- General reduction of the use of hazardous materials at facilities due to safety concerns for operators and the public;
- Security concerns after the terrorist attacks of September 11, 2001;
- Changes in the Western U.S. Uniform Fire Code requiring 100% containment and scrubbing facilities (WEF, 1996);
- U.S. EPA Clean Air Act and Occupational Safety and Health Administration requirements for Risk Management Plans for any facility having more than 1,000 pounds of chlorine gas (Hunter, 2000);
- More restrictive permit limits for receiving waters (e.g., lower concentrations for residuals and disinfection byproducts, and more parameters of concern);
- Increased water reclamation and an increased connection between the Clean Water Act and the Safe Drinking Water Act

As a consequence of driving forces such as these, other disinfection technologies have been developed and POTWs are increasingly evaluating these as possible alternatives to their current disinfection practice. Each alternative has unique advantages and disadvantages with respect to factors such as cost, ability to kill or inactivate a variety of pathogens, availability, safety, penetration of particles, solubility, stability, byproduct formation, and toxicity to humans, aquatic life, and microorganisms. In addition, the effectiveness of each disinfection practice may depend on factors such as contact time, concentration required for effective disinfection, temperature, solubility, and wastewater quality. The knowledge base on each of the alternatives varies widely; some technologies, such as those based on chlorine, have been studied extensively and there are years of experience disinfecting wastewater at the full-scale, while the newest technologies are still being evaluated at the bench scale.

1.1 Objectives and Tasks

The intent of this project was to accomplish two goals: 1) to compile and compare disinfection alternatives in a single document that is easily understood and useable for those in the wastewater treatment industry that are considering making a change in their disinfection practice; and 2) to identify the information or data gaps in our collective knowledge, necessary for adequate comparison of disinfection alternatives.

To accomplish this, the project was organized into three tasks: a literature review, a survey of practice, and a technology assessment. The literature review was conducted to understand the current state of knowledge. It covers peer-reviewed papers, conference proceedings, studies, and reports that have examined the various disinfection alternatives. The survey of practice assessed the demographics of major POTWs utilizing UV and ozone through the use of a number of U.S. EPA and vendor databases, and investigated operator experiences with these technologies through limited surveys. Chlorination was not included in this survey because of the wealth of existing information regarding chlorine disinfection, as documented in several references (e.g., Connell (2002), Water Environment Federation (2006), and Water Environment Federation (1996)) and in past surveys conducted by the Water Environment Federation (WEF) and WEF Disinfection Committee.

The technology assessment compiles the information from the literature review and survey of practice, to compare the advantages and disadvantages of the three mature technologies (chlorine, UV light, and ozone), as well as developing technologies. The basis of comparison and assessment includes historical usage, cost, applicability, effectiveness for pathogen and indicator organism removal and inactivation, ability to meet disinfection standards, reliability and safety issues, residuals, byproducts, and potential impacts to receiving waters. Finally, the compiled information was used to identify gaps in knowledge on each disinfection alternative.

1.2 Scope of Report

The intent of this report is to provide planning level information for wastewater professionals trying to decide whether to make a change in their disinfection practice. The report focuses on the mature disinfection technologies: chlorine added as a gas, hypochlorite, or onsite generation; UV; or ozone. The report also provides the available (but more limited) information on emerging technologies (e.g., peracetic acid and pasteurization).

This report is ***not*** the following:

1. A design manual for the various disinfection technologies; a complete resource for preparing a preliminary design;
2. A listing of all the necessary components for a complete system;
3. An operational manual; a health and safety resource for any of these technologies, or
4. An academic resource that, for example, provides a detailed explanation of the disinfection mechanisms or disinfection byproduct formation reactions.

This report also does not address the sustainability or the energy or global warming gas footprints of these technologies.

1.3 Report Organization

The methodologies used to develop the information for this report are summarized in Chapter 2.0. The mature technologies that are currently installed at POTWs are summarized in Chapter 3.0 (gas chlorine, hypochlorite solution, or onsite generation), Chapter 4.0 (UV), and Chapter 5.0 (ozone). Information for other disinfection technologies is provided in Chapter 6.0. Chapter 7.0 presents the survey of practice for the major POTWs in the 50 United States and Washington, D.C.

CHAPTER 2.0

METHODOLOGY

This chapter summarizes the methodology for this project. Section 2.1 describes the literature review. The survey of practice required several steps: determination of the number of major publicly owned treatment works (POTWs) in the 50 states and Washington, D.C. (Section 2.2); estimation of the number of major POTWs using chlorination, UV, or ozonation as their disinfectant (Section 2.3); a summary of chlorine gas releases (Section 2.4); and the limited survey of owners using UV (Section 2.5) and ozone (Section 2.6) disinfection technologies.

2.1 Literature Review

This project involved an extensive review of peer-reviewed papers, conference proceedings, studies, and reports that have assessed the various alternatives for wastewater disinfection. The literature review focused on studies conducted in the last 10 years and identified >400 peer-reviewed papers and conference proceedings on wastewater disinfection. These references are listed (by disinfection technology) in Appendix D; information on each technology is summarized in Chapters 3.0 through 6.0.

2.2 Determining the Number of Major POTWs

The first step in the survey of practice involved determining the number of major POTWs. Major POTWs have been defined by the U.S. Environmental Protection Agency (U.S. EPA) as facilities that are owned by a state or municipal entity and have an average dry weather design flow of > 0.95 MGD. This portion of the project used three databases: the U.S. EPA Clean Water Needs Survey 2000 (CWNS), the U.S. EPA Envirofacts database, and the U.S. Census.

2.2.1 Clean Water Needs Survey

Each state maintains and updates information in the Clean Water Needs Survey (CWNS) for the POTWs within the state. Fields in this database include the facility name, local or National Pollutant Discharge Elimination System (NPDES) permit number, owner name, dry weather design flow, and unit treatment processes such as the type of primary treatment, secondary treatment, disinfection, tertiary treatment, and solid handling systems. Typically, states spend about 18 months assembling and updating the database, so the data listed in the year 2000 are based on information from around 1998. As required by the Clean Water Act, the U.S. EPA compiles this information every four years and provides the U.S. Congress with a capital requirement update for all U.S. states and territories. For this project, the CWNS 2000 was downloaded in August 2005 from the U.S. EPA website (www.epa.gov/owm/mtb/cwns/index.htm). The database indicated 15,986 POTWs in the 50 states and District of Columbia, and 4,071 major POTWs (flow > 0.95 MGD).

2.2.2 Envirofacts

Envirofacts is a U.S. EPA database that contains information relevant to the NPDES permit program. More than 27,500 NPDES permits in the Envirofacts database use a Standard Industrial Classification (SIC) Code for sewerage systems (4952). In January 2006, the project team used a U.S. EPA Freedom of Information Act request to obtain data from the Envirofacts database for the major POTWs. The information in this database is estimated to be current as of December 2005. The requested data include 4,276 major POTWs with an NPDES permit, and contain information such as the facility name, owner, NPDES permit, flow rate, and unit treatment process at the facility. In general, the field for the unit treatment process was not populated for this database.

2.2.3 U.S. Census

The U.S. Census data were used to find cities that were likely to have a major POTW, given the population size. Based on data maintained by the state of Iowa, the per capita wastewater treatment plant flow is estimated at 173 gallons per day. This number implies that a population of approximately 5,500 people corresponds to the smallest treatment plant that would be considered "major" (average dry weather design flow of 0.95 MGD).

The census database (www.census.gov) was accessed in August 2005, and a list was downloaded of all cities with more than 5,500 people (projected to live in a particular city in 2004) in the 50 states and Washington, D.C. This city list was then used to cross-check the authority field in the two U.S. EPA databases as a quality assurance/quality control step.

2.2.4 Merging of Databases

After cross-referencing among the databases, there were estimated to be 4,450 major POTWs. There were four communities in the U.S. Census that did not have a CWNS or NPDES number. These POTWs were confirmed to exist by checking their respective websites. Table 2-1 summarizes the data sources for the major POTWs and the source for the treatment capacity assigned to each POTW. For the four POTWs without a number, the flow was calculated by multiplying the population served by 173 gallons per capita per day.

Table 2-1. Data Source of Major POTWs in the Fifty States and Washington, D.C.

Data Source	Description of Major POTW	Number	Treatment Capacity Source
CWNS	Has CWNS and/or NPDES No.	4,071	CWNS
Envirofacts	Only has NPDES No.	375	NPDES
US Census	No CWNS or NPDES No.	4	Calculated
Total		4,450	

2.3 Determining Type of Disinfection Technologies Used by Each POTW

Major POTWs use three main types of disinfection: chlorine (in the form of free chlorine or chloramines, and fed as a gas or liquid, or generated onsite), UV, or ozone. The distribution of these disinfection technologies is shown in Table 2-2, and is compared against the values estimated in a 2003 survey of disinfection practices by the WEF Disinfection Committee (WEF, 2004).

For this project, the distribution was determined using the following approach. UV and ozone vendors were contacted to obtain information on their installations. The vendors' lists were cross-checked against the list of major POTWs summarized in Table 2-1. A POTW was assigned UV or ozone only if a link could be confirmed between the two lists. A similar method was used to estimate that ~50 facilities use onsite chlorine generation equipment. This methodology may underestimate the number of facilities using UV, ozone, or on-site generation, because only systems from the largest vendors were identified, and only those plants that could be cross-checked were counted.

The CWNS database was used to estimate the number of major POTWs using no disinfection. The number of POTWs using chlorine was then calculated by subtracting the number of POTWs using ozone, UV, or nothing from the total number of POTWs.

Table 2-2. Project Survey and 2003 WEF Disinfection Committee Survey of Disinfection Practice.

Disinfection Practice	Number	Percent	WEF Disinfection Committee Survey
Chlorine	3,351	75.3%	72 %
UV	918	20.6%	24 %
Ozone	7	0.2%	0 %
No disinfection	173	3.9%	3 %

2.4 Chlorine Gas Release Database

One of the issues prompting the change from gas chlorine to alternative disinfection technologies is the hazard presented by the release of gas chlorine. Regulations require that any release of more than 10 pounds of gas chlorine be reported to the National Response Center. A Freedom of Information Act request was submitted to the National Response Center for all releases of gas chlorine above 10 pounds that occurred from January 1990 through June 2006 for both water and wastewater facilities. All of the obtained data are included in this report, because the database does not identify the facilities with respect to treated flow rates, nor does it differentiate between POTWs and privately owned facilities. Data from water treatment facilities are also included because water treatment facilities are similar to POTWs and the crossover information may also provide some valuable insights to the wastewater industry.

2.5 Limited UV Survey

For the UV survey, operators were asked to fill out a questionnaire on their UV system configuration, performance, cost, and other operational and maintenance issues. Appendix A provides the questions used in the survey. The survey was conducted between June and December of 2006, and approximately 100 responses were obtained from major POTWs using UV as their disinfectant.

The facilities in the survey were selected to cover different plant sizes, geographical distributions, lamp types, and equipment vendors. Plants were separated by average dry weather design flow rate into four categories: 1-2.5 MGD, >2.5-5 MGD, >5-20 MGD, and >20 MGD. These categories were used to determine whether the size of the facility or lamp type were factors in the use of this technology.

2.6 Ozone Survey

The CWNS database was used to develop an initial listing of POTWs that potentially use ozone as their disinfection technology. This approach yielded 32 POTWs. This information was corroborated with information published in the 1980's (U.S. EPA, 1986; Robson and Rice, 1991; Thompson and Leong, 2007). A cross-check of the identified facilities was performed using vendor information and a phone survey (June through November 2006). The ozone phone survey questions are presented in Appendix A.

The ozone phone survey identified five cases of ozone usage. The first was plants that currently use and plan to continue using ozone. The second case was POTWs currently using ozone, but planning to stop. The third case was facilities planning to change to ozone, or currently designing ozone systems. The fourth case was plants that formerly used ozone, converted to an alternative, and are now changing back to ozone. The fifth case was for utilities that converted to ozone and switched to an alternative. The ozone survey contacted 20 POTWs that fit cases one to four. POTWs that covered the fifth case were also contacted, but limited information was obtained, because the move away from ozone may have occurred up to 30 years ago and much of institutional memory of the staff at the facilities was lost.

CHAPTER 3.0

Chlorine Technologies

3.1 Background

3.1.1 Historical Use

Chlorination, or the use of chlorine in various forms, is the most commonly practiced wastewater disinfection technology. In addition to disinfection, chlorine has also been used for control of odors, sludge bulking, foaming, and nuisances (e.g., midge flies). Chlorine compounds are often favored because of their availability, effectiveness, efficiency, relatively low cost, convenience, and the persistence of chlorine residuals.

In 1854, the English Water Commission was the first group to formally recognize chlorine and chlorine compounds for wastewater disinfection, although wastewater in England was not chlorinated until 1884. In the U.S., wastewater disinfection with chloride of lime (a powder containing calcium chloride and calcium hypochlorite) appears to have begun in 1910, with Philadelphia being the first major city to install chlorination (U.S. EPA, 1986). By the late 1950s, approximately 30% of the wastewater treatment plants in the United States were using chlorination of one form or another (White, 1999).

Table 3-1 summarizes data from more recent surveys conducted by the Water Environment Federation (WEF) on the wastewater chlorination practices of its members (Connell, 2002; WEF, 2004). In 1979, the vast majority (95%) of wastewater treatment plants responding to the survey chlorinated their effluents. Since then, the use of chlorine for disinfection at publicly owned treatment works (POTWs) has been decreasing, due to increased awareness regarding the safety in handling, transportation, and storage of chlorine compounds, and concerns about formation of potentially harmful disinfection byproducts.

Table 3-1. Wastewater Chlorination Practice Survey Results. (Modified from Connell 2002; WEF, 2004)

	1979	1989	1996	2003
Chlorination	95%	78%	76%	72%
(Chlorine gas)	(90%)	(74%)	(62%)	(41%)
(Hypochlorite)	(5%)	(4%)	(14%)	(31%)

The survey also indicates that many facilities have moved away from chlorine gas due to the previously mentioned reasons and eliminating potential targets of terrorist attacks. In 1979, approximately 90% of the surveyed facilities used liquefied chlorine gas for disinfection, but only 41% did in 2003. Many of the plants using chlorine gas in 1979 switched to hypochlorite, while others switched to UV disinfection. Of the respondents using chlorine gas in 2003, more than 50% planned to switch to another disinfection process before 2013. Of those planning to switch, more than 60% were planning to switch to ultraviolet (UV) radiation, and the remainder to hypochlorite. On-site generation of hypochlorite is still an emerging technology and was used by only 2% of the POTWs that responded to the WEF survey (WEF, 2004). Discussions with several vendors similarly indicate a small number of installations, primarily at small POTWs. Although the data suggest a decrease in use of chlorine disinfection by POTWs, chlorine compounds are expected to continue to play a significant role in disinfection because of the many advantages offered by this process.

3.1.2 How Chlorine Disinfects

Chlorine is a strong oxidant that reacts rapidly with cellular material to provide fast disinfection. It can inactivate bacteria by damaging the nucleic acids and cell membranes, and inactivate viruses by damaging the protein coat and nucleic acids. It does not inactive protozoa such as *Cryptosporidia* and *Giardia* at practical doses.

3.1.3 Chlorine Chemistry

3.1.3.1 Chlorine Production

The principal compounds that are commonly used as a chlorine source are chlorine gas and sodium hypochlorite solutions, and less frequently, solid calcium hypochlorite. Chlorine does not exist naturally, and must be produced for commercial use. The chemistry of chlorine production for each source is briefly discussed below.

Chlorine Gas The most common process used to produce chlorine gas is brine electrolysis (Eq. 3.1.3.1-1). The produced chlorine is collected, purified, compressed, cooled, stored, and shipped as a liquefied gas under pressure.

$$2\ NaCl + 2\ H_2O + \text{electricity} \rightarrow Cl_2 + 2\ NaOH + H_2 \qquad (3.1.3.1\text{-}1)$$

Sodium Hypochlorite Sodium hypochlorite (also known as liquid bleach) is produced by the reaction of dissolved chlorine gas and sodium hydroxide (Eq. 3.1.3.1-2). The produced sodium hypochlorite is stored and shipped as liquid solutions containing 12.5 to 17% available chlorine.

$$Cl_2 + 2\ NaOH \rightarrow NaOCl + Na^+ + Cl^- + H_2O \qquad (3.1.3.1\text{-}2)$$

Calcium Hypochlorite Calcium hypochlorite, or HTH (high test hypochlorite) is produced by the reaction of sodium hypochlorite and calcium chloride (Eq. 3.1.3.1-3), or the direct addition of chlorine to hydrated lime (Eq. 3.1.3.1-4) to form calcium hypochlorite. The produced calcium hypochlorite is dried and formed into tablets, which are stored and shipped as solid calcium hypochlorite.

$$2\ NaOCl + CaCl_2 \rightarrow Ca(OCl)_2 + 2\ NaCl \qquad (3.1.3.1\text{-}3)$$

$$2\ Cl_2 + 2\ Ca(OH)_2 \rightarrow Ca(OCl)_2 + CaCl_2 + 2\ H_2O \qquad (3.1.3.1\text{-}4)$$

On-site Generation of Chlorine On-site sodium hypochlorite systems use electrolytic cells to produce hypochlorite from sodium chloride (NaCl) salt and water. The salt is charged into a brine tank to form a saturated brine solution, and power is then applied to drive the electrochemical reaction (Eq. 3.1.3.1-5). Sodium hypochlorite is generated as needed.

$$NaCl + H_2O + electricity \rightarrow NaOCl + H_2 \quad (3.1.3.1\text{-}5)$$

3.1.3.2 Free Chlorine

When chlorine gas (Cl_2), is added to water, hydrolysis occurs rapidly to form hypochlorous acid (HOCl), as shown below in Eq. 3.1.3.2-1. HOCl is a weak acid with a pK_a of 7.54 at 20°C, and can dissociate into hypochlorite ions (OCl^-) and hydronium ions (Eq. 3.1.3.2-2). Since both of these reactions release hydronium ions, they therefore tend to reduce the pH and alkalinity of the water.

$$Cl_2 + H_2O \Leftrightarrow HOCl + H^+ + Cl^- \quad (3.1.3.2\text{-}1)$$

$$HOCl \Leftrightarrow H^+ + OCl^- \quad (3.1.3.2\text{-}2)$$

When hypochlorite solutions are added to water, both NaOCl and $Ca(OCl)_2$ form hypochlorous acid (HOCl), as indicated in Eq. 3.1.3.2-3 and Eq. 3.1.3.2-4, respectively. Addition of hypochlorite to water releases hydroxyl ions and, combined with the alkalinity in the hypochlorite solution, tends to raise pH and alkalinity slightly. This may be offset somewhat by hydrolysis of HOCl (Eq. 3.1.3.2-2).

$$NaOCl + H_2O \Leftrightarrow HOCl + Na^+ + OH^- \quad (3.1.3.2\text{-}3)$$

$$Ca(OCl)_2 + 2\ H_2O \Leftrightarrow 2\ HOCl + Ca^{+2} + 2\ OH^- \quad (3.1.3.2\text{-}4)$$

HOCl and OCl^- are often grouped together and collectively called the "free available chlorine" or "free chlorine residual." Both compounds act as disinfectants, but the germicidal efficiency of HOCl is about 40 to 80 times that of OCl^- (Tchobanoglous et al., 2003), so the relative distribution of HOCl and OCl^- is very important. The percentage of HOCl in the free available chlorine is a function of pH and temperature (Eq. 3.1.3.2-3):

$$\%\ HOCl = (1 + K_a \times 10^{pH})^{-1} \cong [1 + (1.4 + 0.06 \times T) \times 10^{(pH-8)}]^{-1} \quad (3.1.3.2\text{-}5)$$

where: K_a is the equilibrium constant (in mole/L) for the ionization reaction Eq. 3.1.3.2-2,

T is the temperature in degree Celsius.

Table 3-2 gives the percentage of HOCl at various temperatures and pH values. The HOCl percentage decreases with increasing pH and temperature. When the decrease is due to temperature, a simultaneous increase in the microbial inactivation rate more than compensates for the change in HOCl composition (see Section 3.2.2.3 for details). However, if pH increases, chlorine dosages may need to be increased to maintain constant levels of microbial inactivation.

Table 3-2. Percentage of HOCl in Free Available Chlorine.

	Temperature (°C)				
pH	10	15	20	25	30
6.00	98%	98%	97%	97%	97%
6.50	94%	93%	92%	92%	91%
7.00	83%	81%	79%	78%	76%
7.50	61%	58%	55%	52%	50%
8.00	33%	30%	28%	26%	24%

3.1.3.3 Combined Chlorine

If free chlorine is added to a wastewater effluent where nitrogen control is not practiced, hypochlorous acid will react with the ammonia present to form chloramines according to the following reactions:

$NH_3 + HOCl \Leftrightarrow NH_2Cl$ (monochloramine) $+ H_2O$ (3.1.3.3-1)

$NH_2Cl + HOCl \Leftrightarrow NHCl_2$ (dichloramine) $+ H_2O$ (3.1.3.3-2)

$NHCl_2 + HOCl \Leftrightarrow NCl_3$ (trichloramine or nitrogen trichloride) $+ H_2O$ (3.1.3.3-3)

Collectively, the chlorine in these chloramines is referred to as "combined available chlorine." The relative amounts of the species depend on factors such as pH, temperature, contact time, chlorine dose, and the ratio of chlorine to ammonia. High pH levels and low chlorine to ammonia ratios favor the formation of monochloramine. Under typical wastewater disinfection conditions, monochloramine (NH_2Cl) and dichloramine ($NHCl_2$) are the dominant chloramine species, and the contribution of nitrogen trichloride (NCl_3) is usually negligible.

The pH has a profound impact on the propotion of monochloramine and dichloramine. At low pH, the following reactions take place:

$NH_2Cl + H+ \rightarrow NH_3Cl^+$ (3.1.3.3-4)

$NH_3Cl^+ + NH_2Cl \rightarrow NHCl_2 + NH_4^+$ (3.1.3.3-5)

The above reactions affect the percentage of monochloramine and dichloramine as a function of pH. Table 3-3 summarizes data from two studies using a series of breakpoint bench scale test under identical conditions, except for pH. Jolley and Carpenter used equimolar concentrations of ammonia and chlorine buffered at different pHs. The Palin work was done from conventional breakpoint curves, starting with 1 mg/L of ammonia-N. The proportionation summarized in Table 3-3 for the Palin work was at the peak of each respective breakpoint curve.

Table 3-3. Percentage of Monochloramine and Dichloramine as Funtion of pH.

pH	NH_2Cl (%)	$NHCl_2$ (%)	Reference
5	13	87	Jolley and Carpenter 1975
6	33	67	**Palin, 1975**
6	57	43	Jolley and Carpenter 1975
7	90	10	**Palin, 1975**
7	88	12	Jolley and Carpenter 1975
8	~100	trace	**Palin, 1975**
8	97	3	Jolley and Carpenter 1975

3.1.3.4 Breakpoint Chlorination

Continued addition of chlorine (increasing the ratio of chlorine to ammonia) results in the complete oxidation of readily oxidizable substances, including chloramine compounds and the remaining ammonia, and the presence of a free chlorine residual beyond the "breakpoint." Figure 3-1 is a generalized curve describing the chloramination process and breakpoint chlorination for a wastewater with 1 mg/L ammonia-N and no chlorine demand. Stoichiometrically, the combined chlorine residual concentration reaches a peak at a chlorine to ammonia nitrogen weight ratio of approximately 5:1. The ratio is approximately 7.6:1 at the breakpoint. However, in wastewater this ratio may approach 10:1 depending on the presence of compounds that may be oxidized by chlorine. Beyond the breakpoint, any added chlorine becomes mostly free chlorine residual, and the process is referred to as "breakpoint chlorination."

Figure 3-1. Generalized Curve of Chloramination and Breakpoint Chlorination. (adapted from Tchobanoglous et al., 2003)

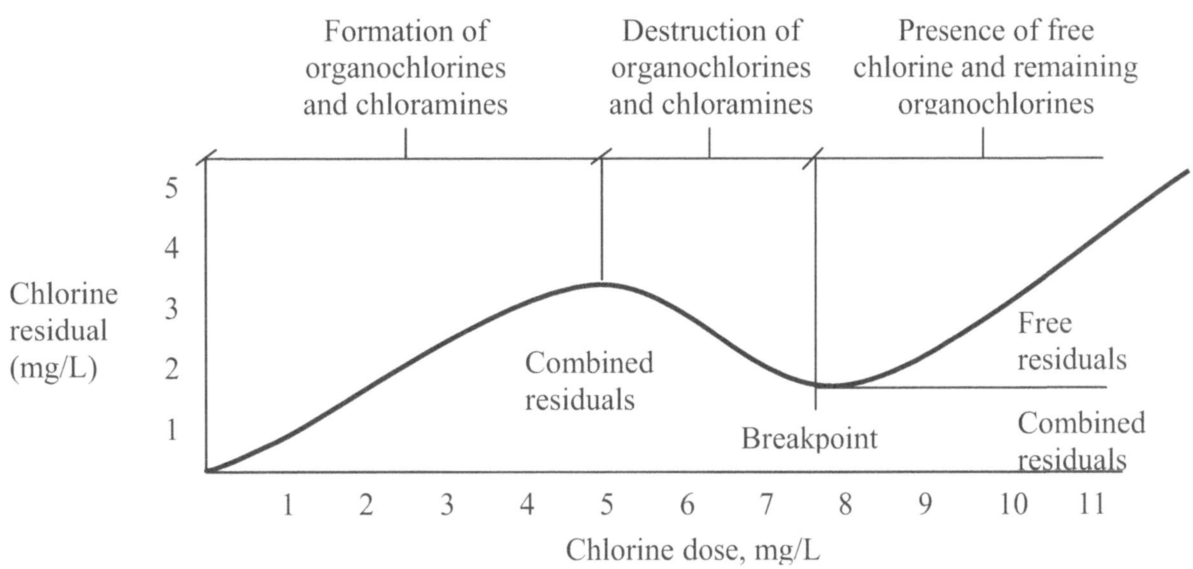

From a practical standpoint, breakpoint chlorination is practiced only when ammonia-N is in the 0.5 – 3.0 mg/L range. Note that at 3.0 mg/L ammonia-N, 23 – 30 mg/L chlorine dosage would be required to reach the breakpoint. These high doses of chlorine make the process less economical, and may also form unacceptable levels of disinfection byproducts. Thus, POTWs that apply breakpoint chlorination must tightly control the levels of ammonia in the effluent and ensure that there is no breakthrough of ammonia.

3.1.3.5 Dechlorination

Dechlorination is most often accomplished by adding chemicals that include sulfur dioxide (gas), sodium bisulfite (liquid), calcium or sodium thiosulfate, and ascorbic acid. Sulfur dioxide (SO_2) and sodium bisulfite ($NaHSO_3$) are the most frequently used dechlorinating agents. Ascorbic acid has the lowest potential to deplete the oxygen in a receiving water when overdosing a dechlorinating agent, followed by thiosulfate, and then bisulfite, and lastly, sulfur dioxide.

Sulfur Dioxide Sulfur dioxide undergoes the following reactions with free and combined chlorine:

$$SO_2 + HOCl + H_2O \rightarrow Cl^- + SO_4^{-2} + 3\,H^+ \qquad (3.1.3.5\text{-}1)$$

$$SO_2 + NH_2Cl + 2\,H_2O \rightarrow Cl^- + SO_4^{-2} + NH_4^+ + 2\,H^+ \qquad (3.1.3.5\text{-}2)$$

$$2\,SO_2 + NHCl_2 + 4\,H_2O \rightarrow 2\,Cl^- + 2\,SO_4^{-2} + NH_4^+ + 5\,H^+ \qquad (3.1.3.5\text{-}3)$$

$$3\,SO_2 + NCl_3 + 6\,H_2O \rightarrow 3\,Cl^- + 3\,SO_4^{-2} + NH_4^+ + 8\,H^+ \qquad (3.1.3.5\text{-}4)$$

Sodium Bisulfite The reactions of sodium bisulfite with free and combined chlorine are shown in Eq. 3.1.3.5-5 to 3.1.3.5-8.

$$HSO_3^- + HOCl \rightarrow Cl^- + SO_4^{-2} + 2\,H^+ \qquad (3.1.3.5\text{-}5)$$

$$HSO_3^- + NH_2Cl + H_2O \rightarrow Cl^- + SO_4^{-2} + NH_4^+ + H^+ \qquad (3.1.3.5\text{-}6)$$

$$2\,HSO_3^- + NHCl_2 + 2\,H_2O \rightarrow 2\,Cl^- + 2\,SO_4^{-2} + NH_4^+ + 3\,H^+ \qquad (3.1.3.5\text{-}7)$$

$$3\,HSO_3^- + NCl_3 + 3\,H_2O \rightarrow 3\,Cl^- + 3\,SO_4^{-2} + NH_4^+ + 5\,H^+ \qquad (3.1.3.5\text{-}8)$$

Note that the dechlorination reactions shown release hydrogen ions, consuming alkalinity and possibly lowering pH.

Sodium Thiosulfate Calcium or sodium thiosulfate is used when oxygen depletion is an issue. Since thiosulfate, regardless of form, is less expensive than ascorbic acid, this is the chemical of choice to evaluate first to determine whether the effluent meets the receiving water requirements. The reactions of sodium thiosulfate with chlorine yield the following:

$$Na_2S_2O_3 + 4HOCl + H_2O \rightarrow 2\ NaHSO_4 + 4\ HCl \qquad (3.1.3.5\text{-}9)$$

$$Na_2S_2O_3 + HOCl \rightarrow Na_2SO_4 + S + HCl \qquad (3.1.3.5\text{-}10)$$

$$2Na_2SO_3 + HOCl \rightarrow Na_2S_4O_6 + NaCl + NaOH \qquad (3.1.3.5\text{-}11)$$

Sodium thiosulfate undergoes the following reactions with monochloramines:

$$4\ NH_2Cl + Na_2S_2O_3 + 5\ H_2O \rightarrow 2\ NaHSO_4 + 4\ NH_3 + 4\ HCl \qquad (3.1.3.5\text{-}12)$$

$$NH_2Cl + Na_2S_2O_3 + H_2O \rightarrow Na_2SO_4 + S + NH_3 + HCl \qquad (3.1.3.5\text{-}13)$$

$$NH_2Cl + 2\ Na_2S_2O_3 + 2\ H_2O \rightarrow Na_2S_4O_6 + 2\ NaOH + NH_3 + HCl \qquad (3.1.3.5\text{-}14)$$

Sodium thiosulfate undergoes the following reactions with dichloramines

$$2NHCl_2 + Na_2S_2O_3 + 5\ H_2O \rightarrow 2\ NaHSO_4 + 2\ NH_3 + 4\ HCl \qquad (3.1.3.5\text{-}15)$$

$$NHCl_2 + 2\ Na_2S_2O_3 + 2\ H_2O \rightarrow 2\ Na_2SO_4 + 2\ S + NH_3 + 2\ HCl \qquad (3.1.3.5\text{-}16)$$

$$NHCl_2 + 4\ Na_2S_2O_3 + 4\ H_2O \rightarrow 2\ Na_2S_4O_6 + 4\ NaOH + NH_3 + 2\ HCl \qquad (3.1.3.5\text{-}17)$$

Dechlorination of chloramines may produce NH_3 or NH_4^+, depending on pH. The reaction produces sodium bisulfate, sodium sulfate and/or HCl. In addition, chloramine reactions produce trace concentrations of ammonium chloride.

Calcium Thiosulfate The reactions of calcium thiosulfate with combined chlorine have not been documented. The reactions of calcium thiosulfate with free chlorine (Hardison and Hamamoto 1998). yields the following:

$$CaS_2O_3 + 2HOCl + H_2O \rightarrow Ca(HSO_3)_2 + 2\ HCl \qquad (3.1.3.5\text{-}18)$$

$$Ca(HSO_3)_2 + 2\ HOCl \rightarrow CaSO_4 + H_2SO_4 + 2\ HCl \qquad (3.1.3.5\text{-}19)$$

In summary

$$CaS_2O_3 + 4\ HOCl + H_2O \rightarrow CaSO_4 + 4\ HCl + H_2SO_4 \qquad (3.1.3.5\text{-}20)$$

In addition, the same authors also report a second reaction between calcium thiosulfate and chlorine.

$$CaS_2O_3 + HOCl \rightarrow CaSO_4 + S + HCl \qquad (3.1.3.5\text{-}21)$$

Ascorbic Acid Ascorbic acid, also known as Vitamin C, is reported to react with chlorine to produce chloride and dehydroascorbate with little or no oxygen depletion observed in a receiving water. With chloraminated water, it is reported to produce insignificant concentrations of ammonium. The reactions with chlorine and chloramine are shown below:

$$C_5H_5O_5CH_2OH + HOCl \rightarrow C_5H_3O_5CH_2OH + HCl + H_2O \qquad (3.1.3.5\text{-}22)$$

$$2\ C_5H_5O_5CH_2OH + 2NH_2Cl \rightarrow 2\ NH_4^+ + Cl_2 + 2\ C_5H_3O_5CH_2OH \qquad (3.1.3.5\text{-}23)$$

Because ascorbic acid is weakly acidic, its addition to water may slighltly decrease the pH. In field applications to neutralize planned releases from drinking water distribution systems vitamin C reduces far more chlorine than an equal weight of sulfur-based compounds (Tikkanen, et al. 2001). On a mass basis, one part of ascorbic acid is reported to neutralize ten parts of chlorine.

3.1.4 Chlorine Demand and Residual

As a strong oxidizing agent, chlorine reacts with reduced compounds in effluents. These compounds are said to exert a chlorine demand. They include ammonia, organic nitrogen, reduced iron and sulfur species, nitrite, and organic matter. The concentration of chlorine remaining in the water after the demand has been satisfied is known as the chlorine residual. This quantity is used to monitor and control the disinfection process, as well as the potential for biological regrowth in distribution lines. Chloramines usually have a slower decay rate than free chlorine, after initial demand is satisfied. Sunlight may affect the residual die-off of both free chlorine and chloramines.

3.1.5 Disinfection with Free or Combined Chlorine

Disinfection can be accomplished using either free chlorine or combined chlorine (chloramines). Because wastewaters contain significant concentrations of ammonia, chlorine disinfection is usually practiced with chloramines. Free chlorine is generally used only when nitrification or other nitrogen control is provided so that breakpoint to a free residual is practical. However, in the absence of ammonia, other sources of chlorine demand (e.g., organic nitrogen) may become significant and reduce the effective disinfectant concentration. As a result, disinfection with a free residual has not been successful with some nitrified effluents with ammonia-N < 0.5 mg/L; to prevent the loss of disinfectant to organic nitrogen and other chlorine demand, ammonia was added to form chloramines and provide adequate disinfection (White, 1999). Ammonia may also be added to fully nitrified effluents before disinfection to form chloramines, which limit the formation of disinfection byproducts such as trihalomethanes.

Free chlorine reacts faster than chloramines, and thus requires much shorter contact times to achieve the same degree of microorganism inactivation. Literature results regarding the efficacy of the mono- and di-chloramine species vary. One study showed that bactericidal properties of dichloramine were better than those of monochloramine, (Wolfe, et al., 1984) while other studies indicate that monochloramine has superior virucidal abilities. Some evidence suggests that solutions of mono- and di-chloramine, in approximately equal concentrations, are stronger disinfectants than those containing only monochloramine or dichloramine (U.S. EPA, 1999). Organochloramines (e.g., N-chloramines and N-chloramino acids) are not effective disinfectants, but are usually included in reported values of combined available chlorine, due to difficulty in differentiating them from inorganic chloramines using current analytical methods.

For facilities without complete nitrification, chloramination is often more appropriate than breakpoint chlorination, because high chlorine dosages would be needed to reach the breakpoint. Due to increasingly stringent requirements on nitrogen discharge, a growing number of plants are implementing biological nitrogen removal. These facilities usually have very low ammonia concentrations in the secondary effluent, and may need to add ammonia to the wastewater to achieve chloramination. For these facilities, free chlorine (after the breakpoint) may be a viable alternative, depending on disinfection requirements. Sections 3.2, 3.5, and 3.7 further discuss the advantages and disadvantages associated with free chlorine versus chloramines, with respect to disinfection performance, byproduct formation, effect on the environment, and costs.

3.1.6 Quantifying the Strength of Chlorine Solutions

The strength of chlorine solutions is quantified using several different units, including molarity, mass concentration, and percentage by weight. Multiple chlorine compounds can be used for wastewater disinfection: chlorine (Cl_2), sodium hypochlorite (NaOCl), and calcium hypochlorite ($Ca(OCl)_2$). To compare the strength of solutions across chlorine compounds, the mass concentrations and percentages are sometimes expressed in terms of available chlorine. For example, the concentration of a NaOCl solution can be given in grams/liter (g/L) of NaOCl, or in g/L of available chlorine. Table 3-4 expresses solution strength in each of the various units for the three chlorine compounds.

Table 3-4. Expressions of Chlorine Solution Strength for 1M Solutions.

	Chlorine (Cl_2)	Pure Sodium Hypochlorite (NaOCl)	Pure Calcium Hypochlorite ($Ca(OCl)_2$)
Molecular weight (g/mole)	70.9	74.5	143
Molarity (M)	1	1	1
Mass concentration (g/L)	70.9	74.5	143
Mass concentration (g/L of available chlorine as chlorine gas)	70.9	70.9	142
Mass concentration (lbs/gal)	0.59	0.62	1.2
Mass concentration (lbs/gal of available chlorine)	0.59	0.59	1.2
Approximate weight percent (%)	6.6	6.9	12.5
Approximate weight percent of available chlorine as chlorine gas(%)	6.6	6.6	12.4

Molarity is measured in moles/liter (M). As shown in Eq. 3.1.3.2-1 to 3.1.3.2-4, one mole of chlorine is electrochemically equivalent to one mole of sodium hypochlorite or half a mole of calcium hypochlorite ($Ca(OCl)_2$). Molar concentrations can be converted to mass concentrations using the molecular weight of the compound (see Eqs. 3.1.6-1 and 3.1.6-2 for examples). Mass concentrations are expressed in grams/liter (g/L) or pounds/gallon (lbs/gal), and can easily be converted to available chlorine as chlorine gas (Eq. 3.1.6-3).

$$1\ \text{M NaOCl} = \left(\frac{1\ \text{mole NaOCl}}{\text{L}}\right)\left(\frac{74.5\ \text{g NaOCl}}{1\ \text{mole NaOCl}}\right) = 74.5\text{g/L as NaOCl} \tag{3.1.6-1}$$

$$\begin{aligned}1\ \text{M NaOCl} &= \left(\frac{1\ \text{mole NaOCl}}{\text{L}}\right)\left(\frac{1\ \text{mole Cl}_2}{1\ \text{mole NaOCl}}\right)\left(\frac{70.91\ \text{g Cl}_2}{1\ \text{mole Cl}_2}\right)\\ &= 70.91\ \text{g/L as available chlorine as chlorine gas}\end{aligned} \tag{3.1.6-2}$$

$$\begin{aligned}1\ \text{g/L NaOCl} &= (1\ \text{g/L NaOCl})\left(\frac{1\ \text{mole NaOCl}}{74.5\ \text{g NaOCl}}\right)\left(\frac{1\ \text{mole Cl}_2}{1\ \text{mole NaOCl}}\right)\left(\frac{70.91\ \text{g Cl}_2}{1\ \text{mole Cl}_2}\right)\\ &= 0.952\text{g/L as available chlorine as chlorine gas}\end{aligned} \tag{3.1.6-3}$$

The approximate weight percent of a solution can be calculated by dividing the mass of the chlorine compound by the total mass of the solution. Eq. 3.1.6-4 shows the calculation for a 74.5 g/L solution of NaOCl. In one liter of solution, there are 74.5 g of NaOCl, and the mass of the solution is approximately 1074.5 g (~1000 g of water in one liter, plus 74.5 g of NaOCl); the exact mass can be calculated using the density of the solution, which varies with the NaOCl concentration. The weight percent as available chlorine can be calculated using the same equation, but with the mass of available chlorine.

$$\left(\frac{74.5\text{ g NaOCl}}{1000\text{ g water} + 74.5\text{ g NaOCl}}\right) = 6.9\%\text{ by weight as NaOCl} \qquad (3.1.6\text{-}4)$$

3.2 Disinfection Efficacy

3.2.1 The CT Concept

In early 1900s, Harriet Chick observed that microbial inactivation is proportional to contact time for a given disinfectant concentration, and Herbert Watson determined the relationship between the inactivation rate constant and disinfectant concentrations. Combining their results yields the Chick-Watson law, which describes the rate of change of microorganism concentrations during disinfection (Eq. 3.2.1-1). The integrated form is given by Eq. 3.2.1-2.

$$\frac{dN}{dt} = -kC^n t \qquad (3.2.1\text{-}1)$$

$$\ln\frac{N_t}{N_0} = -kC^n t \qquad (3.2.1\text{-}2)$$

where: C= disinfectant concentration (or residual)

k = inactivation rate constant

n = an empirically determined constant

N_0 = initial number of microorganisms

N_t = number of microorganisms at time t

t = contact time

Departures from the Chick-Watson law are common, because inactivation rates also depend on wastewater characteristics, including particle size and concentration, temperature, and pH. For example, the Selleck-Collins model (White, 1999) uses the following relationship for disinfection of wastewaters with chloramines:

$$y/y_0 = [Ct/b]^{-n} \qquad (3.2.1\text{-}3)$$

where: C = disinfectant (residual) concentration

b = constant (value of Ct below which no disinfection is observed, i.e., $y/y_0 = 1$)

n = constant (negative of the slope of log (y/y_0) vs. log (Ct/b), when $y/y_0 > b$)

t = contact time

y_0 = initial coliform concentration

y = final coliform concentration

This model suggests that there is an initial lag phase, in which little or no reduction in microorganism levels occurs until a minimum Ct equal to "b" is achieved. This is followed by a second stage in which the rate of microorganism reduction is constant as a function of log (Ct/b). Tchobanoglous et al. (2003) suggest that there is a third stage in which the inactivation rate declines with increasing Ct (a tailing effect attributed to particles in the wastewater).

Eqs. 3.2.1-2 and 3.2.1-3 assume that the chlorine concentration is constant over time, which is generally not true. Despite this, many regulatory agencies control the wastewater disinfection process using the Chick-Watson law with an n value of 1 and a C value corresponding to the chlorine residual concentration (Tchobanoglous, et al., 2003). This variant of the Chick-Watson law gives rise to the "CT" concept, that a given CT value (the multiplication product of disinfectant residual and contact time) provides a constant inactivation rate. In other words, any combination of chlorine residuals and contact times that yields the same CT value will provide the same level of inactivation, within certain limits.

In some states, the CT value and chlorine contact time are specified in regulatory requirements. For example, Florida requires a CT of at least 25 mg-min/L for influent fecal coliform <1,000 per 100 mL, 40 mg-min/L for influent fecal coliforms between 1,000 and 10,000 per 100 mL, and 120 mg-min/L for influent fecal coliform >10,000 per 100 mL at peak hourly flow (Florida Administrative Code, Chapter 62-600.440). California requires a CT value of 450 mg-min/L and a modal contact time of 90 minutes at peak flow for certain reclamation applications (California Code of Regulations, Title 22 Section 60301.230). This CT value is loosely based on the "Pomona virus" pilot studies, which also indicated that for tertiary effluents, chloramine residuals of 5 to 10 mg/L for two hours appear to be as effective as free chlorine residuals of 4 mg/L for the same contact time for virus inactivation (Sanitation Districts of Los Angeles County, 1977). However, because of the variability of wastewaters and indigenous microorganisms, site-specific testing is recommended to establish appropriate chlorine doses and contact times to achieve the desired disinfection target, which must be at least as stringent as the regulatory limit or an equivalent degree of public protection.

3.2.2 Effects of Water Quality on Disinfection

Water quality has a strong effect on chlorine demand and disinfection efficacy. Important water quality parameters affecting chlorination include particles, pH, temperature, organic nitrogen, and other compounds.

3.2.2.1 Particles

Large particles can shield microorganisms from contact with chlorine and can also exert a chlorine demand, thereby reducing residual concentrations. Chauret et al. (2004) tested the effects of three solids (wastewater debris, soil, and iron pipe corrosion debris) on the disinfection of two organisms (*E. coli* O157:H7 and MS2 coliphage) with both free chlorine and chloramines. They found that the particulate matter exerted a demand on the disinfection residuals, especially free chlorine. However, it had no significant effect on bacterial or phage inactivation kinetics (after normalizing for CT) with either free chlorine or monochloramines, with the exception that wastewater debris decreased monochloramine inactivation of *E. coli* O157:H7. Tchobanoglous et al. (2003) report that when suspended solids are present in wastewater, chlorination first inactivates individual free-swimming bacteria and microorganisms in small clumps. After this initial rapid inactivation, the disinfection rate decreases and is a function of the particle size distribution.

3.2.2.2 pH

The germicidal efficiency of hypochlorous acid is much higher than that of the hypochlorite ion. Because HOCl is the dominant species below pH 7.5, free chlorination is expected to be more effective at lower pH. At a given pH, the inactivation efficiency of chlorine when added as a gas or hypochlorite solution is the same. When using gaseous chlorine, the pH will typically drop, while addition of sodium or calcium hypochlorite tends to increase pH. Due to these respective changes in pH, gaseous chlorine is expected to have greater disinfection efficiency than hypochlorite at the same chlorine dose because there will be a greater proportion of HOCl.

For chloramination, pH affects the disinfection efficacy by altering the chloramine species distribution as summarized in Table 3-3. As discussed in Section 3.1.4, the different chloramine species have varying degrees of disinfection efficacy.

3.2.2.3 Temperature

For typical wastewater temperatures, pathogen inactivation by free chlorine or chloramines increases with temperature (US EPA, 1999), even though increased temperatures also decrease the pk_a and thus increase, the concentration of the less effective OCl^- compared to that of HOCl (Gordon and Bubnis, 2000).

3.2.2.4 Chlorine Demand from Organic Nitrogen and Other Compounds

Chlorine residuals decrease in the presence of high chlorine demand, which is caused by organic nitrogen and other compounds (see Section 3.1.4). A higher chlorine dosage therefore may be required to maintain disinfection levels.

3.2.3 Inactivation of Pathogens

The germicidal efficiency of chlorination is typically expressed as the reduction of microorganisms on a logarithmic scale, and is usually given as a function of the CT value, as described in Section 3.2.1. Generally, hypochlorous acid reacts much faster than hypochlorite ions, which in turn react faster than chloramines. As a result, to achieve the same degree of microorganism inactivation, chloramines require much higher CT values than free chlorine. Table 3-5 illustrates this point by providing a relative comparison of the free chlorine at pH ~7 (~80% HOCl, and ~20% OCl^-) and chloramine doses required to achieve a given level of inactivation (Tchobanoglous et al., 2003); absolute CT values will depend upon the water quality of the effluent to be disinfected, the pH, and the temperature. Chauret et al. (2003) found a similar trend: monochloramine provided 2- to 3- log inactivation of *E. coli* O157:H7 with CT values ranging from 20 to 30 mg-min/L, while free chlorine required CT values less than 0.30 mg-min/L.

Table 3-5. Estimated Range of CT Values (mg-min/L) for Inactivation of Bacteria and Viruses. (Tchobanoglous et al., 2003 with permission from McGraw Hill)

	Free Chlorine	Chloramines
Bacteria		
2-log	0.4 – 0.8	12 - 20
3-log	1.5 – 3	30 – 75
4-log	10 – 12	200 – 250
Virus		
2-log	2.5 – 3.5	300 - 400
3-log	4 – 5	500 – 800
4-log	6 – 7	200 – 1,200
Protozoan Cysts		
2-log	35 – 45	70 – 80
3-log	700 – 1,000	1,100 – 2,000

Table 3-5 indicates that bacteria are most susceptible to chlorine disinfection, followed by viruses, then protozoa. Tree et al. (2003) showed similar results; free chlorine was strongest against *E. coli* and *Enterococcus faecalis*, followed by seeded poliovirus, then the F+ specific RNA (FRNA) bacteriophage MS2. Seeded poliovirus was inactivated much more rapidly than indigenous enteroviruses, which had inactivation rates similar to those of natural and seeded FRNA bacteriophage. Cysts of *Entamoeba histolytica* and *Giardia lamblia* and eggs of parasitic worms are very resistant to chemical disinfectants, and are more effectively removed by other methods such as coagulation, settling, and filtration (Connell, 2002).

3.3 System Design and Implementation

This section presents issues related to the design of a chlorination/dechlorination system for wastewater disinfection. The design process must consider site-specific factors including the flow rate, water quality, environmental setting, availability of chemicals, and compliance limits for pathogens and disinfection byproducts. Detailed explanations on system design are beyond the scope of this report; the intent here is to provide an overview of the design issues. For more information, please consult guidance documents, such as the following:

- Municipal Wastewater Disinfection – Design Manual (U.S. EPA, 1986)
- Wastewater Disinfection – Manual of Practice FD-10 (WEF, 1996)
- Alternative Disinfectants and Oxidants Guidance Manual (U.S. EPA, 1999)
- Handbook of Chlorination and Alternative Disinfectants (White, 1999)
- The Chlorination/Dechlorination Handbook (Connell, 2002)
- Wastewater Engineering – Treatment and Reuse (Tchobanoglous et al., 2003)
- Recommended Standards for Wastewater Facilities, also known as the Ten State Standards (Great Lakes - Upper Mississippi River Board of State and Provincial Public Health and Environmental Managers, 2004).
- Review of Municipal Effluent Chlorination/Dechlorination Principles, Technologies and Practices (Environment Canada, 2006)
- Wastewater Disinfection – Training Manual (WEF, 2006)

3.3.1 Maturity of the Technology

The use of chlorine compounds for the disinfection of wastewater is an established and mature technology. It was among the earliest disinfection techniques, and remains the most common method for wastewater disinfection today. As such, it has been studied extensively, and is relatively well characterized for design, operation, and regulatory approval.

3.3.2 Facility Requirements

Chlorine disinfection involves a very complex series of events and is influenced by the type and extent of reactions with chlorine-reactive materials, temperature, pH, suspended solids concentrations, and the viability of target organisms. Therefore, the upstream processes of a treatment plant must be properly designed and effectively operated to minimize the need for excessive disinfection.

In addition, chlorination or chloramination requires a relatively large footprint, in order to provide adequate contact time, especially for reuse. This may be an issue, especially for facilities in urban locations or other areas where acquiring space for chlorine contact basins may be extremely expensive or simply infeasible.

3.3.3 System Components

The elements of a chlorination system are chlorine generation and/or supply, mixing, contacting, dechlorination, and control. The generation/supply subsystem provides a continuous feed of the disinfectant to the system. The mixing element blends the disinfectant into the incoming wastewater. The contacting element provides contact time for the blended wastewater for a sufficient period until the desired microbial inactivation is achieved. The dechlorination subsystem removes the chlorine residual from the disinfected wastewater, to minimize aquatic toxicity, which is associated with chlorine residuals. The control element uses information about the incoming wastewater flow and chlorine residual to adjust the operating parameters of the generation/supply element so as to maintain consistent disinfectant performance (U.S. EPA, 1986).

3.3.3.1 Chlorine Generation/Supply Subsystem

Chlorine gas and liquid sodium hypochlorite are the primary forms of chlorine used in wastewater treatment plants. Solid calcium hypochlorite has been infrequently used, primarily in very small systems. Chlorine chemicals are generally purchased from external sources, but sodium hypochlorite has also been generated on-site. The following sections provide more information on each of these chlorine compounds, as well as on-site generation. Chlorine dioxide, a chlorine derivative that is not widely used for wastewater disinfection, is discussed separately in Section 6.2.

Chlorine Gas Chlorine gas is often referred to as elemental chlorine. It is approximately 2.5 times heavier than air and has a greenish-yellow color and characteristic pungent odor that is noticeable at levels of 0.2 parts per million by volume (ppmv).

Storage facilities for chlorine gas should be equipped with chlorine scrubbers to prevent unwanted emissions to the atmosphere. Chlorine can be withdrawn from storage containers by gas or liquid/external evaporation. Gas withdrawal is not practical in very large plants due to limitations on safe withdrawal of chlorine gas from one ton containers. The chlorine feed into the disinfection system can be done via solution or direct gas. The solution feed process is used in all but very small wastewater treatment plants (less than 1 MGD) or in facilities where chlorine gas is only used on an occasional basis (U.S. EPA, 1986).

Sodium Hypochlorite The use of liquid sodium hypochlorite (also known as liquid bleach) eliminates many of the safety concerns and environmental risks associated with the transport, storage, and feeding of gas chlorine. Commercial-grade sodium hypochlorite is available in solution form, in strengths of 1 to 16% by weight. For wastewater disinfection, hypochlorite is usually delivered in solutions containing 10 to 15% available chlorine. Higher concentrations are not common because the chemical stability of the solution diminishes greatly with increasing strength.

Industry designates the strength of a hypochlorite solution using trade percent, which is defined as the mass of chlorine gas contained in 100 mL of solution. For example, a 10% (trade percent) hypochlorite solution contains 10 mg of chlorine per 100 mL, which is equivalent to 100 mg/L or 0.83 pounds of chlorine per gallon. The weight percent of the solution can be calculated as follows:

$$\text{Weight \% as } Cl_2 = \frac{\text{Available Chlorine (lbs of } Cl_2\text{/gal of solution)}}{\text{Solution Density (lbs of solution/gal of solution)}} \qquad (3.3.3.1\text{-}1)$$

Table 3-6 summarizes available chlorine and hypochloric acid concentrations at several commonly found trade percent levels. Practically speaking, one gallon of 12.5% sodium hypochlorite solution typically contains the equivalent of one pound of chlorine.

Table 3-6. Hypochlorite Solution Strength (Chlorine Institute, 2006 with permission from the Chlorine Institute).

Trade Percent (%)	Available Chlorine ($g\ Cl_2/L$)	Available Chlorine ($lb\ Cl_2/gal$)	Density (lb/gal)	Weight Percent as Available Chlorine Gas (%)
10	100	0.83	9.53	8.7
12.5	125	1.0	9.80	11
15	150	1.2	10.06	12

Hypochlorite solutions are unstable and will decompose during storage; this affects the feed rate and dosage and can produce undesirable byproducts such as chlorate (ClO_3^-, Eq. 3.3.3.1-2), chlorite, and very low levels of perchlorate (ClO_4^-) (Gordon, 1995; Massachusetts Department of Environmental Protection, 2006). The degree of decomposition increases with the length of storage, temperature, exposure to sunlight, impurities in the solution, and decreasing pH. Off-gassing is also a problem for high strength sodium hypochlorite solution (Eq. 3.3.3.1.3). Consequently, hypochlorite solutions should be stored in closed fiberglass tanks or rubber-lined steel containers to prevent corrosive attack and to minimize off-gassing of oxygen (Environment Canada, 2006).

Chlorate Formation

$$3\ NaOCl \rightarrow 2\ NaCl + NaClO_3 \qquad (3.3.3.1\text{-}2)$$

Off Gasing

$$2\ NaOCl \rightarrow 2\ NaCl + O_2 \uparrow \qquad (3.3.3.1\text{-}3)$$

Calcium Hypochlorite Calcium hypochlorite is much more expensive than chlorine gas or sodium hypochlorite (approximately two to three times the cost of sodium hypochlorite on the same active chlorine basis). Therefore it is not widely used as a wastewater disinfectant except for small installations or during emergency situations, as a back-up chlorine source.

Calcium hypochlorite is supplied in a dry form as a powder, granule, disk, briquette, or tablet. Wastewater applications most commonly use the granular or tablet form, which is often supplied in drums of up to 100 lbs, and usually contains 65 to 70% available chlorine. One and a half (1.5) pounds of calcium hypochlorite, with 65% available chlorine, contains the equivalent of one pound of chlorine.

Because of its oxidizing potential, calcium hypochlorite should be stored in corrosion-resistant containers in a cool, dry, well-ventilated location away from other chemicals, especially those that are easily oxidized (U.S. EPA, 1999). The feeding system may directly use solids in proprietary tablet feed chlorinators or may use the calcium hypochlorite in a slurry or solution. Solution feed of calcium hypochlorite has the disadvantage of producing calcium carbonate precipitates that foul surfaces and storage tanks and clog metering pumps, piping, and valves.

On-site Generation of Chlorine On-site generation of liquid sodium hypochlorite is being considered as a competitive chlorine source in some situations. Considerations for on-site systems include cost, concentration of the hypochlorite produced, availability of raw materials, and reliability of the process (U.S. EPA, 1999). The concentration of the sodium hypochlorite solution produced from the system is typically around 0.8%, although recently developed systems may produce solutions as high as 12% NaOCl. The 12% technology utilizes less salt and electricity per pound of hypochlorite produced. Temperature control of the water is important, as low temperatures (less than 15°C or 60°F) may reduce production efficiency and may promote premature electrode coating failure. Because it is difficult to control inlet water temperature at very small remote sites, on-site generation is usually applied at facilities using more than 50 lbs of chlorine per day (WEF, 2006).

This technology generates two waste streams. One is a blow down stream of the electrolytic solution. For the 0.8 % technology the blow down brine is part of the injected hypochlorite while for the 12 % technology the blow down is a separate stream that requires disposal. The other is the formation of hydrogen gas from the electrolytic process. Typically, proper venting addresses the hydrogen gas issue.

The capital cost of an on-site generation system is currently higher than that for a typical chlorine gas or sodium hypochlorite system. In addition, on-site generation may use large quantities of salt, require large storage volumes for salt and/or brine, and increase salt concentrations in the final effluent because of the small blow down stream of brine (Bess, 2003). However, on-site generation enables a facility to produce chlorine compounds at the plant location, rather than relying on shipments of externally generated material. As such, it can provide safer chemical delivery and storage, and eliminate the risk of spill of chlorine gas or 12% hypochlorite into the environment (Environment Canada, 2006). It can also operate unattended at remote sites (U.S. EPA, 1999).

3.3.3.2 Mixing Subsystem

Rapid mixing helps to avoid developing chlorine concentration gradients that may overdose some portion of the wastewater, potentially promoting formation of other chlorinated compounds with little or no germicidal efficiency, and underdose other portions of the wastewater, resulting in insufficient disinfection. Ideally, the mixing device should achieve complete mixing in a few seconds. The primary design options are mechanical mixers that require input of external power such as pumps and motors, and energy dissipation devices (static mixers or hydraulic jumps) that dissipate hydraulic head (if available). Commonly used mechanical devices include turbine agitators, inline mechanical mixers, and submerged vacuum impeller. The use of energy dissipation devices may not be feasible for plants where there is little head difference between the influent and the effluent. However they are inherently more reliable, because their performance is insensitive to power fluctuations or outages and mechanical failures (U.S. EPA, 1986).

3.3.3.3 Contacting Subsystem

The contacting subsystem may consist of a separate contact basin or the outfall pipe. Basins should be designed to promote plug flow, with length-to-width and length-to-depth ratios of at least 40:1 preferred to prevent short-circuiting (White, 1999). As a result, many contact basins employ baffles or serpentine flow configurations to improve effective contact time. In addition, contact tanks should be covered, to reduce chlorine losses to volatilization. Using the outfall pipe reduces the capital costs; however, the pipe must be long enough to meet contact time requirements (U.S. EPA, 1986). Local regulations set the required contact time, which is generally 30 minutes minimum at average flow, or 20 minutes at peak flow. Ten State standards require a 15-minute minimum contact at design peak hourly flow after mixing (Great Lakes - Upper Mississippi River Board of State and Provincial Public Health and Environmental Managers, 2004). In California, reuse regulations require a minimum modal contact time of 90 minutes at peak dry weather flow.

3.3.3.4 Dechlorination Subsystem

The use of dechlorination has increased since the late 1970s, due to concerns about the potentially toxic effects of both free and combined chlorine residuals on fish and other aquatic life. Regulations promulgated by the U.S. EPA and many states require that chlorine residual in effluents be less than 0.05 mg/L, and possibly lower depending on receiving stream classification.

Dechlorination may be accomplished by adding chemicals that reduce chlorine residual to chloride ions, or through activated carbon reactions, aeration, wetlands, or ponding. The chemicals most often used for dechlorination are sulfur dioxide (SO_2) and sodium bisulfite ($NaHSO_3$).

Sulfur dioxide is a nonflammable, corrosive, colorless, liquefied gas stored in steel containers. As with chlorine gas, sulfur dioxide can be fed into the system via gas withdrawal or solution feed; solution feed processes are almost universally used (U.S. EPA, 1986). Gas storage safety considerations for sulfur dioxide are similar to chlorine gas.

Sulfur reducing chemicals that are available in salt or solution form have also been used for dechlorination, especially at sites where hypochlorite is used for disinfection. Besides sodium bisulfite, these chemicals include sodium sulfite (Na_2SO_3), sodium thiosulfate ($Na_2S_2O_3$), and sodium metabisulfite ($Na_2S_2O_5$). Of these compounds, sodium bisulfite is most frequently used for wastewater dechlorination (Connell, 2002).

Other dechlorinating agents used by POTWs include calcium and sodium thiosulfate and ascorbic acid. These alternative dechlorinating agents cause little or no measureable reduction in oxygen levels, even when applied in excess of the chlorine requirement. As a result, they are likely to be used when the dissolved oxygen in the receiving water is a significant concern. In addition, the thiosulfates and ascorbic acid are not considered a hazardous or toxic chemical. These chemicals have advantages in transport, storage, and safety issues due to this classification when evaluating which dechlorination agent to use.

Activated carbon in granular or powder form may also be used for dechlorination, depending on-site-specific conditions. Carbon requirements and contact times are typically determined through on-site pilot testing, with reported powdered activated carbon doses in the range of 30 to 40 mg/L (WEF, 1996). The use of activated carbon for dechlorination alone is not cost effective, but granular activated carbon in contactors that are used for other treatment requirements may also provide dechlorination.

The choice of dechlorinating agent typically depends on availability and economics. The theoretical and practical doses required for dechlorination with each of the sulfur-reducing chemicals are summarized in Table 3-7 (Tchobanoglous et al., 2003 and Tikkanen et al., 2001).

Table 3-7. Dechlorinating Chemical Doses.

Dechlorinating Chemicals	Theoretical Dose (mg/mg Cl_2)	Range in Use (mg/mg Cl_2)
Sulfur Dioxide	0.90	1.0 – 1.2
Sodium Sulfite	1.78	1.8 – 2.0
Sodium Bisulfite	1.47	1.5 – 1.7
Sodium Metabisulfite	1.34	1.4 – 1.6
Sodium Thiosulfate	0.56	0.6 – 0.9
Calcium Thiosulfate	0.53	1.6-2.2
Ascorbic Acid	2.46	2.9-3.3

Chemical dechlorination is generally considered to be fast (seconds), as long as mixing occurs rapidly. The Ten State standards require a minimum of 30 seconds for mixing and contact time at the design peak hourly flow (Great Lakes - Upper Mississippi River Board of State and Provincial Public Health and Environmental Managers, 2004). Consequently, facilities often use the outflow pipe or the outfall structure for dechlorination. However, MacCrehan et al., (2005) found a chlorine residual of up to 0.2 mg/L in wastewater after dechlorination with sodium sulfite. MacCrehan et al. (1998) suggested that slow dechlorination is caused in part by organic nitrogen components (peptides and proteins) in tertiary effluent. According to Bedner et al. (2004), sulfite reduces the most refractory organochloramines too slowly to produce wastewater effluents that meet current ecosystem protection criteria in the U.S. (total chlorine residual less than 0.011 mg Cl_2/L). Reducing agents that were faster than sulfite included dithionite ($Na_2S_2O_4$), thiosulfate, and iodide-mediated sulfite (Bedner et al., 2004). MacCrehan et al. (2005) determined the relative strength of various dechlorinating agents, in order of decreasing strength: metallic iron >> sulfite with iodide ≈ thiosulfate > sulfite alone >> ascorbic acid. Of these compounds, only iron achieved rapid and complete dechlorination; however, use of metallic iron for dechlorination is not being practiced at any wastewater treatment plants.

3.3.3.5 Control Subsystem

The main objective of the control subsystem is to adequately disinfect effluent using minimum chemical doses for chlorination and dechlorination. Effective control of chlorination will also minimize byproduct formation. Many control strategies are available, with the following being the most commonly used:

- Manual or "open-loop" control – This relies on operator intervention.
- Feed-forward – This system is typically flow-proportional.
- Closed-loop feedback control – This relies on continuous sensors for chlorine residual or oxidation-reduction potential (ORP), and is usually used in conjunction with feed-forward flow-proportional control.

In very small plants, where dechlorination is not needed or practiced, manual or feed-forward flow proportional control is generally sufficient. For plants practicing dechlorination, the closed-loop feedback control is a better option, to minimize the potential for overdosing dechlorinating chemicals, which may cause deoxygenation or depress pH (e.g., SO_2) in the final effluent and the receiving water bodies. Feedback control may also be preferred for plants subject to effluent constraints on chlorine residual, or in larger plants where small decreases in chemical dosage result in large cost savings. Although the closed-loop feedback control has the highest capital costs of the three options, the chemical usage and associated costs are smaller when the control system is properly operated (U.S. EPA, 1986).

3.3.4 Ease of Design and Expansion

The wealth of information that is available on chlorine disinfection systems aids in the design process and removes some of the uncertainty on the expected disinfection performance. In addition, the regulatory requirements are generally well established and clearly defined. However, chlorination and dechlorination does require a relatively large footprint, which can make expansion challenging.

3.3.5 Regulatory Acceptance

Because chlorination is a proven technology, the approval process should not be problematic if the design meets all the requirements specified in the relevant regulations. Regulatory agencies limit microbial levels in wastewater effluents and may impose additional water quality requirements, such as turbidity, minimum contact time and/or the CT value to ensure adequate disinfection. The levels of disinfection byproducts and chlorine residuals in chlorinated and dechlorinated effluents must also comply with local regulatory limits, and dechlorination may be necessary if NPDES permits incorporate aquatic toxicity limits.

Areas for chemical storage, handling, and use must follow state and local building and fire codes, published standards, and recommended practices established by the Chlorine Institute, Compressed Gas Association, and American Chemistry Council. Of primary interest to those using chlorine, sulfur dioxide, and anhydrous ammonia are the Building Code, which provides standards and criteria for the storage of toxic or hazardous materials, and the Uniform Fire Code, which was promulgated in 1988 and requires that facilities undergoing expansion install chlorine scrubbers as protection against accidental release of chlorine gas. Utilities should also consult local fire chiefs or marshals regarding changes in chlorine, ammonia, and sulfur dioxide facilities, as these officials have final authority on the installation of these facilities within their jurisdictions.

Finally, utilities using chlorine must develop Risk Management Plans (RMPs) for the Occupational Safety and Health Administration (OSHA) and/or the U.S. EPA. In 1999, OSHA began requiring development and implementation of risk management plans for the accidental release of chlorine at facilities having more than 1,000 lbs of chlorine at a facility (Hunter, 2000). The U.S. EPA implemented RMP requirements under the Clean Air Act, Section 112(r)(7), for facilities that store or handle toxic chemicals. The threshold quantities for triggering RMP requirements for chlorine, aqueous ammonia, and anhydrous sulfur dioxide are 2,500, 20,000, and 5,000 pounds, respectively. The RMP consists of three primary elements: Risk Assessment, Risk Management, and Emergency Response. An RMP should include modeling or forecasting of release scenarios, and the proposed devices, systems, or procedures to be employed to mitigate the impact of these scenarios on the surroundings and the community. The deadline for the first filing was June 21, 1999, and plans must be updated periodically. Some states, such as California and New Jersey, have more stringent reporting requirements and lower threshold levels.

3.3.6 Public Acceptance

Wastewater chlorination has been practiced in the U.S. for approximately a century, and it has been accepted as an effective disinfection technology. However, the public has become more aware of the potential adverse environmental impacts, including the formation of disinfection byproducts and toxic residuals in the effluent as well as environmental risks associated with transport, storage, and handling of the chemicals. It is important to include public input when planning new or expanded chlorination facilities.

3.4 Operation and Maintenance

3.4.1 Ease of Operation and Maintenance

Section 3.3 lists a number of references with information on the operation and maintenance of chlorine disinfection systems. For example, The Municipal Wastewater Disinfection Design Manual describes the careful procedures that must be followed to minimize the potential of chlorine gas leaks during the start-up of chlorination systems, or after a shutdown (U.S. EPA, 1986):

- Check tubing reliability periodically.
- Ascertain minute leakage from lines and fittings from routine inspection and repair.
- Inspect and clean sludge accumulation in evaporator vessels every year or each 200 tons of chlorine delivered.
- Inspect chlorine gas filter semi-annually.
- Clean the chlorine pressure reducing valves and replace the valve spring periodically.
- Disassemble and clean the ejectors and the chlorinator rotameters every six months.

3.4.2 Process Control and On-line Monitoring

3.4.2.1 Chlorine Residual Analyzers

The single most important aspect of controlling the chlorination and dechlorination processes is the accurate measurement of chlorine residuals and the use of instrumentation to control the chlorination and dechlorination chemical feed (WEF, 2002). As a result, the chlorine residual analyzer is probably the system component that requires most careful attention. The following operation and maintenance requirements are essential (U.S. EPA, 1986):

- Inspect the sample lines for solids accumulation and clean filters and screens on a daily basis.
- Clean the electrodes in amperometric systems weekly, or more frequently if erratic or drifting readings occur.
- Check the effluent line pH to see whether the buffer solution strength is appropriate and the solution is properly added.
- Compare the analyzer reading with the results from a manual analysis daily.
- Routinely check the adequacy of potassium iodide (KI) addition, if used.

Chlorine residuals can be measured as free, combined, or total. The total chlorine residual in effluent discharges is generally restricted to a maximum of 0.05 mg/L, although more restrictive (e.g., 0.01 mg/L) standards may be applicable depending on local requirements.

On-line monitoring of chlorine residuals in wastewater is based on techniques that have been developed primarily for potable water. Consequently, many of these monitoring units may not be applicable to wastewater (or may be useful only for high quality tertiary effluents), despite manufacturers' efforts to promote their use with wastewater. Care should be exercised in applying on-line monitoring technology and where possible, a long-term pilot study of the unit should be completed.

On-line amperometric analyzers provide the most accurate indication of chlorine residual for both free and combined chlorine. Units with three-probe sensors, which are designed to adjust to background interferences, provide better accuracy and stability. However, this type of unit is more difficult to operate and requires the highest level of operator intervention. For tertiary wastewater treatment plants and high quality secondary plants, membrane-type residual chlorine analyzers represent the best combination of accuracy and ease of operation. However, these units will quickly foul if used to monitor lower quality effluents, and changes in nitrogen levels will cause significant inaccuracy in the measured residual (Tchobanoglous et al., 2003; Environment Canada, 2006).

Chloramine residuals are usually controlled using conventional chlorine residual or ORP methods, which are not specific for individual chloramine species, but measure combined chlorine residual. Organochloramines, nitrite, manganese, and chromate positively interfere with these classical residual control measurements. Harp (2000) presented a new method of control that is based on a modification of the classical Berthelot analysis for ammonia, and is specific for inorganic monochloramine. This method allows for process control based on the levels of the actual disinfectant, monochloramine, rather than proxies such as total chlorine residual or ORP.

3.4.2.1 Oxidation-Reduction Potential (ORP) Analyzers

Oxidation-reduction potential (ORP) analyzers are suitable for primary and low quality secondary effluents (Environment Canada, 2006), and have been successfully used to control chlorination or dechlorination feeds. In fact, Kim and Hensley (1997) found that the inactivation of coliforms in wastewater correlated better with ORP than with total chlorine residual. One municipal wastewater treatment plant in California replaced conventional methods with a system that automatically modulated chlorine and sulfur dioxide feed by ORP; this reduced chemical use for chlorination and dechlorination by 47% and 62%, respectively, while consistently maintaining compliance with stringent disinfection requirements (2.2 total coliforms/100 mL) and chlorine discharge limits (less than 0.1 mg/L).

Some of the concerns with ORP performance include interferences, sensitivity, and calibration. For example, nitrified effluents seem to increase the millivolt signal of the ORP output, while denitrified effluents seem to depress the signal. Chlorine residual levels >1.0 mg/L and sulfite concentrations <0.5 mg/L can impact the sensitivity of the ORP analyzer. ORP appears to provide more sensitivity in measuring fluctuating free chlorine residual levels than fluctuating combined chlorine residual levels. Ammonia, pH, and temperature effects have been reported to produce inaccurate or erratic measurements. Dissolved chemical compositions, which are site-specific, can affect the background ORP value and ORP levels required for effective disinfection. As a result, operators must determine and account for this characteristic "poise" of their wastewater to achieve reliable performance. Finally, the ORP millivolt signal does not necessarily correlate to the chlorine residual in wastewater; for example, field tests have indicated that effective disinfection may not require elevated ORP levels. Chlorine residual analyzers should be used for monitoring and records, if a chlorine residual reading of the effluent is required (Kopchynski et al., 2001; Connell, 2002).

3.4.3 Operator Safety and Health Risks

Chlorine is highly reactive and is extremely corrosive in the presence of moisture. Chlorine gas is a respiratory irritant and is classified by the U. S. Department of Transportation as a poisonous gas. The effects of chlorine gas exposure include choking, coughing, watery eyes, mild skin irritation, and lung irritation. These effects begin to take effect when chlorine concentration is between 5 to 10 ppmV. The level of chlorine that poses an immediate danger to life or health, issued by the National Institute of Occupational Safety and Health (NIOSH), is 10 ppmV (Chlorine Institute, 1995).

Sodium hypochlorite is a respiratory irritant and is destructive to living tissue. Conditions that may create an extreme hazard, such as excessive heat or mixing with incompatible substances, must be avoided (Marx, 1996). Hypochlorite solutions also release an off-gas, which has a readily detectable chlorine odor and has been identified as chlorine monoxide (Chlorine Institute, 2000). In the presence of moisture, the vapors will corrode metals such as iron and copper. The handling of sodium hypochlorite requires special design considerations because of its corrosiveness and the generation of fumes. Good safety practices call for the use of goggles, gloves, and other protective equipment. Personal protective equipment, in combination with engineering controls, standard operating procedures, and well-maintained feed equipment allow safe handling of NaOCl.

The dechlorinating agent sulfur dioxide is also a corrosive gas and an irritant. Sodium bisulfite should also be treated with care because it can decompose to form corrosive sulfur dioxide, particularly with exposure to acids or heat.

3.4.4 Requirements for Backup Resources

Because pathogen inactivation is one of the main goals for municipal wastewater treatment, facilities should provide redundancy of components to allow for continuous operation during routine maintenance and in the case of un-scheduled repairs (Environment Canada, 2006). For example, emergency back-up power is needed to ensure continuous chlorination and dechlorination during a power outage. Calcium hypochlorite may serve as a back-up chlorine source in some limited situations, where only small quantities (< 1 ton storage) of chlorine are required.

3.4.5 Storage, Inventory, and Replacement Requirements

The amount of on-site storage should be based on the reliability of the delivery method. For plants close to the main suppliers of chemical, storage can be as low as seven days. In locations where winter conditions or institutional constraints (e.g., labor union strikes) can delay shipments, a higher level of storage should be considered. For sodium hypochlorite, excessive storage should be avoided as the high strength solution is not stable (Environment Canada, 2006).

Operators should keep adequate spare parts in inventory for the chlorination and dechlorination systems, to assure short turn-around should the unit fail. Plants should never count on immediate delivery of parts from suppliers as the parts are typically stocked by the manufacturers and extended lead-times are quite common (Environment Canada, 2006).

3.4.6 Operational Impacts on Upstream Processes

As the last process in the wastewater treatment train, the operational impacts on upstream processes should be minimal during typical operations. However, plants using chloramination may have elevated levels of NDMA in the effluent (see Section 3.5.1.1). If this effluent is used to backwash tertiary filters, NDMA-laden water may be returned to the headworks, and potential impacts of this sidestream on the whole treatment plant may need to be evaluated.

3.4.7 Consistency in Performance with Fluctuations in Flow Rate

With proper sizing and design of the chemical feeding system and proper operation of the control system, chlorine doses should be readily adjustable to fluctuations in flow rate and influent water quality. Higher chlorine doses/residuals may be used to compensate for the shorter retention times that occur during high flow conditions. As a result, the performance of chlorination/dechlorination systems should be relatively consistent with fluctuations in flow rate.

3.5 Impacts to the Environment

Potential impacts to the environment from the use of chlorine compounds include the formation of disinfection byproducts, toxicity to aquatic life, and other changes in water quality. There are also potential risks from transport, storage, and use of the various chlorine compounds, as well as regrowth of microorganisms in the final effluent. On the positive side, chlorination may provide incidental benefits. The following sections address these issues.

3.5.1 Changes in Water Quality

3.5.1.1 Disinfection Byproducts

All chemical disinfection processes produce some disinfection byproducts (DBPs). Literature reports over 600 disinfection byproducts for the major chemical disinfectants (chlorine, ozone, chlorine dioxide, chloramines) and their combinations in water treatment (Krasner, et al., 2006). In the case of chlorination, free chlorine or chloramines react with natural organic matter (NOM) to form chlorinated organic DBPs. These compounds are of concern because epidemiological studies have shown an association between DBP levels in drinking water and occurrences of several types of cancer.

The most researched DBPs from chlorination include trihalomethanes (THMs), haloacetic acids (HAAs), nitrosamines including N-nitrosodimethylamine (NDMA), and cyanide. THMs and HAAs are primarily of concern with free chlorine disinfection, while NDMA is mainly of concern during chloramination, and cyanide may be formed with either form of chlorine. These DBPs are discussed in more detail below.

Trihalomethanes and Haloacetic Acids THMs and HAAs represent two major classes of halogenated DBPs formed during chlorination. Factors affecting formation of these halogenated DBPs include the type and concentration of NOM, free chlorine concentration, pH, inorganic and organic nitrogen concentration, temperature, and other water quality parameters; for example, more brominated THMs are created in waters with high concentrations of bromide (e.g., coastal areas). Higher pH, temperature, contact time, chlorine dose, and UV_{254} absorbance tend to produce greater levels of THMs, as do higher levels of dissolved organic carbon (DOC). In addition, for a given wastewater, free available chlorine forms THMs faster than chloramines; consequently, breakpoint chlorination usually generates significantly higher levels of THMs and HAAs than chloramination (U.S. EPA, 1999; Koukouraki and Diamadopoulos, 2001 and 2002; Tchobanoglous, et al., 2003; Yang, et al., 2005). Because of this, some POTWs with fully nitrified effluents choose to add ammonia-N back into the effluent and disinfect with chloramines, rather than free chlorine.

The presence of chloroform and other trihalomethane (THM) species in disinfected drinking water was first reported in 1974, and was linked to chlorination (Singer, 2006). In 1979, the U.S. EPA set the first drinking water standards for THMs. These federal standards have been refined over time, and some states have adopted even more stringent standards to regulate these compounds in wastewater effluent discharges. Depending on the final use (discharge or reuse), POTWs may also need to comply with the federal or local drinking water standards for total or individual THMs and HAAs, as reflected in NPDES or other permits.

N-nitrosodimethylamine (NDMA) Nitrosamines are a group of compounds that are among the most potent carcinogens known (Snyder, 1995). The most studied nitrosamine is NDMA, which can be generated when chloramines react with dimethylamine (DMA) in water and wastewater (Choi and Valentine, 2002; Mitch and Sedlak, 2004). Several studies conducted by the Sanitation Districts of Los Angeles County confirmed NDMA formation during the chloramination process, while free chlorine disinfection (after the breakpoint) did not have a significant effect on NDMA levels (Huitric et al., 2006). NDMA formation can be minimized if preformed chloramines are used as the disinfectant, rather than forming the chloramines from the ammonia already in the wastewater, (Mitch et al., 2005). However, laboratory tests conducted by the Los Angeles County Sanitation Districts suggest that preformed chloramines may provide less effective disinfection (unpublished data), so the ability of preformed chloramines to meet disinfection requirements while minimizing NDMA formation should be verified.

Although there are no federal drinking water standards for nitrosamines, some states, such as California, have established a "notification level" for several nitrosamines in drinking water. These levels can have regulatory consequences for recycled water applications. Due to its high cancer potency, the notification level established for NDMA in California is very low, at 10 nanograms per liter (ng/L) or parts per trillion.

A study of seven POTWs indicated that NDMA occurs in POTW influents due to both domestic and industrial sources. NDMA precursors from domestic sources, industries within the sewage service area, or polymers used within the treatment process also contributed to NDMA levels in the POTW discharge. Secondary treatment (activated sludge or oxygen-activated sludge) removed 0 to 75% of the NDMA, and over 70% of the DMA and other NDMA precursors (Sedlak et al., 2005). Controlling the sources or using a different polymer may help to significantly reduce the effluent NDMA levels.

Cyanide and Cyanogen Chloride Cyanide is a compound that is toxic to aquatic life at elevated levels. It has been detected in the effluent of many POTWs at levels exceeding the influent concentrations. Research has suggested that both cyanide and cyanogen chloride can be formed during chlorination of treated wastewater effluent or drinking water (U.S. EPA, 1999; Kavanaugh, et al., 2003; Na and Olson, 2006). Khoury, et al. (2005) found that cyanide generation was relatively fast under breakpoint chlorination, but was a slower process under chloramination.

From an analytical standpoint, cyanide analysis may be prone to false positive detections, due to sample preparation and preservation methods. The U.S. EPA approved methods for sample preservation include high pH (pH > 12) and a dechlorination agent to prevent oxidation of the cyanide by chlorine. Pandit et al. (2006) found that cyanide levels increased with time, even within the holding times of EPA-approved cyanide analytical methods. Khoury et al. (2005) compared four EPA-approved preservation protocols with immediate analysis to examine the effects of adjusting the pH to ≥12. Relative to samples without pH adjustment, cyanide levels increased slightly when the dechlorinating agent was thiosulfate, and approximately doubled with arsenite. These results imply that EPA-approved preservation protocols may artificially elevate cyanide levels in chlorinated wastewater effluent matrices.

Emerging DBPs The previous paragraphs describe the most widely studied DBPs from chlorination; however, many DBPs remain unidentified. Krasner et al. (2006) conducted a study to evaluate the formation of unregulated DBPs and to identify new DBPs in challenging drinking water sources; overall, only 30% of total organic halogens (TOX) were accounted for by the measured halogenated DBPs, while 70% were unknown. The study involved 12 drinking water treatment plants in the United States: six used only chlorine and/or chloramines, the other six combined either ozone or chlorine dioxide with chlorine and/or chloramines. In addition to currently regulated DBPs, more than 50 DBPs with high potential toxicity were studied. The DBPs included iodinated trihalomethanes, other halomethanes, a nonregulated haloacids, haloacetonitriles, haloketones, halonitromethanes, haloaldehydes, halogenated furanones, haloamides, and nonhalogenated carbonyls. THMs and HAAs represented the two major classes of halogenated DBPs formed on a weight basis, followed by haloacetaldehydes. The researchers also detected 28 previously unidentified DBPs, including brominated and iodinated haloacids, a brominated ketone, and chlorinated and iodinated aldehydes. These results indicate that managing DBPs in wastewater effluents is challenging when using chlorine compounds.

3.5.1.2 Toxicity in Treated Effluent

Low concentrations of chlorine residuals can be toxic to fish and other aquatic life; toxic effects may include mortality and reduced productivity. In general, smaller fish such as fingerlings and minnows are much more sensitive to chlorine toxicity. Studies in California demonstrated that chlorination increased the effluent toxicity in the aquatic environment over non-chlorinated and dechlorinated effluents (Esvelt et al., 1971). Other studies have shown that chlorine residuals of 0.02 mg/L affected aquatic biota, and mortalities of salmon and rainbow trout have been recorded in receiving streams with total chlorine concentrations as low as 0.02

mg/L and 0.026 mg/L, respectively (Environment Canada, 2006). The effluent toxicity may be caused by organochloramines that are not instantaneously neutralized by sulfite (see Section 3.1.3.5) (Blatchley et al., 2005).

3.5.1.3 Chloride, Total Dissolved Solids (TDS), and Hardness

The chlorine disinfectants are all oxidizing agents that produce chloride ions; as a result, the use of these compounds increases the chloride and TDS concentrations of wastewater. This may be important for facilities that are concerned about the chloride and/or TDS levels of the final effluent. Chlorine gas causes smaller increases in chloride and TDS concentrations than sodium hypochlorite solutions, which contain sodium and chloride ions from the production process. A study by the Sanitation Districts of Los Angeles County found approximately 1.5 mg of chloride for every mg of NaOCl in 12.5% sodium hypochlorite solutions (unpublished). On-site hypochlorite generation systems cause the largest increases, because the brine solutions used for on-site chlorine generation are saturated with dissolved inorganic compounds such as sodium and chloride. Finally, the use of calcium hypochlorite as a chlorine source increases the TDS, hardness, and alkalinity of wastewater due to the presence of calcium, chloride, and hydroxide ions.

3.5.1.4 Ammonia

Typically, non-nitrified wastewater effluents contain sufficient ammonia so that chloramines are the disinfectant and, as a result, THMs formation is minimized. POTWs practicing nitrification may add ammonia to the effluent before disinfection to keep the chlorination process under control. Even with nitrification, diurnal ammonia fluctuations in the plant influent may result in episodic ammonia breakthroughs in the secondary effluent, chloramine formation, and difficulty in achieving a reliable chlorine residual. Adding 2-3 mg/L of ammonia-N often helps form a stable chloramine residual, resulting in a more reliable disinfection practice (Calmer et al., 1998, Ryder and Faller, 1993).

Many wastewater treatment plants that practice nitrogen removal also use chloramines for disinfection to minimize THM formation. During dechlorination, the chloramines are converted into chloride and ammonia (see Eqs. 3.3.3.4-2 to 3.3.3.4-4 as examples), thus raising ammonia levels in the effluent. This may be of concern for plants with strict non-ionized ammonia limits for the final effluent. In a study on chloramination of nitrified/denitrified effluents, the Sanitation Districts of Los Angeles County found that approximately 20% of the added ammonia was lost during the chlorination process, and 80% remained in the final effluent after dechlorination (Kuo et al., 2003).

3.5.1.5 pH

The addition of sodium or calcium hypochlorite slightly raises the pH and alkalinity of water due to the release of the hydroxyl ions (see Eqs. 3.1.3.2-3 and 3.1.3.2-4). The addition of chlorine gas slightly reduces the pH by producing hydronium ions (Eq. 3.1.3.2-1). In breakpoint chlorination, the conversion of ammonia-N converted to nitrogen gas consumes alkalinity and reduces pH, unless supplemental alkalinity is added.

3.5.2 Potential Risk

3.5.2.1 Regrowth

Regrowth (also known as after-growth) of microorganisms can occur in receiving water bodies. Regrowth is also a concern in distribution systems for reclaimed water, where biofilms can grow on pipe surfaces; as a result, chlorine residuals may be required in many reuse applications (Tchobanoglous et al., 2003). Ryu et al. (2005) investigated changes in the microbial quality of non-potable reclaimed water distribution systems after chlorination (in some cases in combination with UV treatment) in seven metropolitan areas of the southwestern United States. As the water passed through the distribution pipes, chlorine residuals decreased significantly; this was attributed to bacterial growth and consumption by organic matter in the effluent. Coliform numbers increased, but coliphage numbers did not. *Cryptosporidium* oocysts were detected in 67% of the water samples, and *Giardia* cysts were detected in 43% of the samples. However, no infectious *Cryptosporidium parvum* or enteroviruses were detected. Assimilable organic carbon levels decreased threefold as water passed through the distribution systems, which inversely correlated with bacterial regrowth.

3.5.2.2 Public Safety

The U.S. Department of Transportation classifies chlorine and sulfur dioxide gases as poisonous, and emphasizes prevention of the accidental release from production, transportation, storage, and handling. Sodium hypochlorite and liquid dechlorinating agents reduce the transportation risk and exposure hazard for both the surrounding community and plant operators; however, operators must still receive training on product specifications, receiving and offloading procedures, spill prevention control and countermeasure planning, and vendor materials (Naber, 2003).

3.5.3 Additional Benefits

3.5.3.1 Prevention of Biological Growth in Filters

Prechlorination (i.e., the addition of chlorine upstream of tertiary filters) helps to prevent growth of algae and bacteria on the surfaces of the filter media, and is often done to minimize operational problems associated with biological growth.

3.5.3.2 Incidental Removal of Contaminants

Oxidizing disinfectants such as chlorine may help to remove pharmaceuticals and endocrine disrupting compounds (EDCs), which are of growing concern for the wastewater treatment industry (Scruggs et al., 2005). Snyder et al. (2007) evaluated the EDC/pharmaceutical removal potential of conventional and advanced drinking and reuse water processes. They found that disinfection using free chlorine oxidized approximately half of the target compounds, including all phenolic steroid hormones. Chloramination was far less efficient for contaminant oxidation than free chlorine. Dodd and Huang (2004) found that the antibacterial agent sulfamethoxazole (SMX) reacted directly and quickly with free available chlorine, but more slowly with combined chlorine; other sulfonamide antibacterial agents are also likely to undergo substantial transformation during disinfection with free chlorine. Boyd et al. (2005) observed that the anti-inflammatory pharmaceutical naproxen also reacts readily with free chlorine; more research is needed to identify intermediate and chemical reaction end products and to understand the reaction kinetics of naproxen chlorination for a range of water and wastewater conditions.

3.5.3.3 Other Beneficial Uses in Wastewater Treatment Plants

Chlorine compounds can also be used to control odors and destroy hydrogen sulfide in off-gas (Butts, 2004), control excess growth of filamentous microorganisms causing bulking in activated sludge systems (Caravelli et al., 2003), minimize excess sludge production in the activated sludge process (Chen et al., 2001), and reduce algae and slimes in clarifiers.

3.6 Costs

The cost of chlorination systems depends heavily on the chlorine source (gaseous chlorine, sodium hypochlorite, or calcium hypochlorite). Selection of the chlorine source should be based on factors including dosage, labor costs, availability of chemicals, safety and transportation requirements, size of the plant, and life cycle cost. In the past, larger treatment plants tended to use chlorine gas systems, because the chlorine gas equipment was only a small fraction of the total plant capital cost. Medium-sized plants tended to use chlorine gas or sodium hypochlorite, and calcium hypochlorite was limited to infrequent use in very small plants. However, the balance has been shifting toward sodium hypochlorite in medium or larger plants, due to the safety and potential terrorist issues related to transport, storage, and use of gaseous chlorine.

It should be noted that the 2005-2007 cost climate for chlorine, concrete, steel, stainless steel, and energy has been very dynamic, primarily increasing both capital and O&M costs and making it difficult to use historically based cost curves. Hurricane Katrina temporarily shut down 41% of the chlorine manufacturing facilities (Reisch, 2005). Katrina also affected the railcar distribution network and much of the chlorine demand by such industries as the petrochemical, water, and wastewater treatment facilities. Chlorine costs have been very volatile so cost estimates should be considered very site-specific, and a site-specific project cost estimate is required for any potential project.

In general, it is unclear whether the literature citations used for this section included costs for annual risk assessment surveys and reports that are required by insurers for chlorine, sulfur dioxide and other toxic or hazardous gases or costs for associated liability and insurance. Also, some utilities have to pay more for liability risk insurance for employees that handle toxic and hazardous chemicals. The reduction or increase of these costs should be specifically included in cost analyses of various alternatives when performing a site-specific evaluation.

3.6.1 Capital Costs

Capital costs reflect the initial construction and associated engineering and administrative costs for implementing a disinfection system. Table 3-8 compares the estimated capital costs, taken from several literature sources, of chlorination/dechlorination systems using the various chlorine technologies. These costs have been normalized to a first quarter 2007 *Engineering News-Record* (ENR) 20 cities construction cost index (CCI) of 7870. For convenience, Table 3-8 also includes annual operation and maintenance (O&M) costs (discussed in Section 3.6.2), which have been normalized to a Consumer Price Index of 610.

Table 3-8. Comparison of Capital and O&M Costs for Disinfection with Chlorine Compounds.

Avg. Flow (MGD)	Capacity (MGD)	Chlorine Dosage (mg/L)	Capital ($)[a]	Annual O&M ($/yr)[b]	Source
Gas chlorine disinfection plus sulfur dioxide dechlorination					
1	2.25	5	1,702,000	69,000	Darby et al., 1995
1	2.25	10	1,903,000	83,000	Darby et al., 1995
1	2.25	20	1,918,000	108,000	Darby et al., 1995
5	12.5	10	1,658,000	58,000	Elefritz, 2000
10	20	5	4,739,000	223,000	Darby et al., 1995
10	20	10	5,400,000	319,000	Darby et al., 1995
10	25	10	2,126,000	110,000	Elefritz, 2000
10	20	20	5,965,000	534,000	Darby et al., 1995
15	37.5	10	2,609,000	174,000	Elefritz, 2000
100	175	5	21,661,000	929,000	Darby et al., 1995
			22,303,000		
100	175	10		1,017,000	Darby et al., 1995
100	175	15	22,358,000	1,846,000	Darby et al., 1995
Hypochlorite plus sulfur dioxide dechlorination					
5	12.5	10	949,000	165,000	Elefritz, 2000
10	25	10	1,455,000	331,000	Elefritz, 2000
15	37.5	10	1,898,000	496,000	Elefritz, 2000
240	-	-	22,358,000	1,846,000	Tobin et al., 2005[c]
On-site hypochlorite generation plus sulfur dioxide dechlorination					
5	12.5	10	1,771,000	171,000	Elefritz, 2000
10	22.5	10	2,372,000	343,000	Elefritz, 2000
15	37.5	10	3,606,000	530,000	Elefritz, 2000

[a]Capital costs adjusted to first quarter 2007 (ENR CCI = 7870).
[b]Annual O&M costs adjusted to first quarter 2007 (CPI = 610).
[c]Sodium bisulfite dechlorination.

These cost estimates appear to be planning level costs that have an accuracy of +50/-30%. With the exception of costs from Darby et al. (1995), the costs appear to be "bid" costs (i.e., the facilities costs equivalent to a contractor's bid); Darby et al. (1995) provide total project capital costs that include an additional 20% to account for engineering and administration during facility design and construction. When assessing a project at a specific facility, total project capital costs should also include contingencies reflecting the level of detail in the estimate.

As Table 3-8 illustrates, capital costs increase with plant size and design chlorine dosage. In general, for comparable plant sizes, hypochlorite systems are less expensive than gas chlorine systems meeting new fire code requirements, which in turn are slightly less expensive than on-site hypochlorite generation systems. Table 3-9 shows the capital cost breakdown of gas chlorine systems for chlorination, dechlorination, and Western Uniform Fire Code (UFC) requirements.

Table 3-9. Breakdown of Capital Costs of Gas Chlorine Disinfection Systems. [a]

Avg. Flow (MGD)	Chlorine Dosage (mg/L)	Chlorine Gas Storage Method	Chlorination ($)	Dechlorination ($)	Uniform Fire Code ($)	"Bid" Cost ($)	Total Capital ($)[e]
1.0[b]	5	1-Ton (gas)	619,000	438,000	361,000	1,418,000	1,702,000
1.0[b]	10	1-Ton (gas)	666,000	559,000	361,000	1,586,000	1,903,000
1.0[b]	20	1-Ton (gas)	672,000	565,000	361,000	1,596,000	1,918,000
10[c]	5	1-Ton (gas)	2,725,000	825,000	399,000	3,949,000	4,739,000
10[c]	10	1-Ton (Liquid)	3,098,000	1,003,000	399,000	4,500,000	5,400,000
10[c]	20	1-Ton (Liquid)	3,193,000	1,379,000	399,000	4,970,000	5,965,000
100[d]	5	Bulk	15,303,000	1,557,000	1,190,000	18,051,000	21,661,000
100[d]	10	RR Tank Car	15,495,000	1,900,000	1,190,000	18,586,000	22,303,000
100[d]	20	RR Tank Car	15,518,000	1,923,000	1,190,000	18,631,000	22,358,000

[a]Adapted from Darby et al., 1995 with permission from WERF. All costs adjusted to ENR CCI of 7870.
[b]Peak flow rate of 2.25 MGD.
[c]Peak flow rate of 20 MGD.
[d]Peak flow rate of 175 MGD.
[e]Total capital costs include an additional 20% to account for engineering and administration costs.

Table 3-10 compares the capital costs (from case studies) of the chlorine technology alternatives, for facilities that used gas chlorine and were expanding their disinfection systems. For these plants, expanding capacity with additional chlorine gas facilities has the lowest capital costs.

Table 3-10. Comparison of Capital Costs for Alternative Chlorine Disinfection System Upgrades.

Avg. Flow (MGD)	Capacity (MGD)	Chlorine Gas	NaOCl	On-Site Generation	Source
1.5 – 1.8	2.7	194,000	238,000	498,000	Hanzon et al., 2005
14.7	27.2	0	541,000	2,023,000	Hanzon et al., 2005
16.5	-	2,148,000	2,392,000	4,336,000	Gillette et al., 2003
25	53.5	553,000	553,000	3,099,000	Hanzon et al., 2005
30	61	976,000	2,317,000	2,329,000	Hanzon et al., 2005

3.6.2 Operation and Maintenance Costs

Operation and maintenance (O&M) costs reflect the costs of energy (e.g., electricity), chemicals, labor, maintenance (including replacement parts), and other activities required to operate and maintain chlorination systems. Annual O&M costs are generally estimated based on the annual average flow rate. Actual O&M costs are difficult to quantify accurately (Swift et al., 2000), due in part to the fact that most POTWs do not maintain detailed accounting records of O&M at the process level. Thus, engineering judgment is required when developing project-specific O&M costs, including accounting for unanticipated costs with a contingency.

Table 3-8 compares estimated O&M costs for gas chlorine, sodium hypochlorite, and on-site sodium hypochlorite generation systems that also include dechlorination facilities. As expected, annual O&M costs generally increase as a function of plant size and chlorine dosage.

Table 3-11 shows the breakdown of O&M costs for chlorine disinfection at selected treatment plants. Although absolute costs are site-dependent, chemical costs dominate the O&M costs for all chlorine technologies. For on-site hypochlorite generation, power and maintenance costs are also substantial. On an equivalent chemical basis, chlorine gas is the least expensive, sodium hypochlorite costs 2.5 times more, and calcium hypochlorite costs 8.75 times more (Environment Canada, 2006), which is why calcium hypochlorite is seldom used. As a result, chlorine gas disinfection systems generally have the lowest O&M costs. On-site generation systems generally have lower O&M costs than sodium hypochlorite systems.

Table 3-11. Breakdown of Annual O&M Costs for Selected Chlorine Disinfection Systems. [a]

Avg. Flow (MGD)	Chlorine Technology	Power ($)	Chemicals ($)	Maint./ Labor ($)	Contingency ($)	Other[b] ($)	Total ($)
Baiden et al., 2003							
7.5	Hypochlorite	1,000	249,000	19,000	0	0	269,000
Gillette et al., 2003							
16.5	Chlorine gas	4,000	584,000	71,000	87,000	211,000[c]	958,000
16.5	Hypochlorite	4,000	1,343,000	73,000	142,000	0	1,562,000
16.5	On-site Generation	212,000	604,000	116,000	93,000	0	1,025,000

[a]All costs adjusted to Consumer Price Index of 610 (First Quarter 2007)

[b]Other costs include risk management plans, OSHA compliance, and leak response costs, such as repairs in response to minor leaks, and emergency response and evacuation of employees and nearby residents due to a moderate chlorine gas leak.

[c]Costs include $45,000 for minor leak responses. For a moderate leak response, the leak response costs would increase to $2,212,000 (total "other" costs of $2,378,000) and the contingency costs would rise to $304,000. The total O&M cost would be $3,341,000.

3.6.3 Present Worth (Life-cycle) Costs

Design engineers frequently use net present worth analysis to compare the life cycle economics (combination of capital costs and present value of O&M costs over the project life) of alternative treatment technologies for process selection or to evaluate construction bids. Present worth analysis is project specific due to varying project life and the value of money (i.e., interest rate and inflation). Project life is typically 20 years, but may be as long as 50 years. Interest rates typically range from 3% to 10%.

Table 3-12 compares the net present worth of disinfection systems using various forms of chlorine. The results indicate that chlorine gas is generally the least expensive technology, followed by sodium hypochlorite; on-site hypochlorite generation is the most expensive technology. However, chlorine gas also has disadvantages (see Section 3.7) that may cause a facility to choose other forms of disinfection, even when the costs are equivalent or higher.

Table 3-12. Comparison of Net Present Worth of Disinfection with Chlorine Technologies. [a]

Avg. Flow (MGD)	Chlorine Technology	Net Capital ($)	Present Worth of O&M ($)	Total Net Present Worth ($)
Hanzon et al., 2005 (Case I: Effluent discharge - 20 year life, 4 percent interest rate)				
1.5 – 1.8	Chlorine Gas	194,000	326,000	520,000
1.5 – 1.8	NaOCl	238,000	462,000	700,000
1.5 – 1.8	On-Site Generation	498,000	448,000	946,000
Elefritz, 2000 (Effluent discharge - 10 year life, 5 percent interest rate)[b]				
10	Chlorine Gas	2,126,000	1,344,000	3,953,000
10	NaOCl	1,455,000	2,556,000	4,011,000
10	On-Site Generation	2,372,000	2,649,000	5,021,000
Hanzon et al., 2005 (Case III: Effluent discharge - 20 year life, 4 percent interest rate)				
14.7	Chlorine Gas	0	1,305,000	1,305,000
14.7	NaOCl	541,000	2,229,000	2,770,000
14.7	On-Site Generation	2,023,000	1,726,000	3,749,000
Gillette et al., 2003 (Wastewater reuse – 20 year life; interest rate not specified)				
16.5	Chlorine Gas	2,148,000	20,357,000	22,506,000[c]
16.5	NaOCl	2,392,000	33,169,000	35,561,000
16.5	On-Site Generation	4,336,000	21,743,000	26,079,000
Hanzon et al., 2005 (Case V: Wastewater reuse – 50 year life, 4 percent interest rate)				
22	NaOCl	976,000	16,369,000	17,345,000
22	On-Site Generation	2,317,000	10,118,000	12,435,000
Hanzon et al., 2005 (Case IV: Activated sludge – 20 year life, 4 percent interest rate)				
25	Chlorine Gas	553,000	3,112,000	3,665,000
25	NaOCl	553,000	5,069,000	5,622,000
25	On-Site Generation	3,099,000	3,778,000	6,877,000
Hanzon et al., 2005 (Case IV: Trickling filter/solids contact – 20 year life, 4 percent interest rate)				
25	Chlorine Gas	602,000	4,308,000	4,910,000
25	NaOCl	602,000	7,475,000	8,077,000
25	On-Site Generation	3,369,000	5,423,000	8,792,000

[a]All costs adjusted to first quarter 2007 (ENR CCI = 7870, CPI = 610)

[b]Costs were estimated for 5 MGD, 10 MGD, and 15 MGD facilities. The10 MGD data are provided here, but the trend in relative costs across the chlorine technologies is the same for all three plant sizes.

[c]Includes the cost of a minor leak response. A moderate leak response would increase the life cycle cost to $24,823,000.

3.7 Summary

The use of chlorine compounds is one of the oldest technologies for wastewater disinfection, and is currently the most prevalent at POTWs in the U.S. Most of those facilities use chloramination, but with an increasing number of wastewater treatment plants beginning to practice nitrogen removal, free chlorine disinfection (after the breakpoint) has become a potentially viable alternative to conventional chloramination. This section presents advantages and disadvantages of free chlorine (after the breakpoint) and chloramination, as well as some of the gaps in knowledge that remain.

3.7.1 Advantages

Chlorination is a well-established technology whose inactivation mechanisms are well known. Of the disinfection alternatives, it is often the easiest to apply and least expensive method, regardless of system size. Chlorine chemicals are relatively inexpensive and available in gas, liquid, and solid form. Chlorination can be applied upstream of tertiary filters, thereby reducing biofouling in the filters. It also provides a residual, which is useful for operational monitoring and control, as well as preventing growth in distribution systems during reuse. Finally, it can be easily adjusted for fluctuations in flow rate and water quality.

In addition, each of the chlorination technologies (chloramines and free chlorine) has distinct advantages. Free chlorine is an effective biocide and virucide, and may remove some emerging compounds of concern. It also forms little or no NDMA. Chloramines are more stable and longer lasting than free chlorine, and have much less potential to form THMs and HAAs than free chlorine.

3.7.2 Disadvantages

Chlorine gas is a hazardous and corrosive gas, so special leak containment and scrubber facilities may be required. Some of these disadvantages can be avoided if hypochlorite is used as the chlorine source; however, higher concentrations of sodium hypochlorite are unstable and may produce chlorate as a byproduct. Chlorination is also less effective at high pH and may result in residuals or disinfection byproducts that are harmful to human health or the environment. It increases the total dissolved solids and chloride levels in the effluent. Finally, effluents frequently need to be dechlorinated before final discharge.

Each of the chlorination modes (chloramination and free chlorine after the breakpoint) also has distinct disadvantages. For breakpoint chlorination to a free chlorine residual, high (> 3 mg/L) ammonia-N concentrations in the wastewater would require excessive chlorine doses and is impractical. In addition, free chlorine can form halogenated disinfection products such as THMs and HAAs, as well as oxygenated byproducts that can enhance biological regrowth in water reuse distribution lines if a chlorine residual is not maintained. Chloramines have the disadvantages of being weaker than other disinfectants, and having the potential to generate NDMA. Ammonia addition may be necessary in facilities with nitrification/denitrification, and the added ammonia in the final effluent may enhance biofilm growth in the transmission lines for water reuse.

3.7.3 Gaps in Knowledge

Data gaps identified from the literature review include the following:

- Although many disinfection byproducts have been identified, the majority have not (especially those formed from new or newly discovered compounds).
- Chlorine and chloramines may be able to destroy some emerging chemicals of concern (e.g., naproxen), but the extent and reaction kinetics are largely unknown.
- There are few data on the performance and economics of on-site generation of hypochlorite.
- Regulatory CT requirements should be re-evaluated to account for the chlorine species used for disinfection and the quality of the effluent to be disinfected.
- For free chlorine, methods are needed to reliably measure and maintain residuals in wastewater. In addition, the required operating parameters for wastewater

disinfection are largely unknown and need to be assessed over a range of water qualities and operating conditions. Finally, the toxicity of effluents after treatment with free chlorine needs to be investigated.

- Data on dechlorination kinetics are needed, to ensure that proper contact time is provided.

CHAPTER 4.0

UV TECHNOLOGIES

4.1 Background

Ultraviolet light is the region of the electromagnetic spectrum that lies between the visible light and X-rays (Figure 4-1). The wavelength spectrum of UV radiation ranges from 100 to 400 nanometers (nm), and can be further divided into four regions: vacuum UV (100 to 200 nm), UV-C (200 to 280 nm), UV-B (280 to 315 nm), and UV-A (315 to 400 nm).

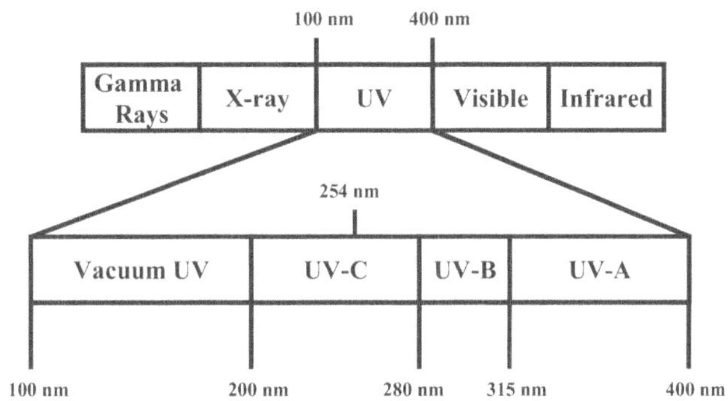

Figure 4-1. UV Light in Electromagnetic Spectrum. (Adapted from U.S. EPA, 2006)

4.1.1 Historical Use

Ultraviolet (UV) light was discovered approximately 150 years ago, and the first commercial UV lamp was made in the early 1900s. Technological advances allowed its first application to water disinfection in 1907 in France, and the first UV system for municipal wastewater was installed in the mid-1970s (Whitby, 2002). In the mid 1980s, the U.S. EPA named UV disinfection as a "best available technology" for wastewater disinfection (Jeyanayagam and Cotton, 2002). Since then, the use of UV disinfection has grown within the wastewater industry, as an increasing number of communities have adopted UV irradiation in place of chlorination. In 1984, only a handful of wastewater treatment plants in the United States used UV systems, and very few of these systems had a design capacity greater than 5 million gallons per day (MGD). At the end of 1999, the total number of installed UV systems for wastewater applications exceeded 500, and the average design capacity of these systems exceeded 20 MGD (Hunter, 2000). By 2006, the number of wastewater treatment plants using UV for disinfection increased to greater than 900. The global market for UV disinfection was valued at around USD $250 million in 2004 and is expected to grow, with estimates of USD $500 to 700 million in 2010 (Stedman, 2004).

A study by Steinberg et al. (2000) indicated that wastewater treatment plants are adopting UV due to the lack of disinfection byproducts, increased operator safety, and concern about meeting regulatory requirements. According to Thomas et al. (2002), concerns over accidental releases of chlorine to the environment, the need to dechlorinate prior to discharge, as well as the introduction and refinement of medium pressure UV lamps have accelerated this transition. Additionally, communities without disinfection are considering UV as the only viable means to comply with tighter microbiological regulations. Based on operational experiences, Amirhor (1999) stated that UV disinfection offers the following advantages: adequate disinfection and compliance with NPDES permit requirements, reliability, safety, elimination of toxic residuals and byproducts, flexibility, simple operation, and no need to comply with specific and costly regulatory requirements.

4.1.2 How UV Disinfects

UV inactivation of microorganisms is considered to be a physical or biophysical process. The germicidal wavelength range lies between 220 to 320 nm, in the regions of UV-B and UV-C. Electromagnetic radiation in this range alters cellular proteins and nucleic acids (i.e., DNA and RNA) through dimerization of the thymine nucleic acids on DNA molecules. The extent of DNA absorbance depends on wavelength, and the peak absorbance occurs around 260-265 nm.

Damage to nucleic acids prevents microorganisms from replicating, and a microorganism that can not replicate will not infect a host. However, it does not prevent the microorganism from performing metabolism and other cell functions. Some microorganisms can repair the damage done by UV light to a limited degree and regain infectivity in some cases (see section 4.5.4 for more details).

4.1.3 The UV Process

In the general arrangement for UV disinfection, wastewater flows through a confined chamber/reactor containing arrays of UV lamps, and the UV radiation from the lamps inactivates microorganisms in the wastewater. A typical UV system consists of a power supply, an electrical system, reactors, lamps, quartz sleeves, a mechanical system to hold the lamps, and a control system (Hunter and Long, 2003). A sensoring system for monitoring UV intensity and a cleaning system for periodical removal of fouling on the quartz sleeves may also be included.

4.1.3.1 Types of UV Systems

There are two basic types of UV systems. In the first type of system, the lamps do not contact the wastewater directly; instead, they are suspended above the water or surround conduits carrying the water (U.S. EPA, 1986). The collimated beam apparatus that is used in bench-scale testing is an example of this type of system, but non-contact systems are otherwise seldom used. In the second type of UV system, the UV lamps are submerged in the wastewater to be treated. The submerged lamps are encased in quartz sleeves to isolate the lamp from direct contact with wastewater and to minimize the cooling effects of the wastewater on the lamps. Submerged systems currently dominate the full-scale applications for municipal wastewater disinfection.

The submerged UV systems can be further classified by the hydraulic design of the UV reactor (closed vessels or open channels), the orientation of the lamps (horizontal or vertical), and the arrangement of the lamps relative to the direction of flow (parallel vs. perpendicular). In one configuration of closed-vessel UV reactor, water enters the unit through an inlet pipe perpendicular to the lamps, redirects, flows parallel to the lamps along the length of the reactor, and finally exits the reactor through an outlet pipe. In other configurations of sealed reactors, the lamps are arranged in a symmetrical array perpendicular to the direction of flow. Baffles at the influent and effluent distribute the flow, and lamps are staggered to encourage turbulence, but the system is arranged with a path that is long enough to promote a plug flow condition (U.S. EPA, 1986). Depending on plant flow rates and redundancy requirements, several reactors in series and/or parallel may be required.

For the open-channel UV systems, the influent flow rate is usually divided equally among a number of open channels in parallel. Each channel typically contains two or more banks in series. A bank is comprised of a specific number of lamp modules, and each module typically contains 2, 4, 8, 12, or 16 UV lamps. The center-to-center spacing between the lamps is usually 3 to 5 inches, depending on the water quality, type and size of the lamps, dose delivery requirement, and headloss considerations. An effluent flow control device such as a weighted flap gate, an extended sharp-crested weir, or an automatic level controller, is often used to regulate the submergence of the uppermost UV lamps in each bank. Modules of lamps are removed occasionally from the channel and cleaned to eliminate deposits (fouling) on the quartz sleeves (Tchobanoglous et al., 2003). Some UV systems offer an automatic wiping and washing system to clean the lamp sleeves.

The pros and cons of vertical versus horizontal lamp arrays in open channel systems have been a topic of debate among suppliers. Both have distinct advantages and any design team should evaluate the potential for each. In some cases, the hydraulics of the plant or the configuration of existing chlorine contact tanks being retrofitted can affect the decision. With vertical systems, individual lamps can be easily replaced while leaving the lamp module in the flow. The water level at the top of the lamps can vary up to a few inches in the vertical lamp system, while the water level in a horizontal system must be kept relatively constant and close to the top of the lamps to avoid short-circuiting of flow; this can lead to ineffective disinfection. However, in the vertical system, a portion of the UV lamps may be uncovered and exposed to the atmosphere at times, and floating materials can be deposited and collected on the upper surface of the quartz sleeve (Tone et al., 1998). The vertical system also requires a deeper channel.

With respect to system performance, Goodman and Mills (2002) compared an open-channel, gravity-flow, medium pressure (MP) system and a closed-vessel MP system and found the two configurations to be equally effective, despite significant differences in design. Ho et al. (1998) evaluated several lamp orientations in parallel pilot tests. Comparing two vertical lamp units and one horizontal lamp unit, they found that the horizontal unit gave a more consistent performance than the vertical pilot units.

4.1.3.2 Types of UV Lamps

UV lamps and fluorescent lamps are similar in design, and both contain mercury vapor. UV light is produced as a result of electrical current flow through the mercury vapor between the electrodes of the lamp. Fluorescent lamps have soft glass with an inside coating of phosphor which converts UV to visible light; germicidal lamps are constructed of UV transmitting glass material (e.g., quartz). UV lamps for wastewater applications can generally be categorized into three groups: low-pressure low output (LPLO), low-pressure high output (LPHO), and medium pressure (MP).

Each lamp type outputs radiation at different wavelengths (Figure 4-2), uses different operating parameters (Table 4-1), and presents different advantages and disadvantages. The following sections provide more details on each lamp type, a brief comparison of the currently available lamp types, and a discussion of some emerging UV lamp technologies. Because technologies are continuously evolving and improving, current manufacturers' literature should be consulted with regard to lamp specifications, especially for the LPHO and MP lamps.

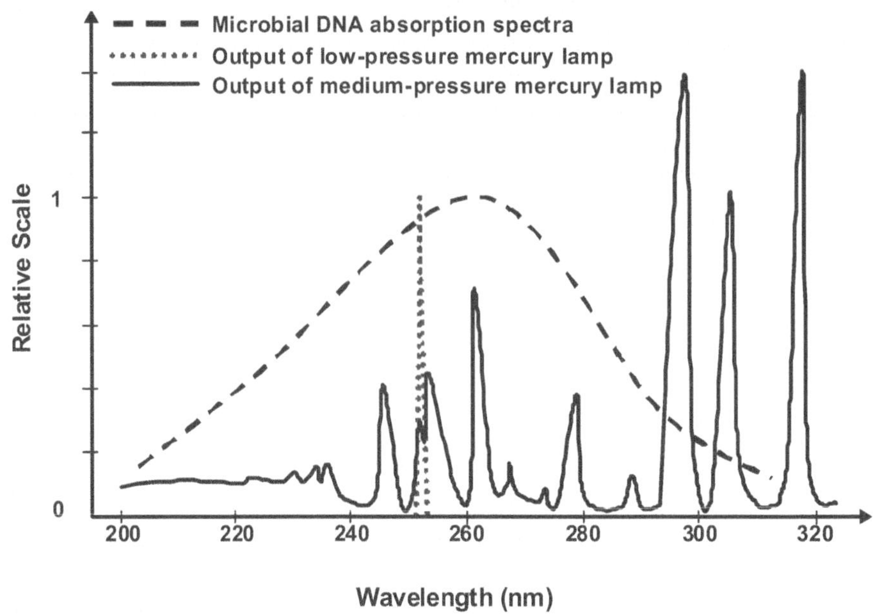

Figure 4-2. Output Spectra of Low-Pressure and Medium-Pressure Lamps and Microbial DNA Absorption Spectra. Not to Scale (From Kuo et al., 2003 with permission from ASCE)

Table 4-1. Comparison of Available UV Lamp Technologies. (Adapted from Jeyanayagam and Cotton, 2002; Tchobanoglous et al., 2003)

	LPLO	LPHO	MP
Spectrum	Monochromatic	Monochromatic	Polychromatic
Input power (W/lamp)	70-90	200-250	1,300-5,000
Germicidal UV (as % input power)	35-40	30-35	15-20
Temperature (°C)	40-60	100-200	600-900
Lamp life (h)	8,000-13,000	8,000-10,000	3,000-5,000
No. of lamps relative to a MP system	10-15	4-8	1
Footprint	Large	Medium	Small

LPLO Lamps LPLO UV lamps operate at an internal mercury pressure of 0.007 mm Hg. Approximately 85% of the total lamp output is monochromatic at 254 nm, very close to the 260-265 nm wavelength range which is most effective for microbial inactivation (Hunter et al., 1998; Jeyanayagam and Cotton, 2002; Tchobanoglous et al., 2003). Although the number of lamps required for a wastewater disinfection project depends on many factors, a rule of thumb for the LPLO system is to use 40 to 60 lamps per MGD flow of wastewater (Hunter et al., 1998). The output of the UV lamp decreases with time due to a reduction in the electron pool within the lamp, deterioration of the electrodes, and aging of the quartz sleeve. The useful life of an LPLO UV lamp varies from 8,000 to 13,000 hours, depending on the number of daily on-off cycles (frequent cycles may shorten the lamp's life). The useful life of the quartz sleeve is about 4 to 8 years (Jeyanayagam and Cotton, 2002; Tchobanoglous et al., 2003).

LPHO Lamps LPHO UV lamps are similar to LPLO lamps with the exception that a mercury-indium amalgam is used in place of mercury, and they operate at an internal mercury pressure of

0.0013 to 0.012 mm Hg. These lamps operate at a higher current discharge, and the use of the mercury amalgam allows greater UV output, typically several times that of an LPLO lamp. The amalgam also maintains a constant level of mercury atoms that can provide longer lamp life and greater stability over a broader temperature range than a MP lamp (Tchobanoglous et al., 2003). The higher output of these lamps reduces the number of lamps required per MGD treatment capacity. Similar to LPLO lamps, the LPHO lamp emission is nearly monochromatic at 254 nm, but its conversion of applied power to UV light is slightly lower than LPLO, at 30 to 35%.

MP Lamps MP UV lamps have been used for more than 40 years in Europe to disinfect drinking water and industrial wastewater. These lamps operate at a pressure of 300 to 30,000 mm-Hg, and they produce a broad spectrum (polychromatic) light. Approximately 27 to 44% of the total energy output is in the germicidal range, but only 7 to 15% of the output is near 254 nm. Consequently, the efficiency of energy conversion to germicidal light for the MP lamps is much lower, at 15 to 20%. However, each MP lamp still generates more UV output than an LPLO or LPHO lamp. The major advantage of the MP system is lower installation cost because fewer lamps are required, as can be seen in Table 4-1, while its major disadvantage is higher operation cost due to lower energy efficiency. Its use is mainly for higher wastewater flows, stormwater overflows, or at space-limited sites (Hunter et al., 1998; Tchobanoglous et al., 2003). The useful life of MP lamps is typically shorter, at 3,000 to 5,000 hours.

Comparison of Current Lamp Types Table 4-2 compares the advantages and disadvantages of low and medium pressure lamps. The main disadvantage of the low pressure lamps is the need for a large number of lamps and correspondingly large footprint; however, the development of the LPHO lamps mitigates this issue to some degree. According to Hunter and Cummings (2002), use of LPHO lamps would reduce the number of lamps per MGD, from 40 to 50 for LPLO systems to 2 to 12 for LPHO systems. The energy demand of LPHO systems is similar to that of LPLO systems, ranging from 1.8 to 3.8 kW input power per MGD for disinfecting secondary effluent to 200 fecal coliform per 100 mL, which is significantly lower than the 6.8 to 14 kW input power per MGD required by medium-pressure UV systems. Some LPHO systems also include automatic sleeve cleaning systems, which makes them more attractive to treatment plant personnel. The study by Hunter and Cummings (2002) indicated that LPHO systems are gaining popularity among utilities that decide to upgrade their disinfection systems.

Table 4-2. Comparison of Low-Pressure and Medium-Pressure Mercury Lamps. (Adapted from U.S. EPA, 2006)

	Low Pressure	Medium Pressure
Comparative Advantages	• High germicidal efficiency; nearly all UV output at 254 nm • Smaller power draw per lamp (less reduction in total dose if one lamp fails) • Longer lamp life	• Higher power output • Fewer lamps for a given application • Smaller reactors/footprint
Comparative Disadvantages	• More lamps needed for a given application • Larger footprint	• Lower electrical to germicidal UV conversion efficiency • Higher operating lamp temperature may accelerate fouling • Shorter lamp life

Emerging UV Lamp Technologies Several emerging UV irradiation technologies are being developed to destroy micropollutants in contaminated groundwater by photolysis or advanced oxidation (UV plus hydrogen peroxide or ozone to form OH• radicals). Among these technologies are pulsed UV, broad-band pulsed xenon, and narrow-band excimers (Hunter et al., 1998). These technologies have shown promise for potential adaptation into conventional wastewater disinfection, but none are currently being used for full-scale wastewater disinfection.

In a pulsed UV system, alternating current is converted to direct current and stored in a capacitor. This energy is then released through a high-speed switch to form a pulse of intense radiation. Each time the switch closes, the lamp is fired and produces ionized gas (plasma) at a temperature of approximately 10,000 K. At this temperature, the plasma emits an intensive broadband (polychromatic) spectrum, including UV (Hunter et al., 1998; Tchobanoglous et al., 2003). The radiation produced through this technology is estimated to be 20,000 times more intense than sunlight at sea level (Asano et al., 2007). Although radiation of germicidal wavelengths constitutes only a small portion of total emissions, the high intensity output still makes the pulsed xenon lamps useful for wastewater disinfection. Marshall (2002) indicated that pulsed UV treatment is effective against fecal coliform, total coliform, and fecal *Streptococcus*, while Weber and Scheible (2003) showed that it can effectively inactivate the coliphage MS2.

Narrow-band excimer lamps produce near-monochromatic radiation like that from low-pressure UV lamps. These lamps use a corona discharge to form excited dimers (dimers are formed by two monomers, e.g., two gas molecules joined together). Gases used for this purpose include xenon, xenon chloride, krypton, and krypton chloride. Once in the excited state, the dimers are termed excimers. The excimers are unstable and will collapse into the normal state by releasing photons. Commercial excimer lamps are available that emit essentially monochromatic light at wavelengths of 172, 222, or 308 nm (Hunter et al., 1998; Tchobanoglous et al., 2003).

The mercury-argon electrode-less UV lamps use microwave energy generated by a magnetron to irradiate the mercury-argon filled lamps. This radiation excites the gas and produces UV light. Because this technology does not use electrodes, the manufacturer claims a longer lamp life.

4.1.3.3 Other Components

Lamp Power Supply and Ballast The power supply and ballast provide the necessary power to energize and operate the UV lamps. Power supplies and ballasts are available in many different configurations and are usually specific to a unique lamp type and application. UV systems may use electronic ballasts, magnetic ballasts, or transformers (U.S. EPA, 2006).

Lamp Quartz Sleeves Each submerged UV lamp is housed within a quartz sleeve to keep the lamp at optimal operating temperature and to protect the lamp from direct contact with water and from breaking. UV lamps are usually centered radially within quartz sleeves using spacers. The sleeve diameter is typically 2.5 centimeters (cm) for LPLO and LPHO lamps, and 5 to 10 cm for MP lamps. Quartz sleeves are typically 2 to 3 millimeters (mm) thick and adsorb some UV light. Quartz sleeves can fracture and foul; their UV transmittance will also decrease as they age (U.S. EPA, 2006).

Cleaning System The quartz sleeves must be periodically cleaned to counteract the effects of fouling. The cleaning method depends on the type of UV system, but in general, manufacturers employ one of two general approaches to clean the lamp sleeves: off-line chemical cleaning (OCC) and on-line mechanical cleaning (OMC).

LPLO and some LPHO systems typically use OCC systems. For OCC in open channel systems, groups of lamps (modules or bundles) are removed from a channel by an overhead crane and transferred to a cleaning tank. Spare modules may be inserted into the channel while cleaning operations are underway. For closed-vessel systems, the UV system is shut down, drained, and soaked with a cleaning solution such as citric acid, phosphoric acid, or a proprietary solution. The reactor is then rinsed and returned to operation.

OMC systems are typically used in MP systems and some LPHO systems. An OMC system is equipped with wipers as a built-in component of the UV reactor. There are two types of wipers used in OMC systems: mechanical and physical-chemical. Mechanical wipers consist of Teflon rings or stainless-steel brush collars that move along the quartz sleeves. Physical-chemical wipers have a collar filled with cleaning solution. With OMC systems, the reactor can remain on-line while the sleeves are cleaned. (U.S. EPA, 2006)

Monitoring UV Transmittance (UVT) The UVT of the wastewater to be disinfected is one of the most important parameters in determining the efficacy of UV disinfection. Therefore, continuous monitoring of UVT is critical to ensure success of a UV disinfection application and operation.

UV Intensity Sensor Analogous to the on-line chlorine residual monitors used to continuously monitor the performance of chlorination systems, UV intensity sensors may be used to indicate dose delivery by providing information related to UV intensity at different points in the reactor. The reading at a specific location responds to changes in lamp output due to lamp power setting, lamp aging, quartz sleeve aging and fouling, as well as changes in UVT of the water being treated (U.S. EPA, 2006).

Temperature Sensor UV systems are often equipped with temperature sensors to monitor the water temperature within the reactors. If the temperature is above the typical operating range, the reactor may need to be shut off to minimize the potential for lamp overheating (U.S. EPA, 2006).

4.1.4 UV Dose

UV dose is a measurement of the UV energy that a surface area receives. In order to inactivate a pathogen, the UV dose must exceed the threshold value for that specific organism; dose-response relationships vary by microorganism. Inactivation is directly linked to the UV intensity acting on the microorganism from all directions and the duration of the UV exposure. In theory, UV dose can be defined analogous to the "CT" concept used in chlorine disinfection, as the product of the UV intensity and the UV exposure time:

$$\text{UV Dose} = I \times t \qquad (4.1.4\text{-}1)$$

where: I = UV intensity, in milliwatts per square centimeter (mW/cm^2)
t = exposure time, in seconds (s)
UV Dose, in $mW\text{-}s/cm^2$ or milliJoules per square centimeter (mJ/cm^2)

The actual UV intensity and exposure time are complex functions of the UV system, operating parameters, and water quality. For example, in order to reach pathogens, the UV radiation must travel through the quartz sleeve, wastewater, and particles (if the microbes are embedded in particles). Consequently, the UV intensity actually reaching the target organisms is lower than that at the surface of the UV lamp, and varies throughout the reactor.

The exposure time is ideally the average hydraulic retention time within the UV reactor (or the reactor volume divided by the flow rate). However, actual exposure times for each target microorganism are a function of reactor volume, flow rate, mixing conditions within the reactor, and extent of short-circuiting. In an ideal plug flow reactor, contact time is uniform for all microorganisms. On the other hand, contact times in an ideal continuous-flow stirred tank reactor range from infinitesimal to infinity. The flow characteristics and residence time distributions of real reactors lie somewhere between these two extremes.

Overall, the UV dose depends on the following factors:

- Lamp energy output – Different types of UV lamps provide different wattages per unit lamp surface area. The higher the energy output, the larger the UV intensity at the lamp surface.
- Wavelength spectra of the UV lamp – Low-pressure mercury lamps emit monochromatic wave while MP lamps generate polychromatic waves. As discussed, the germicidal power varies with wavelength.
- Power setting of the lamp – Newer electronic ballasts allow the lamp output to be modulated. The UV intensity will be lower if the lamp is operated at less-than-maximum power.
- Lamp age – After the initial burn-in period, the lamp output and corresponding UV intensity typically decrease gradually toward the end of the lamp's useful service life.
- Quality of the quartz sleeve – The sleeve is made of quartz to allow UV light to pass through with minimal absorption. The extent of absorption by the quartz sleeve is a function of its age and quality.
- Fouling on the quartz sleeve – Organic or inorganic fouling/scaling can significantly reduce the UV energy entering the wastewater.
- UV transmittance of the wastewater – As UV rays travel through wastewater, their intensity is attenuated continuously because the substances in wastewater absorb some of the UV light. These substances include dissolved organic and inorganic compounds. The UVT is one of the critical water quality parameters determining the UV intensity that will act on the microorganisms.
- Type and size of suspended solids – In addition to dissolved solids, suspended solids in wastewater can absorb and scatter UV light. The concentration of suspended particles and their size distribution will affect the UV intensity that microorganisms receive.
- Particle shielding – Particles can also harbor microbes and protect them from UV radiation. The UV intensity received by a free-flowing microorganism is expected to be greater than that received by a microorganism embedded in a particle at the same location.
- Distance from the center of the lamp – Even without absorption loss, UV intensity decreases with increasing distance from the lamp. This is because the UV energy is distributed over a larger zone as the distance from the lamp increases.
- Hydraulics - Dead space reduces the effective reactor volume and shortens the average hydraulic retention time. Also, without good radial mixing, a microorganism may pass through the UV reactor between lamps (i.e., at the

furthest distance from the lamps) and receive a smaller UV dose than the average value.

Ideally, for the most effective and energy/cost-efficient disinfection, each microorganism would receive a UV dose just above the threshold value for inactivation. However, this is not practically achievable, due to the factors listed above and the variability of the UV dose-response among organisms. For a given average UV dose and a broad dose distribution, some organisms may receive doses much higher than the average value, while others may receive insufficient doses for disinfection. Where possible (e.g., open channel systems), UV systems should optimize reactor hydraulics through inlet and outlet design and use of baffles, and proper upstream treatment should be employed to reduce the concentration of large suspended solids in the wastewater before it enters the UV reactor.

4.2 Disinfection Efficacy

In determining whether UV disinfection is suitable for a given facility, disinfection efficacy is the most important factor. UV treatment is only appropriate if it can inactivate the pathogens of interest in the effluent water at the site. This section discusses water quality parameters that affect UV disinfection, and the efficacy of UV against various pathogens.

4.2.1 Effects of Water Quality on Disinfection

Water quality affects the performance of a UV system by altering the UV intensity within the reactor, and consequently the UV dose received by the organisms within the wastewater. The most important water quality parameters are the UV transmittance (UVT) of the water and the size and number of particles. In addition, dissolved solids may foul the quartz sleeves surrounding the lamps and decrease the effective UV output. As discussed in Section 4.2.1.4, alteration of upstream processes or pretreatment may be necessary to improve the water quality to make UV disinfection effective.

4.2.1.1 UV Transmittance

Substances such as dissolved organics can absorb light, decrease UV intensity, and reduce disinfection efficacy, particularly if the absorption is in the germicidal wavelength range. The extent of UV absorption strongly depends on the types and concentrations of the suspended solids and dissolved matter present. Absorption is characterized by the parameters UV absorbance (UVA) and UV transmittance (UVT) (Eq. 4.2.1.1-1 and 4.2.1.1-2); these two quantities are related by Eq. 4.2.1.1-3. In practice, analytical instruments generally provide UVA or UVT values directly, so light intensities (I_0 and I) are not evaluated independently.

$$UVT_\lambda(\%) = \frac{I}{I_0} \text{ x } 100 \qquad (4.2.1.1\text{-}1)$$

$$UVA_\lambda = \log_{10}\frac{I_0}{I} \qquad (4.2.1.1\text{-}2)$$

$$UVT_\lambda = 10^{-UVA_\lambda} \text{ x } 100 \qquad (4.2.1.1\text{-}3)$$

where: I_0 = initial light intensity, in W/cm^2
I = light intensity after passing through the sample, in W/cm^2

UVA_λ = UV absorbance at wavelength λ, in absorbance units (a.u.)
UVT_λ = UV transmittance at wavelength λ, in %

According to Moreland and Ono (2001), treated wastewaters to be disinfected with UV radiation can be divided into three water quality groups: high UVT_{254} of 90 to 100%, medium UVT_{254} of 40 to 80%, and low UVT_{254} of 0 to 40 %. High UVT_{254} waters have generally received treatment with micro-, ultra-, or nano- filters in combination with reverse osmosis; UV disinfection is relatively easy to apply. Medium UVT_{254} waters are usually secondary effluents and media-filtered secondary effluents, which have well-established UV application histories. Low UVT_{254} waters are primary effluents or any other lower water quality effluents; UV disinfection performance with these waters is difficult to predict.

Waters with low UVT absorb more UV energy and therefore require higher energy input for a specific dose or level of inactivation; consequently, the capital and O&M costs are higher. The lamp spacing may also need to be reduced, which would incur higher headloss across the UV system. These constraints can be mitigated through increased removal of organics in the secondary process and/or tertiary filtration, which increases the UVT of the water.

It is important to assess water quality adequately, so that a successful UV design is provided. Thomas et al. (2002) presented a case study of one community that simply based its design for a major UV construction project on an assumed wastewater UVT. After installation, it was discovered that the apparently clear effluent had a UVT well below 40%. Adherence to recommended design protocols, such as development of a UVT history before design (Melin, 2003), can prevent this type of unpleasant surprise.

4.2.1.2 Particles

Recently, concerns have emerged over the association of microorganisms with particles in wastewater and the resulting effects on UV disinfection performance. Fundamental information on these processes is currently limited, although particles are known to decrease disinfection efficacy by shielding pathogens from UV radiation. In addition, Parker and Darby (1995) found that particle association may cause underestimation of coliform concentrations in disinfected secondary effluents (i.e., overestimation of disinfection efficacies), when the standard multiple tube fermentation test is used for enumeration.

In general, dispersed microbes are easier to disinfect than particle-associated bacteria (Parker and Darby, 1995; Blatchley et al., 2001). This is supported by a linear correlation between the number of particles with associated coliform bacteria, and the residual coliform concentrations after UV treatment; this correlation did not hold when the upstream processes used chemical coagulants (Emerick et al., 1999; Nelson, 2000). Lagoon processes resulted in fewer particles with embedded coliform than other processes such as activated sludge or tricking filters, and these pond effluents were easily disinfected, despite high suspended solids concentrations overall (Emerick et al., 1999; Nelson, 2000). Even in the absence of other particles, bacteria may form UV-resistant aggregates during treatment; a small fraction of these aggregates appears to be essentially unaffected by UV radiation (Blatchley et al., 2001). The dose-response behavior of this highly resistant fraction was variable and difficult to describe quantitatively.

Loge et al. (1999) showed that wastewater solids from trickling filters and activated sludge absorb UV light, rather than scattering it. Radiation could only penetrate highly porous particles and could not pass through solid material; as a result, UV light could not access regions within some particles. Blume and Neis (2004) observed that reducing the concentration of particles >50 microns through sonication improved UV disinfection performance. Smaller particles may also affect efficacy, as Jolis et al. (1996 and 2001) observed decreased kinetic inactivation rates with suspended particles ≥ 7 microns in a tertiary effluent.

4.2.1.3 Fouling Potential

A significant challenge with UV disinfection of wastewater is the accumulation of deposits at the quartz sleeve-water interfaces. Lin et al. (1999a) found that foulants were mostly amorphous inorganics whose composition was site-specific. Species contributing to the formation of the deposits include cations such as iron, aluminum, calcium, magnesium, and sodium, and anions such as carbonate, sulfate, hydroxide, chloride, and phosphate (Lin et al., 1999a; Topnik et al., 2000; Peng et al., 2005). Accumulation of inorganics increased in the longitudinal direction of lamps and fouling kinetics were zero-order, which supported the hypothesis of thermally-induced precipitation (Lin et al., 1999b). Residuals of iron and aluminum, which may be added during upstream treatment processes for enhanced settling and/or phosphorous removal, may significantly accelerate the fouling rate (Lin et al., 1999b). In some tests with LP and MP lamps, UV disinfection was unable to reach the target level of 2,500 coliform/100 mL in waters with high levels of iron (often >3 mg/L) and suspended solids (30 mg/L) and low UVT254 (32%); significant fouling occurred after a few hours (Gehr and Wright, 1998). In the absence of UV radiation and heat from UV lamps, the extent of inorganic fouling was substantially diminished, but organic materials caused rapid fouling; these organics were held only loosely to the quartz surface (Lin et al., 1999b).

Automated chemical and/or mechanical cleaning can help remove most fouling materials. However, some foulants remain permanently on the quartz sleeves after typical cleaning operations. Research indicates that the wipers used in long-term automated cleaning systems damage sleeve surfaces and create scratches or holes with larger surface areas. These defects trap foulants, which attach tightly to the surfaces of the scratches or to each other. For the specific cases studied, chemical plus mechanical cleaning was more effective in removing foulants and avoiding the accumulation of permanent foulants than mechanical cleaning alone. In most cases, UVT of the quartz sleeve was higher than 90% after chemical plus mechanical cleaning, whereas mechanical cleaning alone achieved only approximately 30% UVT of the quartz sleeve after long-term usage. (Peng et al., 2005)

4.2.1.4 Effects of Pretreatment

Unit processes upstream of UV reactors affect water quality and therefore can have significant impacts on the performance of a UV disinfection system by

- removing particles that can shield pathogens from UV radiation,
- increasing UVT by reducing the concentration of UV-absorbing organics and inorganics,
- decreasing UVT by adding chemicals that absorb UV light, or
- altering fouling rates for the quartz sleeve.

For example, residual iron from the use of ferric coagulating agents may decrease UVT and increase fouling of the quartz sleeves, thus negatively impacting disinfection performance. In contrast, increasing the level of pretreatment generally decreases the TSS concentration, improves the UVT, and allows fewer lamps and a smaller reactor size (Brunet et al., 2000). A study at Water Factory 21 in Orange County, California, showed that wastewater treated by microfiltration/reverse osmosis (low suspended solids, low turbidity, and high UVT) required a lower UV dose to achieve the same level of disinfection as conventionally treated wastewater (Norman et al., 1999).

4.2.2 Inactivation of Pathogens

The main mechanism of microbial inactivation by UV treatment is the absorption of UV light by DNA. Consequently, variations in DNA content cause microbes to absorb UV light differently. This contributes to disparities in the susceptibility of different strains of bacteria and viruses to UV disinfection. Even within a single species, different strains behave differently. For example, 2-log inactivation of the pathogenic *E. coli* strains O157:H7 and O26 requires UV doses of 3.0 and 8.1 mJ/cm^2, respectively (Tosa and Hirata, 1999). When evaluating disinfection efficacy, it is important to use microbial assays that measure infectivity, not viability. Early studies based on viability indicated that UV disinfection was not effective for *Cryptosporidium* disinfection. However, later studies based on infectivity showed that low doses of UV light could effectively inactivate *Cryptosporidium* (U.S. EPA, 2006).

4.2.2.1 Relative Resistances to UV Among Organisms

Among the pathogens of interest in wastewater, UV is most effective against *Giardia* cysts, followed by *Cryptosporidium* oocysts, bacteria, and then viruses (Havelaar et al., 1991; U.S. EPA, 2006). Tree et al. (1997) found that the bacterial indicators *E. coli* and *Enterococcus fecalis* were more susceptible to UV disinfection than poliovirus; poliovirus, in turn, was more susceptible than F-plus bacteriophage MS2, which was quite resistant to disinfection. These results imply that the use of bacterial indicators alone would underestimate human enteric virus inactivation (for adenovirus see 4.2.2.2), while the use of F-plus bacteriophage is likely to give a conservative estimate of virus inactivation.

4.2.2.2 UV Disinfection Relative to Chlorine

UV's disinfection capabilities are particularly interesting because conventional chlorine disinfection is weaker against some pathogens, such as spore-forming bacteria and microbes with encapsulating structures such as *Giardia* sp. and *Cryptosporidium* sp., than it is against traditional indicator organisms. UV is often considered more effective than most disinfectants at the maximum allowed concentrations in drinking water, and an excellent alternative for cyst disinfection. For example, UV doses of less than 20 mJ/cm^2 achieved >4-log inactivation of *Cryptosporidium* (U.S. EPA, 2006), and in-vitro infectivity assays of *C. parvum* oocysts showed >3-log inactivation with UV doses of 5 -10 mJ/cm^2 (Bukhari and LeChevallier, 2002). Linden et al. (2002) reported rapid inactivation of >4 logs of *G. lamblia* with a dose of only 1 mJ/cm^2 in buffered saline water.

However, UV is less effective than chlorine against human adenoviruses. A dose of 40 mJ/cm^2 provided only 1-log inactivation, while monochloramine provided up to 2.5-log inactivation with a CT value of 350 mg-min/L, and free chlorine achieved 4-log inactivation with a CT value of 0.22 mg-min/L (Durance et al., 2005). Both of these CT values are below the CT of 450 mg-min/L that is required under California Title 22 for water reuse. A UV dose of approximately 170 mJ/cm^2 was required to achieve 4-log inactivation of adenovirus (Thompson et al., 2003).

4.3 System Design and Implementation

This section presents issues related to the design and selection of a UV system for wastewater disinfection. Detailed explanations on UV system design are beyond the scope of this report; the intent here is to provide an overview of the design approach, methods for determining UV dose, validation of equipment and field systems, and other relevant issues. For more detailed information, the following four guidelines are commonly used in system design and equipment validation of UV disinfection in water/wastewater:

- *Ultraviolet Disinfection Guidelines for Drinking Water and Wastewater Reclamation,* often called the NWRI Guidelines – Published in collaboration with the American Water Works Association Research Foundation, this work has been widely accepted and used in the field of wastewater disinfection (Melin, 2003).
- *Ultraviolet Disinfection Guidance Manual for the Final Long Term 2 Enhanced Surface Water Treatment Rule* (U.S. EPA Report 815-D-06-007) – This U.S. EPA manual provides guidance for drinking water systems on the selection, design, and operation of UV systems to comply with the Long Term 2 Enhanced Surface Water Treatment Rule (U.S. EPA, 2006).
- U.S. EPA Environmental Technology Verification (ETV) Protocol – Water Quality Protection Center, *Generic Verification Protocol for Secondary Effluent and Water Reuse Disinfection Applications* (Scheible and Mignone, 2002). The ETV program was created to accelerate the development and commercialization of better environmental technologies through third-party verification. This protocol was written to closely mimic the NWRI Guidelines for verifying UV systems (Swift et al., 2002).
- U.S. EPA Environmental Technology Verification (ETV) Protocol – Drinking Water Systems Center, *Protocol for Equipment Verification Testing for Inactivation of Microbiological Contaminants* (NSF International, 2003).

4.3.1 Maturity of the Technology

UV disinfection is considered a mature technology and its use in wastewater disinfection is growing. However, the technology is still relatively new and continues to evolve and improve, and to a large extent, applications are vendor and site-specific. The following subsections discuss the current guidelines for design and the steps leading to validation of UV equipment and systems.

4.3.2 Design Approach

4.3.2.1 Characterization of the Facility

Before designing a UV system, the facility and its needs must first be characterized. The characterization process includes choosing the design flow rate, which is often the peak hourly flow rate. However, low flow rates must also be considered to ensure that adequate hydraulics are maintained for all operating conditions. It also involves establishing the water quality profile of the effluent (UVT, suspended solids concentration, turbidity, particle size distribution, fouling potential, etc.), and the compliance requirements (target pathogens, concentration limits, etc.). As explained in Section 4.2.1, water quality can determine the suitability of UV treatment for a given effluent and has a significant impact on the disinfection efficacy, which in turn affects the reactor design requirements. Finally, any restrictions (e.g., space limitations) imposed by the plant must be considered. Alternatively, the ability of the plant to accommodate the needs of the UV system (e.g., space, location, power, and plant hydraulics) should be assessed; the facilities required by the UV system are described in more detail in Section 4.3.3.

4.3.2.2 Design Approach for the UV System

Once the facility has been characterized, the UV system can be designed. In order to expose target microorganisms to a sufficient UV dose and optimize the use of the energy delivered by the lamps, the system should satisfy the UV demand of the wastewater and provide good hydrodynamics. In practical terms, the first step in the design process is to select the types of UV systems, the vendors, and the models appropriate to the site-specific conditions. This involves choosing the reactor type, lamp types and configurations, cleaning system, etc. The automated cleaning system (mechanical only or mechanical/chemical) must be capable of handling the wastewater constituents (e.g., dissolved and suspended solids, other fouling agents, fats, oils and greases).

Every UV equipment model must go through a validation process, which involves collimated beam experiments to generate a dose-response curve, pilot testing of the UV system, and development of a dose-regression model (DRM), which relates disinfection performance to operational parameters (see Section 4.3.4 for more details). In current practice, many facilities simply use the DRM from equipment validation tests to design their systems. However, as discussed in Section 4.2, disinfection performance depends on many site-specific factors, so each facility should consider conducting pilot tests to verify the accuracy and reliability of the DRM's predictions of UV performance and thus provide a more optimum design. The DRM can then be used to size/design the reactors and plant layout (e.g., number of lamps/banks and channels), and develop the operating strategy (flow pacing or dose pacing, etc.). Finally, the NWRI Guidelines recommend field commissioning tests to validate the design and operating parameters for the UV system for the individual site. The results of these tests can be used to optimize the operational UV dose (e.g., maintain adequate disinfection while reducing power demand, the number of reactors/trains on-line, etc.).

The design of the UV system hinges on accurate determination of the UV dose. The NWRI Guidelines define three dose terms for flow-through reactors (Melin, 2003):

- *Delivered UV dose* – The calculated UV dose supplied to a reactor, as determined via collimated beam and bioassay tests (see Section 4.3.4 for more details).
- *Design UV dose* – The delivered UV dose required to meet regulatory limits or achieve the desired level of inactivation of the target microorganism(s).
- *Operational UV dose* – The UV dose applied during operation of the UV system, based on the results of field tests.

The delivered dose is calculated using the dose-response curve (from collimated beam experiments) and data from pilot tests. The design dose is then determined based on regulatory requirements, delivered dose data, and aging and fouling factors. As an example, for media-filtered secondary effluent, the California Department of Health Services requires a minimum design dose of 100 mJ/cm^2 under maximum day flow, after accounting for lamp aging and fouling factors. Minimum dose requirements are 80 and 50 mJ/cm^2, respectively for membrane filtration (e.g., micro- and ultra-filtration) and reverse osmosis effluents.

Reactor Considerations The plant layout should ensure redundancy and flexibility for variations in flow rates, while the reactor design should optimize UV delivery dose and hydrodynamics (e.g., through lamp placement, inlet and outlet conditions, and baffles). Reactors should use properly designed inlet and outlet structures, and eliminate short-circuiting and/or dead zones; otherwise, effective contact times may be shortened and power use may be inefficient (U.S. EPA, 1999; Melin, 2003). Recent studies also indicate that low flow rates should be avoided, because the thicker boundary layers at low velocities reduce the effective amount of radial mixing, yielding non-optimal hydrodynamics and lower disinfection efficacies (Iranpour et al., 1999). The NWRI Guidelines specify that pilot reactors for open channel systems should only be scaled up to full size when velocity profiles can be adequately quantified and matched. Finally, open channel UV systems need appropriate water level control to ensure proper submergence of the UV lamps during all flow conditions (Johns et al., 2002). For closed channel reactors, vendors should provide equivalent assurances.

Lamp Calculations Disinfection is affected by many factors, including the flow rate, wastewater characteristics, variability in the wastewater characteristics, and permit requirements. In calculating the number of lamps required for sufficient disinfection, two approaches can be taken: a probabilistic approach and the conventional, non-probabilistic approach (Loge et al., 1996). Tchobanoglous et al. (1996) reviewed these approaches and identified field data that must be collected to apply each method. The main advantage of the probabilistic approach is that the design of the UV system reflects the variability in the wastewater characteristics, the variability in the field test data, and the statistical nature of the discharge requirements.

4.3.3 Facility Requirements

Before selecting UV treatment as the disinfection technology for a given site, the availability of the necessary facilities (including space, location, power, and plant hydraulics) should be verified. Each of these is discussed in more detail below.

As a physical disinfectant, UV needs less contact time than chemical disinfectants like chlorine. This allows for smaller reactors and lower space requirements than other methods; in fact, chlorine contact tanks may be retrofitted with UV reactors, with space remaining for later expansion (WEF, 1996). In addition to the reactor space, UV systems may require some space or chemical infrastructure for maintenance such as periodic manual cleaning of the lamp sleeves.

In terms of location, most large UV systems in the United States are sited outdoors. However, facilities in regions with potentially adverse weather conditions (e.g., rain, snow, or freezing temperatures) may prefer to locate their UV systems indoors for more convenient maintenance (Tone et al., 1998). High temperatures and/or levels of sunlight may encourage algae growth and bacterial regrowth after disinfection (see Section 4.5.2 for more information on regrowth); providing a covering for the unit or locating it indoors may mitigate these effects. In addition, some vendors require that their power supplies be housed in buildings.

UV disinfection equipment requires a significant amount of electrical power to illuminate the lamps, so accurate assessment of the available capacity of the electrical power distribution system is important. In addition, to ensure continuous operation, the UV disinfection should be connected to emergency power (a generator or uninterruptible power supply) of adequate size (Johns et al., 2002); see Section 4.4.4 for more details on backup power. Transferring a UV system to emergency power under "brown out" or "black out" power scenarios while maintaining 100% disinfection is a design and operational challenge. Because LPHO may need as much as 10 minutes to ramp up to operational conditions (MP lamps may need as much as 30 minutes to warm up), configuring and operating the systems for restarts while maintaining the ability to disinfect wastewater requires careful attention.

Headloss and flow velocities are generally greater across UV disinfection systems than chlorine contact tanks, due to the smaller UV reactor size and the presence of lamps, baffles, etc. Consequently, sufficient head for UV should be verified during the design process, even if the hydraulic head exiting upstream processes is sufficient for chlorination (Johns et al., 2002).

4.3.4 System Design and Validation

4.3.4.1 Determining UV Dose

The methods for determining UV dose are constantly evolving; this section explains some of the practices currently employed. Note that the methods described here ignore the effect of particle shielding, which can have a significant impact on UV dose delivery and distribution.

Theoretical Calculations Wright (2000) presented a method for calculating UV dose for microbial disinfection that includes a measure of the wavelength-dependent factors of UV lamp output, quartz sleeve UV transmittance, microbial action spectra, and wastewater UV transmittance. For MP lamps, he proposed a simplified method that relies only on a measure of the UV transmittance at 254 nm. According to Soroushian et al. (2000), the germicidal efficiency of MP lamps can be estimated using the germicidal effectiveness of each wavelength. For pulsed UV systems, the dose estimation must consider the impact of pulse frequency on disinfection. In addition, calculations must account for differences in lamp arc length, power input, reactor configuration, and dose calculation methods used by UV system manufacturers (Soroushian et al., 2000).

Collimated Beam Bench-scale collimated beam systems generally use a batch reactor, typically a Petri dish, which allows for accurate determination of UV exposure time. The depth-averaged UV intensity within the Petri dish is calculated using the UV intensity at the air-water interface (measured by a calibrated radiometer), and an equation such as 4.3.4.1-1 (Morowitz, 1950). Other equations may also be used, to account for factors including the variability of UV intensity across the air-water interface, and the reflection and deflection of light from water surface (Bolton and Linden, 2003). The UV dose is then determined using the calculated average UV intensity, the exposure time, and Eq. 4.1.4-1.

$$I_{avg}(\lambda) = I_0 \left[\frac{\left(1 - e^{-(\alpha_\lambda)d}\right)}{(\alpha_\lambda)d} \right]$$

Where: $I_{Avg}(\lambda)$ = depth-averaged UV intensity at wavelength λ (mW/cm^2)
I_0 = incident UV intensity at the air-water interface (mW/cm^2)
α_λ = absorption coefficient at wavelength λ in base e (cm^{-1})
d = depth of the sample (cm)

Unfortunately, the dose-response relationships determined in collimated beam tests may vary considerably due to several factors, including the method of calculation, apparatus setup, column dimensions, UV lamp type and output, intensity measurement, shutter type and operation, Petri dish specifications, sample volume and depth of the liquid, mixing conditions, laboratory set up, microbial organism preparation and testing, and water quality (Kuo et al., 2003). As a result, the dose-response curves generated by two different laboratories may be different, even if they use the same sample. Currently, the NWRI Guidelines require that the collimated beam data fall within a range defined by the following two equations:

$$-\log_{10}\left(\frac{N}{N_0}\right) = 0.040 \text{ x } (\text{UV dose in mJ/cm}^2) + 0.64 \qquad (4.3.4.1\text{-}2)$$

$$-\log_{10}\left(\frac{N}{N_0}\right) = 0.033 \text{ x } (\text{UV dose in mJ/cm}^2) + 0.20 \qquad (4.3.4.1\text{-}3)$$

where: N_0 = the concentration of infective MS-2 coliphage before UV exposure
N = the concentration of infective MS-2 after UV exposure

To ensure reproducibility or to have a meaningful comparison of results from different collimated beam tests, a standardized protocol for collimated beam testing and its dose calculation is necessary. Bolton and Linden (2003) presented a detailed protocol for determining UV dose in a bench scale apparatus containing LP or MP UV lamps. This protocol included specifications for constructing the UV testing apparatus, methods for determining the average irradiance in the water, details on UV radiometry, and considerations for microbiological testing.

Flow-through Reactors Calculating the UV dose in a continuous-flow reactor is even more complicated than in a collimated beam test, because flow-through reactors are subject to hydrodynamic irregularities and a variable UV intensity distribution within the reactor, as discussed in Section 4.1.4.1. Each microorganism passing through the UV reactor follows a unique path and is therefore exposed to a unique dose. For example, organisms passing through the UV reactor close to the lamps are exposed to higher UV intensities, compared with organisms traveling near the reactor walls or between lamps. Some microbes move through the reactor quickly while others travel a more tortuous path. A dose distribution depicts the probability that a microorganism passing through a UV reactor will receive a given dose. Figure 4-3 compares dose distributions for two UV reactors with different hydraulics. The narrow dose distribution indicates more ideal hydrodynamic conditions and/or a more uniform UV intensity field. The wider distribution implies less efficient reactor performance; a greater "over-dosing" would be needed to ensure that the threshold UV dose is achieved for the microorganisms at the lower end of the dose distribution (U.S. EPA, 2006).

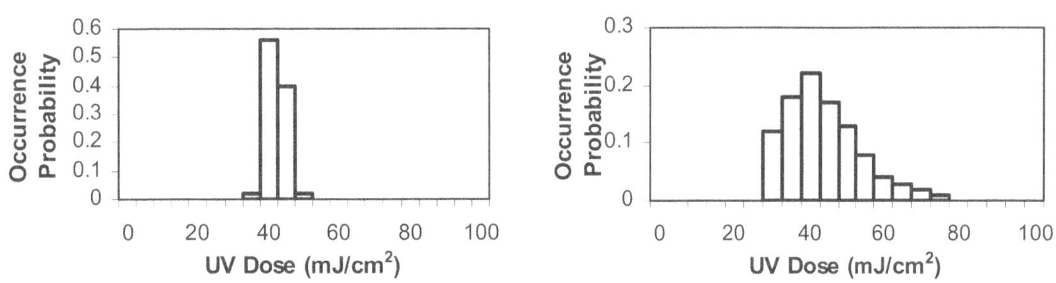

Figure 4-3. Hypothetical Dose Distributions for Two UV Reactors with Different Hydraulics. (Adapted from U.S. EPA, 2006)

In general, four methods have been used to estimate the UV dose in flow-through reactors: the first two are computational models and the last two are empirical. Models need to be validated as they are mathematical representations of the actual UV dose conditions. The first method was commonly used in the design of early UV systems. The average UV dose was derived using the average hydraulic retention time and a spatial average of UV intensity within the irradiated zone. The UV intensity was initially computed mathematically using a procedure known as the point source summation method (U.S. EPA, 1992), and later determined using a model with an infinite number of point sources (Hunter, 2000). However, mathematical modeling has not yet been proven satisfactory for the design of UV disinfection systems, given the many variables that can affect the system's performance. As a result, this approach is now considered obsolete (Tchobanoglous et al., 2003).

The second method uses computational fluid dynamics (CFD) to obtain a distribution of UV doses within a reactor, by integrating velocity profiles and the distribution of UV intensities within the irradiated zone. The CFD model predicts microorganism trajectories through a UV reactor and hence the dose delivered to each microorganism (Blatchley et al., 1995). Although the CFD approach is promising, further work is needed to standardize the methodology and validate a wider range of UV disinfection systems. Also, the reporting of a distribution of UV doses, though accurate, may not easily lend itself to UV system design (Tchobanoglous et al., 2003).

The third method can be used alone or for verification of CFD models. This method involves actinometers, which use compounds such as potassium ferrioxalate or uridine that undergo a chemical reaction when exposed to UV light (Linden, 2000). The extent of the UV-induced reaction is measured and used to calculate the delivered dose. While promising, more work is needed to prove that this approach is practical and accurately predicts UV reactor performance.

The fourth method employs bioassays to determine the delivered UV dose and is currently the most widely used method for design (Qualls and Johnson, 1983). Batch experiments in a collimated beam apparatus first establish a dose-response relationship, by varying the UV dose (calculated as described above) and measuring the corresponding inactivation of an indicator microorganism. These batch calculations of UV dose are applied to a flow-through reactor using biodosimetry, in which the inactivation of a microorganism is measured as it passes through the reactor under a given set of operating conditions. The delivered dose of the flow-through reactor is calibrated by setting it equal to the UV dose that achieves the same level of inactivation in the collimated beam test. More detailed information on this method can be found in the NWRI Guidelines (Melin, 2003).

Dose-Regression Models. Dose regression models (DRMs) have been used as a basis for designing and operating full-scale UV disinfection systems. They relate the delivered UV dose to operating conditions (flow rate and lamp power setting) and water quality (UVT), and typically take the following form (Tang, 2006):

$$DD = a(HL)^b (P)^c (UVT)^d \qquad (4.3.4.1\text{-}3)$$

where: DD = the delivered UV dose in mJ/cm^2, generally divided by approved lamp and fouling factors for design and operation purposes
HL = the hydraulic loading in gallon per minute (gpm) per bank of UV lamps
P = the UV lamp power setting in %
UVT = the UV transmittance of the reactor influent at 254 nm wavelength in %
a, b, c, d = empirically determined coefficients from the regression analysis

However, DRMs should be used with caution; due to their statistical nature, the delivered UV dose calculated from Eq. 4.3.4.1-3 may not match the bioassayed dose under the same operating and water quality conditions. The Sanitation Districts of Los Angeles County (LACSD) conducted pilot-scale UV verification studies at two water reclamation plants, and found that the doses predicted by DRMs were often higher than measured doses (Tang et al., 2006). They also found that the DRM developed for one facility may not accurately predict the delivered doses at other facilities, even if the treatment processes and water quality are similar. Important water quality characteristics affecting UV system performance, such as particle size distribution, may not be fully reflected in UVT, which is the only water quality parameter included in the DRM. Finally, the results indicated that the microbe disinfection across multiple UV banks is not always additive, e.g., the inactivation after two banks in series is less than twice the inactivation after a single bank. This is problematic for equipment validation and full-scale design, because tests are usually conducted with only one bank of UV lamps, and reactors are assumed to be independent and perform identically.

The results from the LACSD study suggest that the validity of the assumptions that dose and log inactivation are additive should be investigated. The results also suggest that the existing practice of relying on DRMs to adjust UV lamp power setting may not always ensure compliance with disinfection requirements. For full-scale UV system design, it is recommended that site-specific pilot testing should be conducted with a pilot plant configuration and number of banks that are the same as the full-scale system.

4.3.4.2 Equipment Validation

The need for effective validation of the dose-delivery performance of commercial reactors is increasing. UV technology validation ensures that the equipment can deliver the target UV design dose, and that the monitoring/control technology modulates the dose appropriately with changes in water quality or operating conditions. In accordance with the NWRI Guidelines, validation tests monitor disinfection over a range of flow rates, UVT values, and power settings.

Determination of UV Dose/Disinfection The bioassay approach for UV reactor validation is recommended over analytical and numerical models. Analytical models, which provide an average dose estimate, have been shown to be inadequate. Numerical models, which utilize computational fluid dynamics and UV light intensity models to predict reactor performance, can be accurate when used by skilled professionals, but require significant validation and/or calibration against bioassay data (Lawryshyn and Cairns, 2003).

Surrogate Microorganisms Selection of the microbe used for dose-delivery validation plays a critical role in the validation process; tests must use organisms that reflect the dose sensitivity of the targeted microbes. MS2 coliphage is widely used for validation of reuse and drinking water reactors at dose levels greater than 30 mJ/cm^2. However, reactors based on these tests may perform poorly at lower dose levels, such as those prescribed for coliform or *E. coli* inactivation in treated wastewaters. Petri and Lawryshyn (2005) found that differences in UV resistance between MS2 and indigenous organisms can significantly impact validation, while differences within the range of responses of indigenous organisms do not affect validation. Scheible and Lawryshyn (2004) recommend using indigenous organisms, or surrogates of similar UV sensitivity, for validation and subsequent sizing of reactors at lower dose levels.

UVT During reactor validation, low UVT water conditions may be simulated by the addition of UV-absorbing chemicals to the feed water. However, caution should be used in applying the results to treatment plant effluents. For a given UVT value, the UV absorbance spectra of the most commonly used compounds (coffee, fluorescein, and lignin sulphonate) differ substantially from that of wastewater treatment plant effluents. This difference impacts dose, especially for MP lamps (Mackey et al., 2004).

Cleaning System To validate the performance of cleaning mechanisms in water and wastewater, Oliver (2002) suggested a protocol that uses a positive control, several criteria to evaluate performance, and long term testing. This protocol successfully distinguished performance between the offline manual, automated mechanical, and automated chemical/mechanical cleaning methods.

4.3.4.3 Field Commissioning

The NWRI Guidelines specify that a field commissioning test of full scale systems should occur before initiating the production of reclaimed water. They recommend testing a number of items, including electrical components, inlet/outlet velocity distribution, water level, flow split between reactor trains, controls and alarms, and instrument calibration. The flow velocities of the system (which are a function of the size, number, and layout of the trains) should fall within the range for which the UV equipment was validated. The test should also verify that the velocity profiles between the full-scale and pilot-scale UV reactors match. If the velocity profiles match, the number of UV lamps in a field-scale UV reactor design may be scaled up by up to ten times the number of lamps in the validated pilot UV reactor (Melin, 2003).

Due to difficulties in matching the velocity profiles between the full- and pilot-scale UV reactors, an alternative method called checkpoint bioassay has been proposed. The checkpoint bioassay uses the same principles as a full-scale bioassay validation with a reduced number of testing conditions, and provides information on disinfection performance. With proper provisions, this method has been accepted by some regulatory agencies (Borroum and Emerick, 2005; Mahar et al., 2006).

4.3.5 Ease of Design and Expansion

UV systems require a much smaller footprint than chlorine contact tanks. Wastewater treatment plants that convert from chlorination to UV disinfection can potentially modify a portion of the existing chlorine contact tanks with a false floor and inter-channel walls to host UV equipment, while the remaining portion can be used for future expansion (WEF, 1996).

4.3.6 Regulatory Acceptance

During the early 1980s, UV disinfection gained some popularity as a result of U.S. EPA funding of the Innovative or Alternative (I/A) technologies program. The funding allowed new technologies to be implemented at wastewater treatment facilities, with an option for complete replacement if the technology could not meet the NDPES permit requirements. As a result, UV technology gained limited acceptance from some state regulatory agencies, although there were also a number of problematic UV systems that resulted in delayed acceptance of the technology in many states until the mid to late 1990s.

In 1993, the National Water Research Institute (NWRI) published guidelines that facilitated California's approval of UV disinfection for specific water recycling applications by setting minimum design requirements for UV disinfection systems (Sakaji, 2000). Testing for compliance with California Title 22 water reuse requirements was conducted during the early 1990s, and results indicated that the UV disinfection systems met the stringent coliform limits for reuse applications (Hunter, 2000). The most recent version of the NWRI guidelines (Melin, 2003) has become an unofficial "standard" in several states with active wastewater reuse initiatives, including Hawaii, Florida, and Texas (Swift et al., 2002). For example, the Florida Department of Environmental Protection can approve UV systems for high-level disinfection if they comply with all requirements of the NWRI guidelines with filtration and chemical feed, or through pilot studies that include pathogen data (enterovirus, *Giardia*, and *Cryptosporidium*) justifying the design and operating parameters. Projects involving low level, basic, and intermediate disinfection are now readily permitted in Florida (York and Potts, 2000).

Other regulatory influences have also positively impacted the growth of UV systems in the US, primarily those increasing the regulatory requirements for chlorine (see Section 3.3.5 for details).

4.4 Operation and Maintenance

4.4.1 Ease of Operation and Maintenance

UV disinfection facilities should be arranged to make operation and maintenance convenient. The Water Environment Federation (WEF) training manual provides a detailed checklist for routine operation of UV systems, as well as trouble-shooting information (WEF, 2006). To ensure that the UV system functions properly, a number of routine maintenance activities should be conducted including the following (Hunter, 2006; U.S. EPA, 2006; WEF, 2006):

- Check and calibrate the on-line UVT analyzers (weekly).
- Check the quartz sleeves and wipers for leaks (monthly).
- Check the cleaning efficiency (monthly). If the UV system is not equipped with an automatic washing system, the quartz sleeves must be cleaned manually, which can range from simple wiping by hand to in-channel cleaning or a separate tank dedicated to cleaning (Mann and Cramer, 1992).
- Check the cleaning fluid reservoir (semi-annually).
- Test the ground fault interrupt (annually).
- Check the accuracy of the flow meters (manufacturer's recommended frequency).
- Replace lamps as necessary, e.g., when a lamp fails or reaches the predetermined service hours, the system does not achieve the required performance, or the on-line UV monitor indicates that the UV intensity is low.
- Replace quartz sleeves by the pre-specified schedule or when damage, cracks, or excessive fouling occurs.
- To maintain or improve performance, plants should also avoid upsets of upstream processes that can lower effluent transmittance, avoid manual operation of the UV

system (because operating at high power levels during low flow can result in more rapid lamp fouling), and experiment with different cleaning solutions (Bush, 2005).

4.4.2 Process Control and On-line Monitoring

The complex interaction of factors affecting UV dose (see Section 4.1.4.1) make disinfection efficacy difficult to calculate or measure directly. However, delivery of the correct UV dose at the correct time is key to ensuring compliance with the discharge permit limits. As a result, good on-line monitoring and process control are essential to successful UV operation.

In order to maintain system control, both manual and automated methods are used. There are two basic types of automated control: flow-paced and dose-paced. Flow-paced control adjusts the number of lamps in service based on the influent flow rate. This type of control is often used alone on LPLO systems, but is typically integrated with other forms of control on LPHO or MP systems (WEF, 2006). Dose-paced control is based on a calculated dose, derived from the flow rate, on-line UVT data, and lamp power (including lamp age and on-line intensity sensor output). This type of control is more commonly used in LPHO or MP systems with either on-line UV intensity or UVT monitors, which allow dose adjustments in response to changing lamp output and water conditions (Wood et al., 2005; WEF, 2006). Automated controls should only be applied over the range of water quality and operational conditions for which the system has been validated.

Other features for successful automated control include proper programming of reporting and failure alarms, effective control logic for lamp bank cycling and flow pacing, UV intensity sensors for monitoring the process, and an operator-friendly control interface that allows easy identification of lamps or ballasts that have failed (Tone et al., 1998). The UV sensor must be fixed at a standardized measuring window on a reference position in the irradiation chamber, and periodically removed during operation for calibration against official specifications.

4.4.3 Operator Safety and Health Risks

Several precautions should be taken to reduce the risks of exposure to UV radiation and other hazards. To prevent over-exposure of skin and eyes to UV radiation, lamps should not be operated in air. This means that UV systems should be equipped with safety interlocks that shut down lamp modules if they are taken out of their operating positions or the water level falls below the top lamps in the reactors. To minimize electrical hazards, ground-fault-interruption circuitry should be included with each module. In addition, protective gloves, eyewear, and footwear are recommended when servicing UV equipment. Finally, operators should be aware of additional hazards from exposure to chemical cleaners, burns from hot lamps, and operation of the overhead crane used to lift lamp modules or banks (Mann and Cramer, 1992; WEF, 2006).

4.4.4 Requirements for Backup Resources

UV lamps are sensitive to power fluctuations, and can lose their arc due to a voltage fluctuation, power quality anomaly, or power interruption. As a result, electrical power supply is a critical component of UV installation, planning, and design. LPLO lamps generally can return to full operation status within 15 seconds after power is restored. However, LPHO and MP lamps may take 4 to 10 minutes to go back to full power if power is interrupted (U.S. EPA, 2006).

In addition to consistent power quality, UV systems require a reliably continuous power supply. An alternative backup power supply may be needed if a backup power supply is not in place or the available load capacity is not sufficient to handle the new load resulting from the UV equipment. In fact, the Colorado Department of Public Health and Environment requires an emergency power supply for all UV systems, although complete redundancy is not necessary (Tone et al., 1998). Selection of the type of backup power supply will depend on the frequency and duration of the power interruption and the consequence of these interruptions on potential off-specification operation of the UV reactors (U.S. EPA, 2006). Existing chlorination systems can also serve as a backup disinfection resource for UV disinfection.

4.4.5 Storage, Inventory, and Replacement Requirements

Because UV disinfection is a physical/biophysical process, it eliminates most chemicals (except the cleaning agents used on the quartz sleeves) and thus the need to transport, store, and handle potentially dangerous chemicals. Operators should maintain adequate spare parts, including ballasts, lamps, and quartz sleeves. Horizontal systems with manual cleaning may require a complete spare module for use when one is removed for cleaning (Tone et al., 1998).

4.4.6 Operational Impacts on the Existing Treatment Train

In manual UV cleaning systems, mild citric acid or phosphoric acid solutions are often used, and the cleaning waste is typically sent back to the head end of the plant. Although the impact to operation may be minimal, plant operators should be aware of this.

4.4.7 Consistency in Performance with Fluctuations in Flow Rate

The number of UV lamps required to achieve a given disinfection efficacy is almost linearly proportional to the flow rate; thus, the design capacity of a UV system is critical in determining the ability of the system to adjust to fluctuations in flow. For design purposes, the peak hourly flow rate is often used. However, without flow equalization, a UV system would experience daily and seasonal fluctuations in flow rate. Redundant UV banks and/or channels can help to deal with the situations with these fluctuations, e.g., wet-weather peak flows. Supplementary chlorine addition is also an alternative for wet-weather peak flows, if the chlorine system is available on-site.

4.5 Impacts to the Environment

4.5.1 Changes in Water Quality

Current literature indicates few if any significant changes in water quality after exposure to typical UV disinfection doses. This includes a lack of disinfectant residuals, which reduces the chance of toxicity in the treated effluent. However, if the water is recycled, an additional chemical disinfectant may be required to provide a residual in the distribution lines to address slimes, algae, and odors.

The data on the effect of UV radiation on organics have been inconsistent, although UV at higher doses has been shown to oxidize organics (see Section 4.5.3 for more details). Soroushian et al. (1996) observed no appreciable differences in the concentrations of volatile and semi-volatile organic compounds between untreated and UV-irradiated waters. However, Awad et al. (1993) found a 2-log reduction in 8 to 16-carbon hydrocarbons, while other studies showed that application of high UV doses to waters with large amounts of reactive organic carbon can increase biodegradable dissolved organic levels (Mofidi et al., 1998; Shaw et al., 2000).

4.5.1.1 Disinfection Byproducts

Formation of disinfection byproducts from UV irradiation would arise either directly from photochemical reactions or indirectly from reactions with the products of photochemical reactions. Photochemical reactions occur when a chemical compound absorbs light, and the resulting excited state decomposes or reacts with others to form a new substance. The resulting concentration of the new species will depend on the concentration of the reactants, the UV dose, and the ambient conditions such as temperature and pH of water (U.S. EPA, 2006).

Compared with chemical disinfectants such as chlorine and ozone, UV disinfection is believed to have little potential for forming harmful disinfection byproducts. For example, UV does not appear to generate THMs, HAAs, or cyanide (Soroushian et al., 1996; Mofidi et al., 1998; Jalali et al., 2005; Linden et al., 2007). Malley et al. (1995) also found no significant DBPs after UV doses of 60 to 200 mJ/cm^2. However, several studies have detected small increases in aldehydes (e.g., formaldehyde, and acetaldehyde), glyoxal, and nitrite (Awad et al., 1993; Malley et al., 1995; Soroushian et al., 1996; Mofidi et al., 1998). Linden et al. (2007) note that byproduct formation depends on the lamp type: unlike LP lamps, MP lamps emit light over a broad range of wavelengths and may form byproducts such as chlorpicrin and nitrite.

4.5.1.2 Toxicity in the Treated Effluent

Several studies have shown that UV radiation does not increase mutagenic activity in water, although one study showed a slight increase in mutagenicity (Wolfe, 1990).

4.5.2 Potential Risks

4.5.2.1 Regrowth and Repair

Some microbes may not be adequately inactivated during the disinfection process. These organisms can reproduce, causing regrowth and an increase in the concentration of pathogens or indicator organisms. For example, Hancock and Davis (1999) found that *Daphnia sp.* (small crustaceans 0.2 to 0.5 mm in length) ingested and shielded fecal coliforms, *E. coli*, fecal *Streptococci* and *Enterococci* in wastewater; this shielding was a major contributor to increased bacterial concentrations, after UV treatment and normal *Daphnia sp.* die-off. Within the regrowth population, they identified representative fecal coliform colonies to be *E. coli*, which were only a small fraction of the original fecal coliform population (Hancock and Davis, 1999). In contrast, Baron (1997) observed no relationship between regrowth and effluent suspended solids concentrations that ranged from 4 to 12 mg/L.

Some microbes may be inactivated but not destroyed by the disinfection process. These organisms may repair the damage from disinfection and regain pathogenicity. As shown in Table 4-3, repair increases the concentration of infective organisms and can lead to regrowth. For these sources, the counts after repair/regrowth are lower than the pre-UV levels. Because both repair and regrowth cause increases in organism concentrations, they are difficult to differentiate and the literature often does not distinguish between these two mechanisms.

Table 4-3. Recovery of Pathogens After UV Disinfection.

Organism	Water	UV Dose (mJ/cm²)	Incubation Time (hr)	Recovery	Source
Thermotolerant coliform	Secondary effluent	40-60	2-24	Up to 2 log	Baron, 1997
E. coli	Secondary effluent	34.5-90	3	0.1-0.6 log	Baron and Bourbigot, 1996
E. coli O157:57	Buffer	1.1-3.3	3	Up to 0.3 log	Tosa and Hirata, 1999
E. coli O26	Buffer	3.9-13	3	0.4-3 log	Tosa and Hirata, 1999
Fecal *Streptococcus*	Secondary effluent	40-60	2-24	Up to 1 log	Baron, 1997
Enterococcus	Secondary effluent	34.5-90	3	None	Baron and Bourbigot, 1996
Salmonella	Secondary effluent	60	24	None	Baron, 1997
Viruses	Secondary effluent	60	2	None	Baron, 1997

The literature reports that repair can occur under dark conditions (dark repair) or under light (photoreactivation). Dark repair was discovered in the 1960s, and appears to occur through two mechanisms: excision repair and recombination repair. In excision repair, enzymes cut out DNA that was damaged by UV, while in recombination repair, enzymes use the complementary strand to restore the damaged DNA. Viruses lack the enzymes needed for dark repair, but can use the host cell's enzymes to mend their DNA. Dark repair is slow, relative to photoreactivation. (Asano et al., 2007)

Photoreactivation was discovered in 1949, and is a two-step process involving the enzyme photolyase (Asano et al., 2007). In the first step, which does not require light, the enzyme binds to the damaged site on the DNA. When exposed to light, photolyase mends damage to the chemical bonds in the DNA. Viruses lack photolyase but can use the host cell's enzymes to mend their DNA. Photoreactivation depends on a number of factors, such as environmental conditions (e.g., the visible light intensity after disinfection), UV operational parameters (e.g., the UV wavelength and dose) and the microbe species and strain.

Kashimada et al. (1996) observed apparent photoreactivation after UV treatment for heterotrophic bacteria, coliform bacteria, and fecal coliform in raw sewage, but not *E. coli*; the photoreactivation rate increased with increasing visible light intensity. Tosa and Hirata (1999) found that visible light photoreactivation occurred with enterohemorrhagic *E. coli* O26, but was minimal with *E. coli* O157:H7 after disinfection with LP lamps. They then conducted a second round of UV disinfection and found that the UV dose required for 1-log inactivation of *E. coli* O26 was 12.0 mJ/cm^2, much higher than the UV dose of 5.4 mJ/cm^2 used for the initial 1-log inactivation (Tosa and Hirata, 1999). This may be problematic if it implies that the bacteria present after disinfection and repair are more UV-resistant. Some reports indicate that regrowth can be minimized if UV-treated water is kept in the dark for approximately three hours after disinfection (Martin and Gehr, 2007), while Kalisvaart (2004) reported that polychromatic UV radiation from MP lamps prevents the recovery of microorganisms. Oguma et al. (2004) found less regrowth of *E. coli* after UV disinfection with MP lamps than with LP lamps. However, they found that *Legionella pneumophila* showed photoreactivation after disinfection, regardless of lamp type; organism concentrations increased 2.7-log after exposure to LP lamps and 2.5-log after exposure to MP lamps. Clearly, more research is needed to better understand the factors affecting repair and regrowth.

4.5.2.2 Mercury Release

Because UV lamps contain mercury, their disposal is covered by federal and state regulations. Lamps may be disposed of by a local hazardous waste hauler, or sent to a recycler who removes the mercury from the lamps and separates the components for reuse. If a UV lamp is broken, it can release mercury to the water (except in the case of LPLO and LPHO mercury amalgam lamps, where the mercury is contained in a solid phase when the lamp is not lit). Most breakage occurs due to carelessness and excessive force during installation. Misaligned wiping mechanisms caused by faulty assembly can also be a problem, so wiping mechanisms should be tested before the lamps are ignited.

4.5.3 Other Benefits

Higher UV dosages can act as a barrier to micropollutants through photolysis, as well as provide disinfection, in wastewater treatment plants (Williamson 2002). In particular, UV photolysis has received attention for the ability to treat N-nitrosodimethylamine (NDMA), and is one method currently used to remove NDMA from recycled water (Cutler et al., 2002). However, as shown in Table 4-4, the doses required to meet potential regulatory limits for NDMA are likely to be high. Thus, only incidental removal of NDMA is likely to be achieved by UV systems designed for disinfection.

Table 4 -4. NDMA Removal Using UV Radiation.

System	UV Dose (mJ/cm²)	NDMA Log Removal	Source
Collimated beam	100	0.3	Jalali et al. (2005)
Collimated beam	550	1	Soroushian et al. (2001)
Pilot system	100	0.22	Jalali et al. (2005)
Pilot system	>2000	1	Soroushian et al. (2001)
Pilot system	1000	1	Mitch et al. (2003)

4.5.4 Reliability and Compliance History

Just as the issues affecting the water quality and UV system design are site-specific, the factors that affect the ease and reliability with which a UV system meets compliance standards vary by facility. These factors include the relative size or design flow of the wastewater treatment plant, the variability of the flow and water quality, presence of treatment units to limit that variability, the amount of maintenance performed on the system, the ability of the system alarms to notify the operator of a power outage or system failure, and the operator's ability to respond and return the system to service.

Hunter et al. (2005) surveyed 19 wastewater treatment plants with design flow rates >100 MGD. The plants differed on water quality requirements and discharge limits. The survey indicated that these UV systems were reliable and met NPDES requirements, even with changes in flow rates and effluent quality. However, one municipal wastewater treatment plant in the Northwest experienced problems disinfecting effluent to the required standards for the first few months following expansion of the facility. The problems were attributed to premature scaling of the quartz sleeves and constituents in an industrial discharge from a printed circuit board manufacturer (Swift et al., 2001). The results of the survey conducted in this study yielded similar results (see Chapter 7.0): most treatment plants have found that UV disinfection reliably meets compliance standards, but a few facilities have had difficulties meeting performance objectives (often due to poor water quality).

To improve the reliability of UV disinfection systems, guidelines such as those from the Division of Water Quality of the State of North Carolina (2001) and the NWRI (Melin, 2003) are available as references for the design (and design review) of UV systems. Requirements include reliable feed water quality (i.e., proper design and consistent performance of upstream processes), safe and proper design of the UV system and its components, good operational and maintenance practices, a reliable power supply, redundant units or a backup disinfection system, and prompt detection of and response to system failure.

4.6 Cost

When assessing UV-based disinfection systems, the design engineer should estimate the capital costs, annual operation and maintenance (O&M) costs, and life cycle costs (net present worth) associated with the UV systems under consideration. It should be noted that the 2005-2007 cost climate for chlorine, concrete, steel, stainless steel, and energy has been very dynamic, primarily increasing both capital and O&M costs and making it difficult to use historically based cost curves. Cost estimates should be considered very site-specific, requiring a site-specific project cost estimate for any potential project.

In general, it is unclear whether the literature citations used for this section needed or included additional costs for annual risk assessment surveys and reports that are required by insurers for this technology. Some utilities have to pay more for liability risk insurance for employees that handle toxic and hazardous chemicals. It is not clear whether mercury from catastrophic lamp failures or potential exposure to UV light change liability or workers insurance premiums. The reduction or increase of these costs should be specifically included in cost analyses of various alternatives when performing a site-specific evaluation.

Many factors affect the cost of UV disinfection, including the following (Cotton et al., 2001):

- Water quality – As water quality improves (e.g., higher UVT, fewer and smaller particles), fewer UV lamps are needed and the system costs decrease.
- Microbial inactivation goals – The UV dose required to inactivate microorganisms varies by species, and the degree of inactivation increases with increasing UV dose. More stringent effluent or reuse standards require higher UV doses and a larger UV system, which results in higher capital and operation and maintenance (O&M) costs.
- Types of UV lamps/equipment – The UV lamp type significantly affects the O&M costs for operating a UV system; in general, LPLO lamps are the least expensive and MP lamps are the most costly. The design life of the LP lamps is approximately twice that of MP lamps. However, fewer MP lamps are required for a specific flow and UV dose.
- Design flow rate – The size of the UV system is almost linearly proportional to the flow rate of water to be treated. Selection of an appropriate design flow rate (usually peak hourly flow) has a significant impact on the capital cost of a UV system.
- System redundancy requirement – Higher degrees of system redundancy require larger systems with more parts, and the associated capital and maintenance costs.
- Retrofitting constraints – If a UV system is to fit into an existing plant, issues such as the plant hydraulic profile, existing plant configuration, equipment redundancy,

and space constraints must be considered. In terms of hydraulics, flow through a UV system causes headloss, which depends on the flow rate, types and configurations of lamps, reactor configuration, baffle configuration, associated piping and valving, and ancillary equipment (e.g., flow meters and automatic washers). If the existing plant does not have sufficient hydraulic head for gravity feed to the UV system, it will need dedicated pumps for the UV reactors.

- Energy costs – These are a major component of the UV O&M costs. The type and number of lamps determine the power requirement to operate a UV disinfection system.

4.6.1 Capital Costs

Capital costs reflect the initial construction and associated engineering and administrative costs for implementing a disinfection system. The capital costs for UV disinfection for a given water quality increase as the system size or the dose increases. Actual costs will depend on the number and type of UV lamps and type of ancillary facilities needed as well as site-specific conditions, geographical location, and water quality. More specific information for capital costs of equipment can be obtained from individual vendors of UV systems.

Table 4-5 compares estimated capital costs from several literature sources for UV disinfection systems at several wastewater treatment plants. These capital cost estimates appear to be planning level costs that have an accuracy of +50/-30%. The capital costs have been normalized to a first quarter 2007 *Engineering News-Record* (ENR) 20 cities construction cost index (CCI) of 7870. With the exception of costs from Darby et al. (1995), values appear to be "bid" costs (i.e., the facilities costs equivalent to a contractor's bid); Darby et al. (1995) provide total project capital costs that include an additional 20% to account for engineering and administration during facility design and construction. When assessing a UV project at a specific facility, total project capital costs should also include contingencies reflecting the level of detail in the estimate.

Table 4-5 Capital and O&M Cost estimates for UV Disinfection.

Average Dry Weather Flow (MGD)	Capacity (MGD)	UV System Type	Capital Cost ($)[a]	Annual O&M ($)[b]	Source
Effluent Discharge					
1	2.25	LPLO	174,000	11,000	Darby et al. 1995
-	2.7	Unspecified	421,000	19,000	Hanzon et al. 2005
-	3	LPLO	317,000	23,000	Hanzon and Vigilia 1999
5	12.5	LPLO	936,000	42,000	Elefritz et al. 2000
5	12.5	MP	838,000	48,000	Elefritz et al. 2000
10	20	LPLO	1,263,000	95,000	Darby et al. 1995
-	22.5	LPHO	2,106,000	98,000	Baiden et al. 2002
-	22.5	MP	2,106,000	176,000	Baiden et al. 2002
10	25	LPLO	1,655,000	73,000	Elefritz et al. 2000
10	25	MP	1,455,000	94,000	Elefritz et al. 2000
-	27.2	Unspecified	1,976,000	68,000	Hanzon et al. 2005
15	37.5	LPLO	3,210,000	182,000	Elefritz et al. 2000
15	37.5	MP	1,835,000	154,000	Elefritz et al. 2000
-	53.5	Unspecified	3,210,000	182,000	Hanzon et al. 2005
-	61	Unspecified	5,054,000	266,000	Hanzon et al. 2005
100	175	LPLO	9,136,000	727,000	Darby et al. 1995
Reuse or Low Coliform Limit					
1	2.25	LPLO	938,000	60,000	Darby et al. 1995
5	12.5	LPLO	3,304,000	158,000	Elefritz et al. 2000
5	12.5	MP	2,689,000	106,000	Elefritz et al. 2000
10	20	LPLO	6,830,000	516,000	Darby et al. 1995
10	25	LPLO	6,421,000	315,000	Elefritz et al. 2000
10	25	MP	4,476,000	210,000	Elefritz et al. 2000
15	37.5	LPLO	8,958,000	402,000	Elefritz et al. 2000
15	37.5	MP	5,947,000	317,000	Elefritz et al. 2000
100	175	LPLO	50,247,000	3,493,000	Darby et al. 1995
240	-	MP	54,840,000	3,301,000	Tobin et al. 2005

[a]Capital costs adjusted to first quarter 2007 (ENR CCI = 7870),
[b]O&M costs adjusted to First Quarter 2007 (Consumer Price Index of 610).

The estimated capital costs are clearly site-dependent, but the relative costs can be compared across UV technologies. In general, capital costs are higher for LPLO UV systems than for MP UV systems. However, as discussed in Section 4.6.2, the operating costs for MP lamps are sometimes higher than for LPLO technologies, so the overall costs may be competitive (see Section 4.6.3). Note that more stringent discharge or reuse limits (e.g., lower coliform densities) greatly influence the cost of the UV systems.

The capital costs for selected UV systems are broken down into components in Table 4-6. Data from Samel et al. (2003) indicate that stricter effluent requirements for reuse (14 CFU/100 mL fecal coliform monthly average, 80 mJ/cm^2 minimum UV dosage) result in higher capital costs than for effluent discharge (200 CFU/100 mL fecal coliform monthly average, 30 mJ/cm^2 minimum UV dosage). Systems with MP lamps incurred slightly higher construction costs than systems with LPHO lamps. This resulted in higher total capital costs when the system met the discharge requirements, but for reuse, higher LPHO equipment costs offset the increase in MP construction costs, and the total capital costs were similar.

Table 4-6. Breakdown of Capital Costs of UV Disinfection Systems.[a]

Capacity (MGD)	UV Lamp Type	Equipment ($)	Other Construction ($)	Total ($)	Source
Effluent Discharge					
3	LPLO	83,000	161,000	317,000	Hanzon and Vigilia, 1999
30	LPHO	521,000 – 756,000	197,000 – 263,000	784,000 – 977,000	Samel et al., 2003[b]
30	MP	700,000	555,000	1,255,000	Samel et al., 2003
40	LPHO	437,000 – 654,000	395,000 - 435,000	872,000 – 1,060,000	Samel et. al., 2003[b]
40	MP	423,000	774,000	1,197,000	Samel et al., 2003
Reuse or Low Coliform Limit					
30	LPHO	894,000 – 1,212,000	464,000 – 617,000	1,499,000 – 1,802,000	Samel et al., 2003[b]
30	MP	794,000	918,000	1,712,000	Samel et al., 2003

[a]Costs normalized to first quarter 2007 (ENR CCI = 7870)
[b]Costs estimate ranges are for three different LPHO UV equipment vendors

4.6.2 Operation and Maintenance Costs

Operation and maintenance (O&M) costs reflect the costs of energy, lamps and parts, chemicals, labor, and other activities required to operate and maintain the UV system. Annual O&M costs are generally estimated based on the annual average flow rate. Actual O&M costs are difficult to quantify accurately (Swift et al., 2000), due in part to the fact that most POTWs do not maintain detailed accounting records of O&M costs at the process level. Although O&M costs can be obtained from UV manufacturers, they may be generic and/or incomplete and may require revision to reflect site-specific conditions.

Table 4-5 lists annual O&M cost estimates from the literature for several wastewater treatment plants. Table 4-7 shows the breakdown of O&M costs, in 2007 dollars per million gallons treated ($/MG), for selected wastewater treatment plants for power, lamps and parts, chemicals, maintenance and labor, and other categories. These costs appear to be planning level O&M cost estimates, which have been adjusted to first quarter 2007 using the consumer price index. However, the basis of the costs is not provided in most cases.

As expected, annual O&M costs generally increase with increased plant capacity and with more stringent treatment goals (reuse vs. typical effluent discharge limits). For effluent discharge, O&M costs for MP systems are generally higher than for LP systems, but this may not be the case for reuse applications. Total O&M costs range from about $8/MG to $72/MG. In general, O&M costs are dominated by the costs of electrical power and lamp and part replacement. Electrical costs comprised the largest percentage of total O&M costs for systems with MP lamps, while lamp and part replacements comprised the largest percentage of total O&M costs for LPHO lamps. Overall O&M costs for LP lamps were generally lower than for MP lamps, due in part to the lower power requirements/costs, longer lamp life, and lower cost per lamp of the LP lamps.

Table 4-7. Breakdown of O&M Costs (in $/Million Gallons of Wastewater Treated) for UV Systems.[a]

Avg Flow (MGD)	UV Lamp Type	Original Electricity Unit Cost ($/ kwh)	Power ($/ MG)	Lamps Components ($/ MG)	Chemicals ($/ MG)	Maintenance and Labor ($/ MG)	Testing ($/ MG)	Total ($/ MG)	Source
Effluent Discharge									
0.4	MP	0.06	42.24	16.91	0.40	12.81	0.00	72.36	Swift et al., 2000[b]
1.8	LPLO	0.06	12.04	12.44	3.59	8.99	0.00	37.07	Swift et al., 2000[b]
2.0[c]	LPLO		5.57[c]	4.75	1.97	12.85	6.96	32.11	Hanzon & Vigilia, 1999
3.0	LPLO	0.06	8.37	5.61	1.08	3.51	0.00	18.57	Swift et al., 2000[b]
3.6	MP	0.06	32.94	21.46	0.09	10.01	0.00	64.50	Swift et al., 2000[b]
5.28	MP	0.06	37.92	17.16	0.06	3.92	0.00	59.06	Swift et al., 2000[b]
7.25[d]	LPHO	0.06	12.40	17.77	1.65	4.13	0.00	35.59	Baiden et al., 2003
7.25[d]	MP	0.06	43.80	14.88	1.65	4.13	0.00	64.46	Baiden et al., 2003
15.0	LPHO[e]	0.042	2.60 – 4.52	5.12 – 5.74	0.00 – 0.09	0.30 – 1.21	0.00	8.10 – 11.24	Samel et al., 2003[g]
15.0	MP	0.042	14.23	4.59	2.30	0.09	0.00	21.20	Samel et al., 2003[g]
20.0	LPHO[e]	0.042	2.74 – 4.29	5.33 – 5.38	0.00 – 0.07	0.28 – 1.21	0.00	8.51 – 10.89	Samel et al., 2003[g]
20.0	MP	0.042	10.79	4.79	2.30	0.09	0.00	17.96	Samel et al., 2003[g]
Reuse or Low Coliform Limit									
15.0	LPHO[e]	0.042	7.11 - 11.44	13.30 – 14.49	0.00 – 0.06	0.69 – 3.85	0.00	21.96 – 29.56	Samel et al., 2003[g]
15.0	MP	0.042	28.45	11.72	4.59	0.21	0.00	44.98	Samel et al., 2003[g]

[a]All costs adjusted to Consumer Price Index of 610 (First Quarter 2007).
[b]Original costs based on actual field measurements
[c]Original electricity unit cost basis not specified
[d]Estimated (33.3% of design flow)
[e]Costs estimate ranges are for three different LPHO UV equipment suppliers.
[g]Original costs from competitive construction bids

Within each lamp type, there was considerable variability, due to design and site-specific factors such as upstream treatment processes, water quality, electricity rates ($/kwh), maintenance procedures, and compliance requirements. Liberti et al. (2003) demonstrated the effects of upstream processes by testing filtered secondary effluent and secondary effluent in a pilot-scale system in West Bari, Italy; O&M costs were 17.5 euro and 35 euro/1000 m^3 ($54.15 and $108.30/MG), respectively, to achieve 2 cfu/100 mL of total coliforms. The data from Samel et al. (2003) indicate that the effluent requirements influence the cost strongly; O&M costs associated with meeting reuse requirements more than doubled the cost of meeting discharge requirements.

Facilities may reduce energy costs by reducing lamp intensity and/or turning off lamps when not needed, either based on flow-pacing (adjusting for the water flow rate) or dose-pacing (maintaining a dose set point) (Swift et al., 2000). Tone et al. (1998) also noted that flow pacing can significantly reduce power costs; however, this savings must be balanced with the shorter lamp and ballast life caused by frequent on-off cycling.

4.6.3 Present Worth (Life-cycle) Costs

Design engineers frequently use net present worth analysis to compare the life cycle economics (combination of capital costs and present value of O&M costs over the project life) of alternative treatment technologies for process selection or to evaluate construction bids. Present worth analysis is project-specific due to varying project life and the value of money (i.e., interest rate and inflation). Project life is typically 20 years, but may be as long as 50 years. Interest rates typically range from 3% to 10%.

Table 4-8 compares the net present worth of alternative UV disinfection systems competitively bid for two wastewater treatment plants, one with two treatment goals of discharge and reuse (Samel et al., 2003). Consistent with the capital and O&M costs, MP systems are generally more expensive than LPHO systems and stricter effluent (reuse) requirements increase net present worth. Despite the higher costs, MP systems may be the best alternative in some cases, e.g., for facilities with limited space (because MP systems generally have smaller footprints).

Table 4-8. Comparison of Net Present Worth for Four Alternative UV Systems.

Design Flow (MGD)[a]	Vendor	UV System Type	UV Dosage (mJ/cm^2)	Capital Cost ($)	Annual O&M ($)	Present Worth[b] ($)
Effluent Discharge						
30	A	LPHO	30	872,000	52,000	1,407,000
30	B	LPHO	30	1,060,000	62,000	1,698,000
30	C	LPHO	30	875,000	44,000	1,328,000
30	D	MP	30	1,198,000	116,000	2,392,000
40	A	LPHO	30	784,000	64,000	1,443,000
40	B	LPHO	30	977,000	80,000	1,800,000
40	C	LPHO	30	788,000	62,000	1,426,000
40	D	MP	30	1,255,000	131,000	2,603,000
Reuse						
30	A	LPHO	80	1,510,000	120,000	2,745,000
30	B	LPHO	80	1,802,000	162,000	3,409,000
30	C	LPHO	80	1,499,000	124,000	2,775,000
30	D	MP	80	1,712,000	246,000	4,244,000

Source: Samel et al., 2003. Costs adjusted to first quarter 2007 (ENR CCI = 7870)
[a]Peak flow rate (average flow rate 50% of peak)
[b]Project life of 20 years and interest rate of 7.375%.

4.6.4 Comparison with Chlorine Technology Costs

Costs are a key factor for POTWs that are considering changes to their disinfection systems due to plant expansions or new effluent limits, but are very site-specific. Cost comparisons of UV and chlorine disinfection systems at nine wastewater treatment plants were developed from six literature sources. All costs have been adjusted to first quarter 2007 dollars.

Table 4-9 provides information on the plant size, the original disinfection system, the disinfection system objective, and the reference for each plant. The disinfection system sizes range from 1.5 MGD to over 240 MGD. Five of the POTWs had existing gas chlorine systems, three had sodium hypochlorite (NaOCl) systems, and one was a new plant. The disinfection objective was low coliform limit and/or reuse at four POTWs, effluent discharge limit at four POTWs, and wet weather discharge at the other POTW.

Table 4-9. Characteristics of POTWs with Cost Comparisons of Disinfection Systems.

Plant ID	Avg Flow MGD	Design Flow MGD	Original Disinfection System	Disinfection Objective	Source
1	0.3	1.5	NaOCl	Reuse	Hamby, 2006
2	1.5 – 1.8	2.7	Chlorine gas	Effluent Discharge	Hanzon et al., 2005
3	16.5	18	Chlorine gas	Low coliform limit	Gillette et al., 2003
4	-	22.5	Chlorine gas	Effluent discharge	Baiden et al., 2002
5	14.7	27.2	Chlorine gas	Effluent discharge	Hanzon et al., 2005
6	25	53.5	Chlorine gas	Effluent discharge	Hanzon et al., 2005
7	20	61	None (new POTW)	Low coliform limit, reuse	Hanzon et al., 2005
8	25	65	NaOCl	Wet weather discharge	Keller et al., 2005
9	240	-	NaOCl	Low coliform limit	Tobin et al., 2002

Table 4-10 summarizes the estimated capital costs for the nine POTWs and indicates the reason for each evaluation. Seven analyses were for expansions and/or upgrades, one involved a disinfectant change, and the other was for a new facility. Only four of the analyses considered gas chlorine as an alternative. Most of the references did not specify whether the capital costs were total project or "bid" cost estimates, but all technologies appear to be treated the same in each analysis, so comparable costs were provided.

The estimated capital costs are clearly site-dependent, but tend to increase as a function of flow rate and more stringent disinfection objective (low coliform limit or reuse vs. effluent discharge). Comparing the relative costs across technologies, capital costs are generally higher for UV systems than for gaseous chlorine and NaOCl systems.

Table 4-10. Comparison of Capital Costs for Disinfection with UV and Chlorine.

Plant ID	UV	NaOCl	Chlorine Gas	Comments
1	\$1,103,000	\$994,000		Expansion and upgrade
2	\$421,000	\$238,000	\$194,000	Upgrade
3	\$18,177,000 [b]	\$2,392,000	\$2,148,000	Upgrade
4	\$2,106,000	\$2,407,000		Disinfectant change
5	\$1,976,000	\$541,000	\$0	Upgrade
6	\$3,210,000	\$553,000	\$553,000	Upgrade
7	\$2,329,000	\$976,000		New facility
8	\$6,355,000	\$1,983,000	\$3,925,000	Expansion and upgrade
9	\$54,146,000	\$31,273,000		Upgrade

[a]All costs adjusted to first quarter 2007 (ENR CCI =7870).
[b]Value does not include the cost of infrastructure needed for residual and prevention of secondary growth.

Table 4-11 compares estimated annual O&M costs for UV and chlorine disinfection systems for the nine POTWs. The details for the O&M costs were generally not provided in the reference. For two plants, these costs were estimated from the present worth of the O&M costs.

Annual O&M costs are site-dependent, and tend to increase with plant size and more stringent disinfection objective, in a manner similar to capital costs. In general, the UV systems have lower O&M costs than the chlorine systems.

Table 4-11. Comparison of Annual O&M Costs for Disinfection with UV and Chlorine. [a]

Plant ID	UV	NaOCl	Chlorine Gas	Comments
1	$125,000	$147,000		
2	$19,000	$34,000	$24,000	
3	$465,000[b]	$1,562,000	$958,000	UV costs estimated[b]
4	$98,000[c]	$269,000		
5	$69,000	$164,000	$96,000	
6	$182,000	$373,000	$229,000	
7	$415,000	$1,204,000		
8	$92,000[d]	$234,000[d]	$130,000[d]	No Annual costs provided
9	$3,301,000	$2,073,000		

[a]All annual costs adjusted to first quarter 2007.
[b]Annual O&M costs not provided; estimated from present worth of O&M costs using calculated present worth factor for NaOCl and chlorine gas alternatives.
[c] With LPHO lamps; estimated annual O&M cost with MP lamps is $176,000.
[d]Estimated from present worth of O&M assuming 6% discount rate.

Table 4-12 compares the net present worth of UV and chlorine disinfection systems at the eight POTWs that provided estimates. Methodology differed among references, with a 4% discount rate used in four analyses, 6% rate in two analyses, and no discount rate information provided for two analyses. One analysis also considered a 3% inflation rate.

The results are mixed, with UV sometimes being more expensive and sometimes less expensive than chlorine. Therefore, care should be taken to develop site-specific costs so that the appropriate comparison of costs is provided for the individual POTW.

Table 4-12. Comparison of Net Present Worth of Disinfection with UV and Chlorine[a].

Plant ID	UV	NaOCl	Chlorine Gas	Project Life (Years)	Comments
1	-	-	-	-	No costs provided
2	$680,000	$712,000	$523,000	20	4% discount rate
3	$21,607,000	$32,017,000	$20,228,000[b]	20	No discount rate provided
4	$2,800,000[c]	$4,800,000		20	6% discount rate
5	$2,905,000	$2,764,000	$1,303,000	20	4% discount rate
6	$5,688,000	$5,616,000	$3,677,000	20	4% discount rate
7	$7,969,000	$17,344,000		50	4% discount rate
8	$5,209,000	$3,389,000	$3,819,000	20	No discount rate provided
9	$100,329,000	$60,314,000		20	6% discount rate; 3% inflation

[a] All costs adjusted to first quarter 2007.
[b] Includes the cost of a minor leak response. A moderate leak response would require increase the present worth cost to $22,440,000.
[c] With LPHO lamps; estimated net present worth with MP lamps is $3,035,000.

4.7 Summary

4.7.1 Advantages

UV disinfection of wastewater provides effective inactivation of most viruses, spores, and cysts. It is a physical process rather than a chemical disinfectant; this eliminates the need to generate, handle, transport, or store toxic/hazardous or corrosive chemicals. This also eliminates any effect from disinfectant residuals that can be harmful to human or aquatic life, and is user-friendly to operators. UV can operate effectively with a shorter contact time than other disinfectants, and uses a correspondingly smaller footprint. UV does not increase the TDS of the treated water unlike chlorination/dechlorination. Finally, it may provide incidental removal of emerging contaminants such as NDMA.

4.7.2 Disadvantages

UV disinfection also presents some potential disadvantages. It is less effective against adenovirus than other disinfectants such as chlorine, and low dosages may not effectively inactivate some viruses, spores, and cysts. UV disinfection is also less effective for wastewater with low UV transmittance (UVT); turbidity and suspended solids in the wastewater usually require very high UV doses making this technology less attractive. Microorganisms can sometimes reverse the destructive effects of UV through a repair mechanism, known as "photo-reactivation" under sunlight, or "dark-repair" in the absence of light. The lack of a disinfectant residual can result in regrowth and may cause problems (e.g., biofouling of the distribution pipeline for water reuse). Finally, on-line monitoring of delivered dose can be challenging, and a preventative maintenance program is necessary to control fouling of sleeves.

4.7.3 Gaps in Knowledge

Although UV technologies have become well established for wastewater disinfection, UV manufacturers are continually updating their equipment and systems. Future challenges include:

- Identification and modeling of all of the important parameters that affect disinfection efficacy, such as water quality, flow conditions, and biological characteristics (initial concentration of organisms, disinfection kinetics of organisms, etc.)
- A better understanding of the effects of water quality and operational conditions on regrowth
- Development of standardized method(s) to measure, calibrate, or certify how well equipment works before or after installation
- Improved understanding of hydraulic limitations and scale-up of closed channel systems
- An accurate method to scale up and validate, from pilot test results to full scale (e.g., Is disinfection performance by individual UV banks additive?)
- Reliable determination of in-situ UV intensity (which varies with lamp age, fouling, distance from lamps, flow paths, temperature, and location along the length of the lamp)
- Development of better control systems (integration with plant SCADA systems, more user-friendly interface, etc.)

- Understanding and optimization of the destruction of emerging compounds of concern; although this requires high doses, it may be an important decision criterion for some plants
- Investigation of disinfection byproducts: although UV does not use chemicals, it does apply large amounts of energy to the water (especially medium pressure lamps) and creates radical and excited species that may form byproducts that have not yet been identified
- Improve means to provide continuous UV disinfection after a shut down due to lack of flow or power due to power outage to UV system resulting in lamp restart conditions
- Improved energy efficiency of MP lamps
- Evaluation of the performance of developing UV technologies such as pulsed UV, broad-band pulsed xenon lamps, and narrow-band excimer lamps

CHAPTER 5.0

OZONE TECHNOLOGIES

5.1 Background

Under ambient conditions, ozone (chemical formula, O_3) is a blue gas with a pungent odor. To form ozone, an energy source is used to dissociate oxygen molecules (O_2) into oxygen atoms (O), which collide and react with oxygen molecules to form ozone. Ozone is relatively unstable and rapidly decomposes in water to form free radicals, including the hydroxyl radical (OH•). Both ozone and the resulting free radicals are stronger oxidizers than chlorine and can oxidize many organic and inorganic compounds in water and wastewater, thus making ozone a very effective disinfectant as well as a very effective oxidant. It is effective against bacteria and viruses, and provides some protection against cysts and eggs (Paraskeva and Graham, 2002).

Ozone can also be used for odor control at POTWs, for removal of soluble refractory organic compounds in advanced wastewater treatment, effluent oxygenation (Tchobanoglous et al., 2003), and to reduce solids production (Camacho et al., 2007). Recent research indicates that ozonation effectively removes a large number of endocrine disrupting compounds (EDCs) and pharmaceuticals and personal care products (PPCPs) that survive biological treatment processes; these data have caused renewed interest in the use of ozone for wastewater disinfection and for oxidation (Ternes et al., 2003; Huber et al., 2005; Snyder et al., 2007).

5.1.1 Historical Use

Ozone was first used for drinking water disinfection in 1893 in the Netherlands, and has been used since 1906 at a water treatment plant at Nice, France. In contrast, ozone was first used for wastewater disinfection in the United States in 1975 at Indiantown, Florida. Since then, almost 50 major POTWs have installed ozonation equipment, mostly during the 10 years between 1975 and 1984 (Thompson and Leong, 2007); see Section 7.4 for more details. Most of these POTWs have stopped using ozone for a variety of reasons: changes in legislation, operational and maintenance problems (a major issue with first-generation ozone facilities), inability to attain performance objectives without major modifications, and high ozone dosage requirements at some facilities that raised the cost of ozonation relative to other alternatives (U.S. EPA, 1986; Paraskeva and Graham, 2002).

In 2006, seven POTWs were using ozone, with a combined capacity of 155 MGD; at least four major POTWs were in planning, design, or construction to change their disinfection technology to ozonation (Thompson and Leong, 2007). These seven POTWs represent approximately 0.2% of all major POTWs in the United States; in contrast, 25% of the drinking water in the U.S. is treated with ozone. However, advances in ozone generation and solution technology developed by the drinking water industry have made ozone more economically competitive during the past 10 to 15 years, so ozone use at POTWs may increase, particularly with increasing concerns over microconstituents such as PPCPs.

5.1.2 How Ozone Disinfects

Inactivation of bacteria by ozone is attributed to the oxidation of cell membrane components, including glycoproteins, glycolipids, or certain amino acids such as tryptophan. In addition, ozone can disrupt bacterial enzymatic activity by acting on the sulfhydryl groups of certain enzymes. It may also attack both purines and pyrimidines in nucleic acids within the cell (U.S. EPA, 1999a).

For virus inactivation, ozone appears to modify and break the protein capsid sites that the virus uses to fix on the cell surfaces. This liberates and inactivates the viral ribonucleic acid (RNA) or deoxyribonucleic acid (DNA). With the tobacco mosaic virus, ozone was found to attack both the protein coat and RNA (U.S. EPA, 1999a).

For cysts, ozone is hypothesized to damage the cyst exterior, thereby enabling inactivation. For example, it can destroy the mucoid plugs covering the pores of *Naegleria gruberi* cysts. For *Giardia muris*, ozone is speculated to make the cyst wall more permeable to aqueous ozone, which can then damage the plasma membranes and affect the nucleus, ribosome, and other ultrastructural components (U.S. EPA, 1999a).

5.1.3 Ozone Dosage and Residual

A number of parameters are used to monitor ozone disinfection: applied ozone dosage, transferred ozone dosage, and ozone residual. Sometimes referred to as simply "dosage," the applied ozone dosage is the amount of ozone added to the water. The transferred ozone dose (TOD) is the amount of applied ozone that dissolves into the water. The TOD is defined as the difference in the ozone applied in the feed gas and the ozone remaining in the contactor off-gas (exhaust) per unit volume of wastewater treated, as shown in Eq. 5.1.3-1. The ozone residual, which is sometimes referred to as the "dose," is the concentration of dissolved ozone in the water (i.e., the ozone available for disinfection) at a given time.

$$\text{TOD (mg/L)} = \{([O_3]_{gas\ in} * Q_{gas\ in}) - ([O_3]_{gas\ out} * Q_{gas\ out})\} / Q_{water} \qquad (5.1.3\text{-}1)$$

where: $[O_3]_{gas\ in}$: ozone concentration in the feed gas (mg/L)
$[O_3]_{gas\ out}$: ozone concentration in the off-gas (mg/L)
$Q_{gas\ in}$: feed gas flow rate (e.g., m^3/day)
$Q_{gas\ out}$: off-gas flow rate (e.g., m^3/day)
Q_{water}: wastewater flow rate (e.g., m^3/day)

5.2 Disinfection Efficacy

Analogous to chlorine disinfection, ozone disinfection efficacy likely depends on the ozone residual and the reaction time. Although some early studies indicated that TOD is the most important factor governing the extent of ozone disinfection and that contact time has relatively little effect (U.S. EPA, 1986; Janex et al., 1999), these studies generally did not include information on the ozone residual concentrations at points within the contactor. Consequently, it is unknown whether the ozone residual persisted for the full contact time, and is not possible to assess the disinfection CT conditions under which these data were developed.

5.2.1 Effect of Water Quality on Disinfection

The ability to provide, maintain and monitor the dissolved ozone concentration is critical for compliance with disinfection regulations. Consequently, factors that accelerate ozone decomposition are undesirable because the ozone residual dissipates faster, requiring a corresponding increase in the applied ozone dosage, and thus increasing the ozone system's operating cost (U.S. EPA, 1999a).

The rate of ozone decomposition is a complex function of pH, temperature, and concentration of organic and inorganic constituents in water. This section describes the effects that pH, temperature, carbonate species, suspended matter, and other wastewater constituents have on the reaction rate of ozone and pathogen inactivation.

5.2.1.1 pH

Ozone decomposes faster at higher pH, forms hydroxyl radials more rapidly, and forms various oxidants with differing reactivities. This means that for a given TOD, a higher pH level will result in a lower level of ozone residual available for disinfection. For a given level of dissolved ozone residual measured, the reported effects of pH on pathogen inactivation varied from negative, to no effect, to positive (U.S. EPA, 1999a).

5.2.1.2 Temperature

The impact of temperature can be estimated from the regulatory requirements of USEPA's Surface Water Treatment Rule (SWTR) and Long-Term 2 Enhanced Surface Water Treatment Rule (LT2SWTR) (U.S. EPA, 2006). In the rule's *Giardia lamblia* CT tables, the required CT doubles for every 10ºC decrease. The LT2SWTR disinfection CT tables for *Cryptosporidium parvum* require a four-fold increase in CT for each 15ºC decrease in the water temperature. These requirements provide an indication of the change in disinfection kinetics to be expected with changes in wastewater temperatures.

5.2.1.3 Carbonate Species

Carbonate and bicarbonate alkalinity scavenge the radicals produced by ozone (U.S. EPA, 1999a; Tchobanoglous et al., 2003), but have essentially no effect on ozone decomposition. These scavenging reactions may partially explain the decreased disinfection efficacy at higher pH, because higher alkalinities are typically associated with higher pH.

5.2.1.4 Suspended Matter

Suspended solids in water may increase ozone demand and shield embedded bacteria (Tchobanoglous et al., 2003), but literature results are conflicting. Ozone inactivation of viruses and bacteria contained in aluminum flocs was not reduced at turbidity levels of 1 and 5 NTU; this implies that the microorganisms received no protection from the flocs (Walsh et al., 1981). Similar results have been obtained for poliovirus 1, Coxsackie virus A9, and *E. coli* associated with bentonite clay. However, adsorption of the f2 bacteriophage at 1 and 5 NTU of bentonite clay did retard the rate of inactivation by ozone (Boyce et al., 1981).

5.2.1.5 Other Wastewater Constituents

Many wastewater constituents react with ozone, thereby exerting an ozone demand that must be satisfied to ensure sufficient ozone residual is available for disinfection. Both organic and inorganic compounds can exert an ozone demand. For organic compounds such as humic materials or greases, the degree of interference depends on their functional groups and chemical structures. Reduced forms of inorganic compounds such as nitrite, sulfide, iron and manganese can be oxidized by ozone (U.S. EPA, 1999a; Tchobanoglous et al., 2003).

Literature on the effects of organic matter (typically measured as COD, BOD, or TOC) and color on ozone disinfection vary, with some reports indicating no strong effect on the overall effectiveness of ozonation treatment (e.g., Paraskeva, 1998). However, White (1999) presented the following empirical equations to estimate total and fecal coliform concentrations as functions of water quality and ozonation parameters for unfiltered secondary effluent:

$$\log_{10} TC = 0.96 \log_{10} TC_o - 0.89 + 0.012\ TCOD + 0.6\ NO_2\text{-}N + 0.013\ TSS - 4.024 \log_{10} TOD - 0.57\ R^{1/2} \quad (5.2.1.5\text{-}1)$$

$$\log_{10} FC = 4.06 + 0.020\ TCOD + 0.37\ NO_2\text{-}N + 0.012\ TSS - 3.94 \log_{10} TOD - 0.59\ R^{1/2} \quad (5.2.1.5\text{-}2)$$

where:
FC = fecal coliform concentration (count/100 mL)
TC = total coliform concentration (count/100 mL)
TC_o = initial total coliform concentration (count/100 mL)
TCOD = total COD concentration (mg/L)
TOD = transferred ozone dosage (mg O_3/L)
NO_2-N = nitrite concentration (mg N/L)
TSS = total suspended solids concentration (mg/L)
R = ozone residual concentration (mg/L)

5.2.2 Inactivation of Various Pathogens

Ozone is very effective against bacteria, and can inactivate total and fecal coliforms as well as enteric viruses in secondary effluent. The ozone dosages typically range between 2 and 15 mg/L, with contact times between 1 and 30 minutes (Paraskeva and Graham, 2002). In one study, *E. coli* levels in water were reduced by 4 logs (99.99%) in less than 1 minute with an ozone residual of 0.009 mg/L at 12°C. Bacteria spores are more resistant to ozone disinfection, but ozone can easily inactivate all forms at relatively low levels (U.S. EPA, 1999a).

Viruses are typically more resistant to ozone than vegetative bacteria, but an ozone dosage of 5 mg/L with a contact time of 5 minutes was sufficient to achieve 5-log reduction of the highly resistant MS2 bacteriophage (Lazarova et al., 1998). It should be noted that the information in this study did not differentiate between the applied ozone dose (ozone demand + ozone residual) and transferred ozone dose (TOD), which is the effective portion causing the disinfection.

Little information exists regarding the performance of ozone with other microorganisms and parasites (e.g., protozoan cysts and nematodes) in wastewater effluents. However, protozoan cysts, such as *Giardia* and *Cryptosporidium*, are known to be much more resistant to ozone than vegetative forms of bacteria and viruses. Table 5-1 compares the disinfection efficacy of ozone against several organisms. Ozone was effective against *Pseudomonas aeruginosa*, moderately effective against *Giardia lamblia*, and relatively ineffective against *Cryptosporidium parvum* in secondary and tertiary effluents treated with an ozone dosage of 15 mg/L with a 10 minute contact time. Because of the low levels of *Cryptosporidium parvum* in the water, the effectiveness of ozone against this microorganism was not conclusive. With respect to these microorganisms, ozone is still much more effective than chlorine, chloramines, or chlorine dioxide.

Table 5-1. Effectiveness of Ozone on Selected Pathogens. (Liberti et al., 2000)

	Log Removal	
Microbe	**Secondary Effluent**	**Tertiary Effluent**
Pseudomonas aeruginosa	1.8	2.0
Giardia lamblia	0.4	0.5
Cryptosporidium parvum	0.1	0.0

5.3 System Design and Implementation

5.3.1 Maturity of the Technology

There is more than 30 years of design experience in the use of ozone at both POTWs and water treatment plants (WTPs) in the U.S. Although the use of ozone for wastewater disinfection has decreased significantly and been limited to relatively few major POTWs during much of the past 20 years, most of the technology advances in ozone system design in the water industry are applicable to the wastewater industry. Therefore, some of the design information in this chapter is based on relevant experience in the water industry.

Ozone generation is an established process, but ozone generation and solution technologies are still evolving. There have been significant increases in ozone generator capacity as well as advances in ozone generator efficiency, oxygen gas (supply) quality, ozone contactor design and transfer systems during the past 25 years. These ozone system advances have been complemented by advances in both the instrumentation used to monitor the ozone gas concentration and the dissolved ozone concentration in water as well as in control systems. Although these technical advances have been driven by the use of ozone in the water industry, the wastewater treatment industry can make use of most, if not all, of these advances.

5.3.2 Design Approach

The design of an ozone disinfection process includes selecting the capacity of the ozone system equipment to economically meet the disinfection objectives over the anticipated range of operating conditions. Important design factors include the energy efficiency of the process, effluent disinfection requirements, wastewater quality, and wastewater flow fluctuations (U.S. EPA, 1986). Typical design steps are listed below and examples can be found in U.S. EPA (1986) and WEF (1996):

- Determine the initial ozone demand and slope of dose-response curve
- Estimate the ozone-transfer dose and efficiency
- Select the number and contact time for the ozone contactor(s)
- Determine the required applied ozone dosage and associated ozone system capacity
- Select the number and capacity of the ozone generators

- ♦ Determine the number and capacity of the liquid oxygen (LOX) tank(s) and/or of the on-site oxygen concentration unit(s)

Both the applied ozone dosage and TOD are needed for design. The applied ozone dosage determines the ozone production capacity required. The TOD is used to establish the relationship between the applied ozone dosage and ozone transfer efficiency. When designing an ozone disinfection system, disinfection performance should be evaluated with respect to the TOD and/or the product of dissolved ozone concentration (as a function of time) and the available contact time. Table 5-2 presents typical TOD values to achieve different effluent disinfection standards for various types of wastewater for a 15-minute contact time (Tchobanoglous et al., 2003). The numbers serve only as a guide for initial estimation; site-specific TOD values should be derived from pilot-scale studies.

Table 5-2. Typical Transferred Ozone Dosages to Achieve Effluent Coliform Standards. (Tchobanoglous et al., 2003 with permission from McGraw Hill)

		Ozone Dosage (mg/L)			
	Initial Coliform	Effluent Coliform Standard, MPN/100 mL			
Wastewater Type	**(MPN/100 mL)**	**1000**	**200**	**23**	**<2.2**
Raw wastewater	10^7 - 10^9	15 - 40			
Primary effluent	10^7 - 10^9	10 - 40			
Trickling filter effluent	10^5 - 10^6	4 - 10			
Activated-sludge effluent	10^5 - 10^6	4 - 10	4 – 8	16 - 30	30 - 40
Filtered activated-sludge effluent	10^4 - 10^6	6 - 10	4 – 8	16 - 25	30 - 40
Nitrified effluent	10^4 - 10^6	3 - 6	4 – 6	8 - 20	18 -24
Filtered nitrified effluent	10^4 - 10^6	3 - 6	3 – 5	4 -15	15 -20
Microfiltration effluent	10^1 - 10^3	2 - 6	2 – 6	3 - 8	4 -8
Reverse osmosis effluent	None				1 -2

The required TOD depends on the quality of the wastewater to be treated, effluent discharge limits, and the anticipated ozone contactor performance. Due to the variables involved, selection of an appropriately, but not excessively, conservative value for the TOD is probably the most challenging consideration in process design. Generally, two approaches can be taken to select the design TOD. One is to review the applied ozone dosages and transfer efficiencies at similar existing wastewater treatment plants and conduct the design based on these values. Because only a few POTWs currently use ozone, the recommended approach is to conduct a pilot-plant evaluation on the wastewater to be disinfected to establish the site-specific relationship between organism inactivation and TOD. The U.S. EPA (1999b) indicates that pilot-testing and calibration of ozone disinfection systems is critical prior to installation, to ensure that the disinfection system can meet discharge permit requirements for each specific site.

5.3.3 Facility Requirements

At a minimum, secondary treatment should be provided prior to ozone disinfection. Generally, ozone treatment of secondary effluents has been found to be capable of achieving an effluent quality equivalent to the European Union bathing waters and World Health Organization (WHO) agricultural standards as well as other comparable standards. However, to meet more stringent standards such as the California Title 22 regulations, tertiary filtration may be necessary before ozonation (Paraskeva and Graham, 2002).

Due to safety concerns, ozonation system equipment should generally be isolated in its own room, with separate exterior entrances, separate heating and ventilation system, noise control, etc. Ozone generators and each ozone generator's associated power supply unit (PSU) should be located indoors for protection from the environment. Operation and maintenance staff and electrical components should be protected from possible ozone gas exposure that could occur with leaks. Ventilation should be provided to provide a suitable temperature and to exhaust ozone-rich air from the room in case of a leak (U.S. EPA, 1986; U.S. EPA, 1999a).

The earliest generation of ozone generators used higher voltage electrical power than the current generation, but the electrical wires still requires special insulation (U.S. EPA, 1986). The electrical energy required to produce ozone is typically between 4 and 6 kilowatt-hours per pound (kWh/lb) of ozone for the ozone generator and 0.5 to 5 kWh/lb of ozone for other system requirements. Consequently, the electrical energy required to produce ozone will typically range between 5 and 10 kWh/lb ozone.

Due to the strong oxidizing power of ozone, all construction materials downstream of the ozone generator must be resistant to ozone oxidation (stainless steel 316 or 316L, glass, ceramic, Teflon, Hypalon, or polytetrafluoroethylene). The ozone gas pipeline connections should be welded or flanged. The ozone gas delivery pipeline should include a pipe loop that rises at least 5 feet above the maximum possible water level inside the ozone contactor to prevent water from flowing back into the ozone generator, which can have expensive and dangerous consequences. On the positive side, the size of an ozone contactor required for disinfection can be significantly smaller and less costly than a chlorine contactor with the same disinfection goals (Thompson and Leong, 2007).

5.3.4 Ozone System Components

Ozone is an unstable gas that decomposes to elemental oxygen very rapidly after generation, and must therefore be generated on-site for use. Major ozone system components are shown in Figure 5-1 and typically include:

- Oxygen supply system, associated oxygen gas vaporizers (for LOX systems), and gas quality monitoring instruments, including dew point analyzers;
- Ozone generators, and the associated power supply units (PSUs) and ozone gas concentration monitors;
- Ozone contactors, and the associated ozone gas transfer system and dissolved ozone concentration analyzers;
- Ozone contactor off-gas handling and residual ozone gas destruction system, and ozone gas monitors;
- Ambient oxygen gas and ozone gas concentration monitoring system;
  Control system

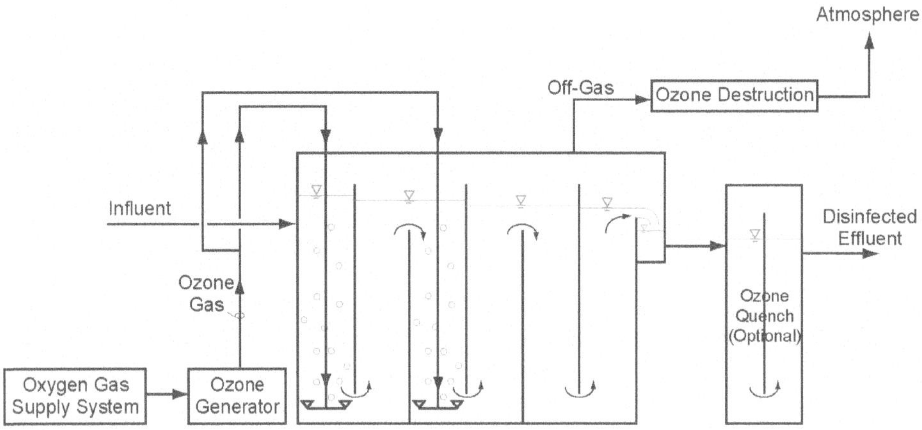

Figure 5-1. Schematic Diagram of an Ozone Disinfection System.

5.3.4.1 Oxygen Supply Systems

The current ozone system designs for both POTWs and WTPs are dominated by the use of high-purity oxygen gas supplies delivered to the site as liquid oxygen (LOX) or concentrated on-site from atmospheric air using molecular sieve technology. The LOX is stored on-site in a cryogenic tank and converted to gaseous oxygen (GOX) in a LOX to GOX vaporizer prior to delivery to the ozone generator. One of the major advantages of storing and using LOX on-site is that the oxygen gas supply flow rate to the ozone generators can be changed relatively rapidly in response to a change in the required/desired ozone production rate.

Although some of the early ozone systems included the ability to reuse the oxygen-rich off-gas from the ozone contactor(s) as part of the ozone generators' oxygen gas supply, operational problems associated with the high moisture content in the off-gas have eliminated use of this design approach. However, it should be noted that some POTWs were designed to recirculate the oxygen-rich off-gas from the ozone contactor(s) back to the activated sludge process as the primary, or as a supplemental, oxygen source.

5.3.4.2 Ozone Generators

The ozone generators and their associated power supply unit (PSU) systems that are currently available are essentially the third generation of ozone generator technology. Early ozone generators typically had the capacity to produce up to about 250 pounds of ozone per day at up to 1% by weight from dry air or up to about 4% by weight from high-purity oxygen gas. In comparison, the current state-of-the-art ozone generators can produce as much as 4,000 pounds of ozone per day at 10 to 12% by weight from high-purity oxygen gas.

The advances in generator technology have significantly reduced power and chemical costs, relative to earlier generations. Due to the improvements in power supply technology, 10 to 12% by weight ozone concentrations can be produced using the same amount of electrical power per pound of ozone (kWh/lb) that was required to produce 1% by weight ozone concentrations 25 years ago. In addition, the ability to produce ozone at higher ozone gas concentrations significantly reduced the amount of LOX (10 versus 100 pounds when the gas concentration increased from 1% by weight to 10% by weight) needed to produce one pound of ozone, and hence, the production cost due to purchased LOX.

5.3.4.3 Ozone Contactors

The ozone contactor typically consists of a series of downflow ozone transfer chambers and upflow reactor chambers (see Figure 5-1), where ozone gas is dissolved into the water and given time for disinfection and reaction. The design of the contactors and the associated ozone transfer systems has undergone significant advances during the past 25 years. Ozone contactor design changes have been due to both the increase in the ozone gas concentration produced by newer generations of ozone generators, as well as responses to regulatory requirements.

Ozone Transfer Technology The technology used in the ozone transfer chambers has evolved with improvements in ozone generators over the years. As ozone concentrations increased, less gas was needed and gas flow rates decreased, which allowed the use of more efficient gas transfer equipment. The first generation technology produced 1 to 2% by weight ozone gas from clean dry air, and used large ceramic diffuser rods to transfer the ozone gas to the water. The shift in oxygen source to LOX (which produces 4 to 6% by weight ozone gas concentration) allowed the use of more efficient ceramic "dome" or "disk" style diffusers, while the current generator technology (which produces 10 to 12% by weight ozone gas concentration) and/or lower ozone dose required for some treatment goals has permitted the use of eductors to transfer the ozone gas into the water.

Ozone Contactor Design The changes in ozone transfer equipment have affected contactor design. Early contactors had large ozone transfer chambers to provide enough space for the diffuser rods and sufficient time for dissolution, and relatively small reactor chambers. As transfer technology moved to smaller and more efficient diffusers around 20 years ago, the size of ozone transfer chambers decreased to approximately the same size as the subsequent reactor chambers.

Several water treatment plants use pipelines as the ozone contactor. High-efficiency gas eductors are recommended to efficiently transfer ozone from the gas phase into a dissolved condition in the water when a pipeline is used as the contactor. This application may be evaluated on a case-by-case basis and could reduce capital cost by eliminating the larger reinforced concrete ozone contactor structure(s) and reducing O&M cost by possibly eliminating the need for re-pumping.

On-line Monitors for Dissolved Ozone Concentrations in the Contactor Almost none of the POTWs that currently apply ozone for disinfection use on-line monitors to measure dissolved ozone concentrations in water as it flows through the ozone contactor. However, on-line monitoring is recommended. Without data on the aqueous ozone concentrations (residuals), the true disinfection contact time (i.e., the time period until the residual disappears) is unknown, and a CT model cannot be used to predict disinfection.

5.3.4.4 Off-Gas Destruction System

The ozone gas concentration in the ozone contactor off-gas will typically be between 0.1 and 1.0% by volume (between 1,000 and 10,000 ppmv), depending on the ozone dose, ozone demand, gas transfer system efficiency, contactor design, and other factors. The current OSHA time-weighted 8-hour exposure limit for ozone is 0.1 ppmv and the 15-minute exposure limit is 0.3 ppmv. Therefore, the ozone contactor off-gas must be treated to reduce the ozone concentration to less than 0.1 ppmv prior to venting to the atmosphere. The off-gas destruction ("destruct") system generally uses a catalyst to convert ozone back to molecular oxygen.

5.3.5 Ease of Expansion

The size of an ozone system is proportional to the flow rate; therefore an increase in the POTW design flow rate would require adding more ozone generator(s), possibly increasing the oxygen supply system capacity, and constructing additional disinfection contactor capacity.

5.4 Operation and Maintenance

5.4.1 Ease of Operation and Maintenance

Because the ozonation system is relatively complex, routine regular maintenance is essential (WEF, 1996). This section describes some recommended maintenance activities.

5.4.1.1 Oxygen Gas Supply System

The high purity sources of oxygen (LOX, on-site cryogenic oxygen, or on-site molecular sieve oxygen) that are used to generate ozone at most plants are clean and dry with a dew point below -60°C (-76°F). However, source gas with a high dew point (above -60°F) will require more frequent maintenance of the ozone generators because ozone reacts with moisture in the feed gas to generate a very corrosive condensate that contains nitrogen oxides such as nitric acid. On all systems, particle filters should be replaced periodically; the replacement cycle is site-specific, but inspection intervals for the filters should not exceed three months.

Each of the high purity oxygen systems has specific maintenance requirements. LOX systems require specialized maintenance by appropriately trained technicians. It is recommended that the LOX supply contract include a maintenance service agreement. Cryogenic oxygen distillation systems require annual maintenance, including significant shutdown and start-up time prior to and after conducting the maintenance tasks. Cryogenic systems should include a LOX system as both a backup and supplemental oxygen supply. On-site molecular sieve oxygen concentration units require annual maintenance, and their associated air compressors also need annual maintenance.

On the older ozone systems that use ambient air, the gas pressurization units (air compressors) can contaminate the feed gas and should therefore be checked regularly. After-coolers require annual inspection for scale accumulation, corrosion, or metal deterioration. Refrigerant dryers should be inspected at 3-month intervals or more frequently. Finally, desiccant dryers need weekly inspection with an annual maintenance that includes disassembly of the unit and inspection of the desiccant media.

5.4.1.2 Ozone Generator

Ozone generators and on-site oxygen gas concentration systems consume a lot of electrical power, so power usage should be optimized with respect to disinfection performance. Because most wastewater treatment plant staff are unfamiliar with ozone generation equipment, maintenance contracts with the equipment suppliers should be used, to provide operation and maintenance training and emergency assistance during the first few years of operation.

There must be no leaking connections in or surrounding the ozone generation system, because a very small leak of ozone gas from the generator can cause a dangerous ambient ozone concentration (see Section 5.4.3 for details). The ambient ozone and oxygen gas concentration monitoring equipment should be tested and calibrated as recommended by the manufacturer. Ozone generation system components including air compressors, coolant levels, air supply, gaskets, and dielectric assemblies should be regularly monitored to ensure that they are in good working condition and serviced according to the manufacturer's recommendations.

5.4.1.3 Ozone Generator Power Supply Unit and Other Auxiliary Equipment

Each ozone generator's power supply unit (PSU) should be inspected and tested annually. Flow meters, temperature and pressure sensors, analyzers for ozone gas and dissolved ozone concentrations, and other instrumentation and controls should be inspected and maintained regularly to ensure accuracy of data and to provide data needed to optimize system operation.

5.4.1.4 Ozone Contactor and Off-Gas Destruction System

The ozone contactor(s) should be inspected annually for both structural and the functional integrity of all components. The ozone contactor(s) should be well-ventilated and internal air quality tested before personnel are permitted to enter. Gaskets in the diffusers are exposed to the highest concentrations of ozone and should be replaced annually. Other gaskets in the ozone system should be checked annually and replaced if their condition has deteriorated.

Each component of the ozone destruct system should be inspected regularly. The ozone destruction catalyst will typically provide satisfactory service for 3 to 5 years, and should be replaced when the ozone gas concentration in the off-gas discharged to the atmosphere increases to 0.1 ppmv. If chlorine is used as a backup disinfectant in the ozone contactor, the off-gas destruction system should not be operated, because chlorine will poison the catalyst.

5.4.2 Ozone Production Process Control and On-line Monitoring

The key process control parameters are ozone dosage, transfer efficiency, mixing, ozone half-life, and contact time. For most ozone systems (i.e., those that use LOX as the oxygen source to produce ozone), the main objectives of process control and automation are to minimize energy use and oxygen consumption. Older ozone systems that use clean, dry air as an oxygen source focus on optimizing electrical power consumption.

Ozone production rates can be regulated by adjusting the ozone generator(s)' power setting. Three approaches are commonly used for ozone production process control. The simplest control method is to provide a constant applied ozone dosage. This method works best when the wastewater flow rate, wastewater quality and associated ozone demands, and the disinfection goals performance do not vary considerably. However, the ozone dosages needed for adequate disinfection typically vary on a routine basis. The second control method uses ozone residual control, which is effective in theory because disinfection performance is typically closely related to ozone residual. However, this method is generally more difficult in practice because it requires maintaining calibration of the dissolved ozone residual monitors' sensors, which are vulnerable to fouling associated with contaminants in the treated, disinfected effluent as well as changes in water characteristics, and time-delays between the residual monitor signal and the control system. The third option is off-gas ozone concentration control, which was a relatively new approach in the mid-1980s (U.S. EPA, 1986), but has not been widely adopted.

During operation, ozone concentrations should be analyzed at a number of locations The ozone concentration in the product gas stream of the ozone generators should be monitored to determine the applied dosage. The ozone content of the off-gas from the contactors should be monitored to determine the ozone transfer efficiency; high ozone concentrations may indicate problems with the ozone transfer system. The ozone concentration in the off-gas after the destruct unit(s) should be monitored to ensure the safety of staff and detect any problems with the ozone destruct system. The residual ozone concentration in the disinfected water leaving the first ozone transfer chamber and leaving at least one subsequent chamber in the ozone contactor should be monitored to determine the current ozone demand, the ozone half-life, and to verify that the desired CT condition is being achieved.

5.4.3 Operator Safety and Health Risks

The primary safety concern for ozone systems is the ozone itself. Ozone is a hazardous gas and should be handled accordingly. Gaseous ozone is reported to be explosive when its concentration reaches about 240 mg/L. If inhaled, ozone causes dryness of the mouth, coughing, and irritation to the nose, throat, and chest at a significantly lower concentration. Ozone exposure may also cause labored breathing, headache, and fatigue. The characteristic sharp, irritating odor is readily detectable at concentrations as low as 0.01 to 0.05 parts per million by volume (ppmv), or 2×10^{-5} to 1×10^{-4} mg/L. Ozone concentrations of 0.1 ppmv inhaled during an eight-hour period in a work area is the maximum limit set by regulations from the Occupational Safety and Health Administration (OSHA).

POTWs that generate ozone from high purity oxygen gas are vulnerable to hazards from both elevated ozone gas and elevated oxygen gas concentrations. Ozone gas will remain in the generation equipment and the gas delivery pipeline system for prolonged periods of time. Therefore, these units must be carefully purged before being opened, and at least two self-contained breathing apparatus (SCBA) should be kept nearby. Entry into an ozone contactor should be carefully planned and carried out for maximum personnel safety, due to possible oxygen deficiencies at the bottom of the enclosed tanks (WEF, 1996; WEF, 2006).

In addition, ozone or oxygen gas concentrations could rise to hazardous levels in work areas if a leak occurs in an ozone gas pipeline, ozone generator, or the oxygen gas supply system, including the high-purity oxygen gas pipelines. All ozone systems should have ambient ozone monitors to measure the ozone gas concentration at locations within the plant that could potentially be contaminated by ozone; systems that use high purity oxygen should also monitor ambient oxygen gas concentrations. The ozone system controls should be able to automatically shut down the ozone generator(s) and the oxygen supply system if the ambient ozone gas or ambient oxygen gas sensors indicate that there may be a gas leak. The ozone generator room should also include a "panic button" to permit the operating staff to effect an emergency shutdown of the ozone generators if an unsafe condition is apparent (U.S. EPA, 1986). In addition, ozone or oxygen gas concentrations could rise to hazardous levels in work areas if a leak occurs in an ozone gas pipeline, ozone generator, or the oxygen gas supply system, including the high-purity oxygen gas pipelines. All ozone systems should have ambient ozone monitors to measure the ozone gas concentration at locations within the plant that could potentially be contaminated by the ozone contactor off-gas, which will typically contain ozone gas at concentrations well above the OSHA-permitted levels. This off-gas must be treated to convert the ozone to molecular oxygen prior to discharge to the atmosphere, because ozone poses worker health risks and is a regulated air pollutant.

One final concern is the electrical risk posed by ozonation equipment. The systems are complex and use high-voltage power, so only properly-trained and qualified electrical and mechanical technicians should be permitted to do any equipment maintenance or repair tasks.

5.4.4 Operational Impacts on Existing Treatment Trains

Ozone addition may have a number of impacts on a POTW's existing infrastructure. For example, ozone oxidation of iron and manganese generates insoluble oxides that may increase loading on tertiary filters and hence, filter backwash frequency, if ozone is added before filtration. Ozone also reacts with other oxidants such as chlorine and chloramines; however, using ozone ahead of chlorine disinfection reduces the chlorine demand of the treated water and allows a stable chlorine-compound residual to be maintained at a much lower level. This process train may also result in lower chlorinated DBP formation.

The use of ozone may convert some of the refractory organic matter in effluents to more biodegradable forms that can increase biological growth, which may increase the BOD in the treated effluent. Ozonation can also increase the dissolved oxygen (DO) levels in the final effluent, particularly when ozone is produced from high-purity oxygen rather than from clean dry air. This may improve effluent quality and/or satisfy discharge requirements. An elevated DO concentration in the effluent may promote biological growth, if disinfection is incomplete. However, biological regrowth has generally not been observed following ozonation (see Section 5.5.2) so this may not be a serious problem in practice.

5.4.5 Storage, Inventory, and Replacement Requirements

Since ozone is generated on-site and applied to the wastewater directly, there is no storage of ozone gas at the site. However, materials used for ozone generation may have storage requirements. For example, the storage of LOX is subject to regulations in building and fire codes. These regulations will impact the space requirements and may dictate the construction materials of adjacent structures if certain setback requirements cannot be met (U.S. EPA, 1999a).

5.4.6 Consistency in Performance with Fluctuations in Flow Rate

Ozone generation capacity is designed to achieve a specified applied ozone dosage at the design flow rate, which is typically two to three times the average daily dry weather flow rate. Stand-by ozone generation capacity is typically not provided for peak conditions beyond the design flow rate. Where peak flows are frequent and unpredictable, back-up or stand-by equipment should be provided (U.S. EPA, 1986).

In addition to possible concerns over high flow rates, ozone generators and ozone gas transfer systems often perform less efficiently when the ozone production rate is less than 25% of the maximum production rate and when the ratio of gas flow rate (in SCFM) to transfer basin volume (in cubic feet) is less than 0.01 minutes^{-1}. This issue can be resolved by including a sufficient number of ozone generators to permit removing one or more generators and/or contactors from operation when the combination of ozone dosage and plant flow rate is low.

5.5 Impacts to the Environment

5.5.1 Changes in Water Quality

Ozone is a strong oxidant and can react with constituents in effluents, thereby affecting the chemical composition and overall quality of the water (Hoigne, 1998). Studies of ozone treatment of secondary effluent have generally found that ozonation results in decreasing concentrations of COD, turbidity, color, UV absorbance, and anthropogenic compounds such as detergents, pesticides, phenols, aromatic compounds, unsaturated fatty acids, polysaccharides, and amino acids (see Section 5.5.3.2 for more details). Reductions in effluent color and UV absorbance (at 254 nm) are strongly correlated with ozone dose. The effect of ozonation on TOC and BOD_5 is not always consistent; some studies reported insignificant changes in these parameters following ozonation, while others reported increasing or decreasing concentrations. Wert et al. (2007) found an increase of threefold or more in the concentration of assimilable organic carbon (AOC) with ozone dosages above 3.6 mg/L. Finally, because ozone decomposes to oxygen, the dissolved oxygen (DO) level in the treated effluent typically increases following ozonation. This may be a benefit to POTWs that discharge to receiving water with a minimum DO requirement (Paraskeva and Graham, 2002).

5.5.1.2 Disinfection Byproducts

Ozone is capable of reacting with a wide range of organic and inorganic compounds in the water, leading to the formation of reaction intermediates and stable byproducts. Byproducts generated from ozonation include short-chain aldehydes (e.g., formaldehyde, acetaldehyde, glyoxal, methyl glyoxal), short-chain fatty acids (e.g., acetic acid, and formic acid), carboxylic acids (e.g., glyoxylic acid, pyruvic acid, ketomalonic acids, oxalic acid, succinic acid), alcohols, alkanes, ketones, and free amino acids. Ozone can also react with bromide, if present, to form bromate and brominated organics (bromoform, brominated acetic acids, bromopicrin, brominated acetonitriles, cyanogen bromide) (U.S. EPA, 1999a; Paraskeva and Graham, 2002).

Among the byproducts generated by ozonation, bromate (a regulated DBP) has received the most attention. It is a potential carcinogen and has a drinking water standard, or maximum contaminant level (MCL) of 10 μg/L. Another disinfection byproduct is monobromamine, which occurs when ammonia reacts with hypobromous acid (that is formed when ozone oxidizes bromide). The rate of the bromamine reaction is several orders of magnitude greater than the reactions that form bromate or bromoform. Both bromate and monobromamine formation appear to be correlated with the product of ozone residual concentration and ozonation time (C_RT), as well as the bromide and ammonia concentrations in the water; bromate and monobromamine concentrations increase as C_RT increases (U.S. EPA, 1999a; Paraskeva and Graham, 2002).

Bromate formation can be reduced by using a combination of ozone and hydrogen peroxide: O_3/H_2O_2 dosages of 2.1/1.0 mg/L and 3.6/2.0 mg/L yielded no bromate formation, and an O_3/H_2O_2 dosage of 7.1/3.5 mg/L yielded 13-16 μg/L bromate, approximately half of the concentration formed by the same dosage of ozone alone. However, the addition of hydrogen peroxide also increased biodegradable DBPs by 15-53%, and generally reduced disinfection efficacy; consequently, the use of ozone alone is recommended over the combination of ozone and hydrogen peroxide (Wert et al., 2007). A combination of ozone with hydrogen peroxide has been shown to increase conversion of bromide to bromate for some operating conditions at water treatment plants. Site-specific studies should be conducted to evaluate this risk.

5.5.1.3 Toxicity in the Treated Effluent

Ozonation may produce some toxic mutagenic or carcinogenic intermediate compounds, but these compounds may be unstable. Data are inconsistent and indicate that ozone treatment can either increase or decrease toxicity and mutagenicity, and the results are site-specific and seasonal (Rein et al., 1992; Blatchley et al., 1997; Monarca et al., 2000).

Ozone itself can be acutely toxic to many forms of aquatic life. Tolerance to ozone varies with the type and age of organism, period of exposure, and residual ozone concentration. Paraskeva and Graham (2002) found that between 20 seconds and two minutes were required for

0.2 to 1.0 mg/L of residual ozone to decay to nondetectable concentrations. Drury et al. (2006) found that ozone dosages between 2 and 9 mg/L usually decreased to residual concentrations less than 0.5 mg/L in 5 to 8 minutes, and decayed below the detection limit in 10 to 20 minutes. The relatively rapid ozone dissipation rate suggests that ozone residuals will normally not be detected by the time an effluent is discharged into receiving waters.

Finally, although residual ozone has been shown to be toxic to aquatic life, it is, by comparison, no more so than residual chlorine. In addition, the dissolved ozone residual disappears rapidly in the environment. At dosages typically used in wastewater disinfection, any ozone residual would be short-lived and should pose few problems to the receiving water bodies (WEF, 1996).

5.5.2 Potential Risk

Microorganism reactivation after ozonation is unlikely to occur (U.S. EPA, 1986; Paraskeva and Graham, 2002); Lazarova et al. (1998) did not observe a measurable regrowth of *E. coli* in seawater after ozonation.

In terms of risk to the surrounding community, there are fewer safety issues associated with shipping and handling than for chlorine, because ozone is generated onsite, and the quantity of ozone present in the ozone generator(s), gas conveyance pipelines and contactor(s) is typically much less than 1% of the ozone system's daily production capacity.

5.5.3 Additional Benefits

5.5.3.1 Dissolved Oxygen, Color, and UV Transmittance

Ozonation increases the dissolved oxygen concentration and reduces color in effluents. Ozone may also significantly increase the UV transmittance of wastewater, especially for secondary effluents.

5.5.3.2 Incidental Removal of Contaminants

As described in Section 5.5.1, the use of ozone can improve a number of water quality parameters, such as COD, turbidity, color, and UV absorbance. Ozone can also convert or destroy certain refractory organic substances such as humic acids and the insecticide malathion (White, 1999). In recent years, ozone has received attention for its ability to reduce the concentrations of many trace organic contaminants such as hormones, pharmaceuticals and personal care products (PPCPs), many of which are classified as EDCs. These trace organics include the following:

- Antibiotics: clarithromycin, erythromycin, roxithromycin, sulfamethoxazole, and trimethroprim
- Antiepileptics: carbamazapine, dilantin
- Antiphlogistics: acetaminophen, diclofenac, ibuprofen, indomethacin, naproxen
- Betablockers: atenolol, celiprolol, metoprolol, propranolol, sotalol
- Lipid regulators: clofibric acid, fenofibric acid, gemfibrozil
- Other pharmaceuticals: diazepam, fluoxetine, hydrocodone, pentoxifylline
- Hormones: androstenedione, estradiol, estriol, estrone, ethynylestradiol, progesterone, testosterone
- Musk fragrances: AHTN (Tonalide), HHCB (Galaxolide)

- Other: atrazine (pesticide), caffeine, phenols, combined amino acids and polysaccharides

Most of these compounds were reduced to below detection limits in secondary or filtered secondary wastewater effluents, with 9 to 15 mg/L of ozone and an 18 to 24 minute contact time. In addition, ozone reduced estrogenicity by more than 90% at a dosage of 5 to 9 mg/L and a 24 minute contact time. At lower ozone doses of 2 to 5 mg/L, twelve of the 34 compounds studied were incompletely removed and were detected: atenolol, caffeine, clofibric acid, dilantin, estriol, estrone, fenofibric acid, HHCB (Galaxolide), ibuprofen, metoprolol, pentoxifylline, and sulfamethoxazole. Other compounds were recalcitrant to ozonation and remained in the water even after treatment at ozone dosages of 9 to 10 mg/L: DEET (insect repellent), meprobamate (pharmaceutical), oxybenzone (sunscreen), trimethoprim (antibiotic), TCEP (flame retardant), and several iodinated x-ray contrast media (iomeprol, iopamidol, iopromide, and diatrizoate). Adding hydrogen peroxide or UV following ozonation significantly improved the removal of three of the four x-ray contrast media tested (iomeprol, iopamidol, and iopromide), but the diatrizoate was especially recalcitrant (Ternes et al., 2003; Snyder et al., 2007).

Although there are no current discharge or reuse limits on most trace contaminants because the potential health effect caused by these chemicals is not known, the ability of ozone to remove many of these chemicals may become more important if future regulations require more stringent limits for water reuse or discharge. Information provided by wastewater regulators in California indicates that unlike chlorine, there is currently no discharge limit for dissolved ozone in treated effluent (Carlton, 2007).

5.5.3.3 Using Ozone to Reduce Solids Production

Recent studies on the use of ozone for cell lysis in waste activated sludge (WAS) indicate that ozone is significantly more effective than heat (Camacho et al., 2005). Use of ozone to lyse cells could also increase the biogas produced in digesters.

5.6 Cost

The cost of ozone treatment depends on-site-specific factors, and cost information in the literature is limited. The following cost estimates assume a properly designed and operated secondary treatment system meeting a discharge standard of 30 mg/L BOD_5 and 30 mg/L TSS. It should be noted that the 2005-2007 cost climate for concrete, steel, stainless steel, and energy has been very dynamic, primarily increasing both capital and O&M costs and making it difficult to use historically based cost curves. Cost estimates should be considered very site-specific, requiring a site-specific project cost estimate for any potential project.

The costs in this section do not address costs for the associated insurance premiums and the annual risk assessment surveys and reports that may be required by insurers for LOX and accidental exposure to ozone. Also, some utilities have to pay more for liability risk insurance for employees that work in these environments. These costs have not been considered in this section of the report. The reduction or increase of these costs should be specifically included in cost analyses of various alternatives when performing a site-specific evaluation.

5.6.1 Capital Cost

Table 5-3 summarizes the estimated capital costs of ozone systems, based on information from WTPs and POTWs that recently added or upgraded an ozone system. Costs include the oxygen supply and ozone generation systems, the associated ozone transfer system, ozone contactor(s), off-gas destruct system, and ozone system controls. Each cost assumes a maximum applied ozone dosage between 5 and 7.5 mg/L and a minimum ozone contactor contact time of at least 10 minutes during peak flow conditions. Both the 1 and 10 MGD capacity systems include one ozone contactor, and the 100 MGD system includes two 50 MGD capacity ozone contactors.

Table 5-3. Capital Cost (in Millions of Dollars) for Ozone Disinfection Systems at POTWs.

	1.0 MGD	10 MGD	100 MGD
Transfer System and Ozone Contactor(s)	\$0.20 – 0.35	\$1.0 – 1.2	\$5.0 – 6.0
Oxygen Supply and Ozone Generator System	\$0.80 – 1.00	\$2.0 – 2.5	\$10.0 – 15.0
Total System Cost	\$1.00 – 1.35	\$3.0 – 3.7	\$15.0 – 21.0
Cost per MGD of Capacity	\$1.00 – 1.35	\$0.30 – 0.37	\$0.15 – 0.21

These costs have been normalized to a first quarter 2007 Engineering News-Record (ENR) 20 cities construction cost index (CCI) of 7870. The estimates are based on either planning level costs that have an accuracy of +50/-30%, or on construction bid costs adjusted by an additional 20% to account for other project costs, including engineering planning and design, construction management, and project administration. Cost estimates include an 8% mark-up for tax on materials, a 25% planning level contingency, 20% for contractor mark-ups, and 20% for design, construction administration and legal costs associated with the project.

The estimates in Table 5-3 are in agreement with a reported project cost of \$8.7 million for an ozone generation system and two ozone contactor stages at a 60 MGD capacity wastewater reclamation facility, for two construction phases that were bid in 1998 and 2002 (Spivey, 2007). After adjusting this total to include an additional 20% for other project costs (see previous paragraph for details), and normalizing to a first quarter 2007 ENR 20 cities CCI of 7870, the project cost is approximately \$14.2 million, or about \$240,000 per MG of treatment capacity.

5.6.2 Operating Cost

Table 5-4 provides estimated O&M costs for labor, power, and parts and supplies. These estimates are based on information provided by WTP managers in May 2007 for ozone systems with capacities to produce between 500 and 5,400 pounds of ozone per day. Costs have been normalized to a Consumer Price Index of 610. The costs assume an average applied ozone dose of 5.0 mg/L and a peak dose of 7.5 mg/L, labor cost of \$50/hour (including benefits), and electrical power costs of \$0.10 per kWh. The power consumption is based on an average applied ozone dose of 5 mg/L and an average flow that is 1/3 the treatment capacity. The cost per MG of water treated assumes that the plant operates at the design capacity.

Table 5-4. Estimated O&M Cost for Ozonation Disinfection Systems at WTPs.

Item	Annual Cost per MGD of Treatment Capacity	Cost/MG of Water Treated
Labor	$1,250	$3.40
Electrical Power Cost (Based on Power Use of 76.0 MW-hr)	$7,600	$20.80
LOX	$5,800	$15.90
Parts and supplies	$100	$0.30
Total	$14,750	$40.40

Table 5-4 indicates that electrical power and LOX comprise most of the O&M cost associated with ozonation. The total O&M cost (~$40/MG of water treated) is lower than that reported by Liberti et al. (2000). Based on pilot plant tests, they estimated O&M costs of €37.3/1000 m^3, or approximately $180/MG of water treated. This discrepancy may be due to a variety of factors, including issues with scaling costs from pilot to full scale, the assumption in Table 5-4 of operating at design capacity, or local variations in disinfection requirements or costs for power, chemicals, labor, etc. Costs may vary widely, and a site-specific cost estimate should be calculated for each individual facility.

5.6.3 Comparison with Other Disinfection Technologies

In terms of capital costs, Thompson and Leong (2007) indicate that the ozone generator and high-purity oxygen supply system is generally significantly more expensive than gas chlorine or hypochlorite solution systems, but this may be balanced by the significantly lower cost of an ozone contactor relative to a chlorine contactor and the elimination of a dechlorination system. Drury et al. (2006) estimated that the overall capital costs of ultrafiltration and microfiltration membranes with ozonation would be approximately 30% higher than the existing sand filters with UV disinfection.

As discussed in Section 5.6.2, the O&M costs associated with ozonation are dominated by electrical power and LOX costs. On a mass basis, the chemical cost of ozone is relatively high compared to the cost of chlorine gas and about the same as chlorine supplied as hypochlorite solution. However, it is more appropriate to compare ozone and chlorine on a molar basis, with 48 grams of ozone providing the same oxidation as 71 grams of chlorine: the relative cost of ozone per milliequivalent (meq) of electron donor is about 1.4 times more expensive than chlorine gas, but about 20% less than the cost of hypochlorite.

The ozone O&M costs in Table 5-4 and Section 5.6.2 ($7 to $180/MG of water treated) are similar to literature reports of UV O&M costs of $5 to $75/MG of water treated (see Section 4.6.2 for details). However, as discussed in Section 5.6.2, costs are very site-specific and are difficult, if not impossible, to compare across studies. Therefore, any facility comparing disinfection technologies should conduct its own cost estimates, based on the site requirements and conditions.

5.7 Summary

Ozone has been used as a disinfectant at major POTWs for over 30 years, but the number of facilities using ozone has significantly declined. Concerns of emerging contaminants, color, indirect water reuse, and reoxygenation of receiving waters have been drivers for POTWs that have installed or are planning to install ozone. Many improvements to ozone technology have been developed in the drinking water industry, where ozone use is common. Most of these improvements in equipment and design can also be applied to POTWs considering the use of ozone.

5.7.1 Advantages

The advantages of ozonation for wastewater disinfection include the following:

- Effective disinfectant for inactivating bacteria, viruses, parasites, and parasite cysts
- Relatively short contact times (approximately 10 to 15 minutes) at peak flow rates
- Rapid decomposition, so no harmful residuals to remove after ozonation
- Little increase in TDS due to disinfection process
- No chlorinated disinfection byproduct (THMs and HAAs) formation
- No regrowth of microorganisms except for those shielded by particulates
- Relative to chlorine, fewer safety problems associated with shipping and handling because ozone is generated onsite, and the quantity of ozone present in the ozone generator(s), gas conveyance pipelines and contactor(s) is typically much less than 1% of the ozone system's daily production capacity
- Increases the dissolved oxygen concentration of the effluent
- Usually reduces effluent COD, and sometimes BOD and TOC concentrations
- Helps with removing color, odor, and iron and manganese
- Can oxidize some emerging organic pollutants such as PPCPs and other EDCs

5.7.2 Disadvantages

Disadvantages of ozonation for wastewater disinfection include the following:

- Requires complicated equipment to provide high transfer efficiency and efficient contact for disinfection
- Relatively complex to operate and maintain; more suitable for medium to large POTWs
- Ozone is very reactive and corrosive, especially to elastomers and gasket materials, and therefore requires rigorous maintenance. Ozone systems must use special thicker low permeability covers of concrete to protect reinforcing steel in structures, or use of corrosion-resistant material such as stainless steel. These requirements increase capital costs.
- Ozone in off-gases from ozone contactors must be destroyed (converted back to O_2) to protect workers from an unsafe exposure to high levels of ozone, which is extremely irritating and toxic at relatively low levels
- Low dosages may not effectively inactivate some viruses, spores, and cysts
- No sustainable disinfection residual; chlorine may still be needed if disinfection residual is required
- Can generate disinfection byproducts, some of which (e.g., bromate) are potentially harmful to human health
- Requires at least secondary treatment processes that reduce high levels of suspended solids and natural organic matter because these increase ozone demand and reduce ozone effectiveness
- Higher capital costs than a hypochlorite solution system assuming there is an existing chlorine contact basin.

5.7.3 Gaps in Knowledge

Data gaps identified from the literature review and technical assessment of ozone include the following:

- The mechanisms of ozone byproduct and intermediate product formation, fate of these chemicals, and toxicological relevance of these compounds resulting from ozonation of wastewater need to be better understood.
- Limited data are currently available on the effectiveness of ozonation to inactivate specific pathogenic organisms (e.g., protozoan cysts) in wastewater effluents.
- Regrowth potential of these microorganisms following ozonation is also not well characterized.
- Results from studies conducted to date on the overall effect of ozone on effluent toxicity are variable.
- The effect of ozone on TOC removal, especially refractory organic matter not removed by biological treatment processes, should be further studied.
- One of the key challenges for ozonation is the perception that ozonation is significantly more expensive than other disinfection alternatives; however, this is based on data from previous generations of ozone generators and contactor equipment. More cost data are needed that reflect the past 20 years of advances in ozone technology.
- For new plants, cost data are needed to compare ozone to gas chlorine and hypochlorite solution, including the cost of chlorine contact basins and the dechlorination system.

CHAPTER 6.0

OTHER DISINFECTION TECHNOLOGIES

6.1 Overview

Chapters 3.0 through 5.0 discuss the most commonly used wastewater disinfection methods; this chapter covers other alternatives that have been investigated and/or are under development. Section 6.2 deals with chlorine dioxide, which has been used mainly for disinfection of drinking water but rarely for wastewater. The rest of the chapter focuses on alternatives that are currently being studied for wastewater disinfection; many have been tested at the pilot scale, but only limited data are available on water quality effects, operations, cost, etc. These technologies are split into three sections. Section 6.3 covers methods that rely on the addition of chemicals for disinfection: peracetic acid, ferrate, and brominated compounds. Section 6.4 discusses physical methods that rely on the input of energy for disinfection: pasteurization, ultrasonic cavitation, electron beam, and photocatalysis. Section 6.5 discusses alternatives that combine disinfection technologies, such as the use of UV with chlorine. Section 6.6 provides a brief summary of the technologies.

6.2 Chlorine Dioxide

6.2.1 Background

Chlorine dioxide gas (ClO_2) is a proven bactericide and a much more effective virucide than chlorine. It has been used for disinfection of potable waters, especially those containing phenols or other taste and odor producing compounds; however, experience with wastewater disinfection is very limited. Few pilot studies have been conducted, and no known major POTWs in the U.S. were identified as currently using chlorine dioxide for wastewater disinfection in this project's survey of practice. Several factors have discouraged the use chlorine dioxide as a wastewater disinfection alternative: it must be generated on-site due to chemical instability and safety concerns, chemical costs are high, and the overall system is complex to operate and maintain, relative to chlorination (U.S. EPA, 1986).

Chlorine dioxide is a neutral compound, with a chlorine oxidation state of +4. When reduced, it generally forms chloride (Cl^-) or chlorite (ClO_2^-). It is a stronger oxidant than chlorine and has highly selective reactivity due to the unique, one electron transfer mechanism that reduces it to chlorite (ClO_2^-). Although the disinfection mechanisms are not well understood, they appear to vary by the type of microorganism and involve specific oxidation reactions between chlorine dioxide and biomolecules, rather than gross physical damage to organisms. Chlorine dioxide can react with amino acids and free fatty acids, and possibly viral ribonucleic acid (RNA). Through these reactions, chlorine dioxide may damage peripheral structures (e.g., alter the virus protein coating, or increase the permeability of the outer membrane) or may disrupt internal functions such as protein and/or RNA synthesis (U.S. EPA, 1999). The ability of chlorine dioxide to adsorb to viral protein coatings may explain its strength as a virucide (Tchobanoglous et al., 2003).

6.2.2 Disinfection Efficacy

Water quality parameters that may affect the performance of chlorine dioxide disinfection include suspended solids, temperature, and pH. As with all disinfection technologies, suspended solids may shelter pathogens from disinfectants; clumps of cysts and viral aggregates are more resistant to disinfection than free-floating cysts and single state viruses. Like chlorine, the disinfection efficacy of chlorine dioxide decreases as temperature decreases. Chlorine dioxide is less sensitive than chlorine to pH changes between 6.0 and 8.5 for the inactivation of viruses and cysts; however, pH does have some effect. Increasing pH generally increases inactivation of protozoa such as *Crytosporidium*, *Giardia*, and *Naegleria*; data on the effect of pH on *E. coli* inactivation are conflicting, with one study showing increased inactivation at higher pH and another showing no effect between pH 6.0 and 10.0 (U.S. EPA, 1999).

Chlorine dioxide is generally equally or more effective than chlorine for the inactivation of bacteria, viruses, and cysts, particularly in higher pH solutions (U.S. EPA, 1999). Roberts et al. (1981) compared chlorine dioxide and chlorine disinfection in laboratory experiments with secondary effluent. For the inactivation of poliovirus and indigenous coliphages, chlorine dioxide was the more effective disinfectant in both filtered, nitrified wastewater and effluents from a conventional activated-sludge process. For total coliform, chlorine dioxide generally required lower CT values to achieve the same degree of inactivation; for a given contact time, chlorine dioxide required 2 to 70 times lower residual concentrations. In some cases, water quality affected disinfection efficacy. Chlorine dioxide demand was lower than chlorine demand in the nitrified, filtered secondary effluent, and the required mass dose of chlorine was 2 to 10 times greater than that of chlorine dioxide, depending on the effluent coliform standard. In contrast, chlorine dioxide demand was higher than chlorine demand in the conventional activated sludge process effluent, and the required mass doses of chlorine dioxide and chlorine were approximately equal: 2 mg/L to comply with a standard of 1,000 total coliforms/100 mL, 2.5 mg/L for 200 total coliforms/100 mL, and 8 mg/L for 2.2 total coliforms/100 mL.

6.2.3 System Design and Implementation

Because chlorine dioxide gas is extremely unstable and explosive, transportion is not recommended; consequently it must be generated on-site before application. Generators typically form chlorine dioxide from the reaction of sodium chlorite ($NaClO_2$) with gaseous chlorine, aqueous hypochlorite (OCl^-), or hydrochloric acid (HCl) (U.S. EPA, 1999):

$$2\ NaClO_2 + Cl_{2\,(g)} \rightarrow 2\ ClO_{2\,(g)} + 2\ NaCl \qquad (6.2.3\text{-}1)$$

$$2\ NaClO_2 + HOCl \rightarrow 2\ ClO_{2\,(g)} + NaCl + NaOH \qquad (6.2.3\text{-}2)$$

$$5\ NaClO_2 + 4HCl \rightarrow 4\ ClO_{2\,(g)} + 5\ NaCl + 2H_2O \qquad (6.2.3\text{-}3)$$

In commercial generators, several of the above reactions may take place simultaneously; some favor low pH levels and others are independent of pH, so operating conditions may differ, even with the same chemical feed stocks. The ratio of the feed reagents affects the purity of the final product, as chlorite ions may not react or may form ClO_2 or chlorate (ClO_3^-). Generators are operated to obtain the maximum yield of chlorine dioxide, while minimizing free chlorine residual or other residual oxidant formation. The specific yield for generators is typically greater than 95%; the measurable excess chlorine should be less than 2% by weight (U.S. EPA, 1999).

Chlorine dioxide generators are simple mixing chambers, often filled with inert media such as Teflon chips, ceramic or raschig rings to create hydraulic turbulence for mixing. The hydraulic retention time of the generator is typically about one minute. In addition to the generator, basic components of the chlorine dioxide generation system include storage and feed systems for the chemical reagents (sodium chlorite, gaseous chlorine, aqueous hypochlorite, and/or hydrochloric acid) and chlorine dioxide feed piping and dispersion equipment (U.S. EPA, 1999; Tchobanoglous et al., 2003).

6.2.4 Operation and Maintenance

Operators should be well informed of the potential hazards and safety requirements related to the storage and handling of chlorine dioxide and other chemical reagents used in the disinfection process. Chlorine dioxide solutions cause severe eye irritation, and sprays or mists irritate the mucous membranes of the nose and the throat. There is no established threshold limit value (TLV) for chlorine dioxide. Because of these health impacts, systems must be completely flushed before any service or maintenance activities, and personnel should wear protective equipment. Chlorine, which can be used for the generation of chlorine dioxide, also has a number of safety issues that are discussed in Sections 3.3.5 and 3.4.3. Contact with hydrochloric acid, another potential reagent, can cause burns to skin or mucous membranes, so appropriate precautions, procedures, and equipment should be used (Connell, 2002; WEF, 2006).

Safe storage and handling of chemical reagents also avoids the unintentional and uncontrollable release of high levels of chlorine dioxide. For example, accidental acidification may produce explosive or flammable concentrations. Lower temperatures and/or higher concentrations may cause the formation of crystals that plug pipelines, valves, and other equipment. Sodium chlorite that crystallizes and dries to a powder becomes a fire hazard and can ignite if in contact with combustible materials (U.S. EPA, 1999).

Process monitoring and control are used to produce chlorine dioxide only as needed, thus minimizing storage requirements. Process sensors (e.g., residual analyzers or flow meters) regulate the rate of chlorine dioxide generation, typically by varying reagent flow rates. Residual analyzers may be particularly useful for control, because chlorine dioxide yields may fluctuate as sodium chlorite feed solutions stratify (particularly below 4°C) or degrade (particularly in sunlight or at high temperatures). When determining the concentrations in wastewater, chlorine dioxide must be differentiated from chlorite and free and combined chlorine; both the DPD and the amperometric methods can be modified to do this (U.S. EPA, 1986).

6.2.5 Impacts to the Environment

Chlorine dioxide residuals are believed to degrade more rapidly than chlorine residuals, and are therefore likely to pose less of a threat to aquatic wildlife (Tchobanoglous et al., 2003). Pure chlorine dioxide does not generate THMs, and generally produces few organic byproducts. However, chlorine dioxide disinfection does generate some byproducts, including chlorite, chlorate, and organic DBPs. For example, it may generate non-halogenated byproducts similar to those produced during ozonation. In addition, the chlorine dioxide production process uses chlorine and sodium chlorite, which may generate organic chlorine compounds similar to those caused by chlorination.

Chlorite and chlorate are potentially toxic, and are produced to varying degrees as end products during generation, treatment, and subsequent degradation of chlorine dioxide. Factors affecting the concentrations of chlorine dioxide, chlorite, and chlorate in the final effluent include the ratio of the applied dosage to oxidant demand, blending ratios of sodium chlorite and chlorine during chlorine dioxide generation, exposure to sunlight, and chlorate levels in sodium chlorite feedstock (U.S. EPA, 1999). Because ~70% of the chlorine dioxide consumed during disinfection reverts to chlorite, concentrations of these disinfection byproducts in the effluent may limit the applied chlorine dioxide doses. Chlorite has a maximum contaminant level (MCL) of 1 mg/L; chlorate has no current MCL, but a maximum residual discharge limit of 0.8 mg/L (U.S. EPA, 1999; WEF, 2006).

The addition of chlorine dioxide also increases the chloride and total dissolved solids concentrations of the treated effluent; this may be an issue for some water reuse applications. On the positive side, the application of chlorine dioxide does not produce THMs in wastewater. In one study, the measured amount of total organic halogen (TOX) formed by chlorine dioxide disinfection was almost statistically insignificant; chlorination formed at least 10 to 20 times more undesirable halogenated byproducts, measured as TOX (Roberts et al., 1981).

6.2.6 Costs

This section presents costs for the addition of chlorine dioxide to a conventional treatment plant. Both capital, and operation and maintenance (O&M), costs are considered. The WEF Wastewater Disinfection Training Manual (WEF, 2006) states that the capital costs for chlorine dioxide disinfection are low, but the operating costs are high. According to Roberts et al. (1981), wastewater disinfection with chlorine dioxide is at least as expensive as, and generally more expensive than, disinfection with chlorine.

6.2.6.1 Capital Cost

The capital costs of a chlorine dioxide disinfection system include chlorine feed and storage facilities, chlorine dioxide generation and feed systems, and contact basins. Cost estimates presented herein are based on planning level costs that have an accuracy of +50/-30% and include an additional 20% to account for other project costs, such as engineering planning and design, construction management, and project administration. The capital costs have been normalized to the first quarter 2007 Engineering News Record (ENR) 20-cities construction cost index (CCI) of 7870.

Changing from Chlorine Gas to Chlorine Dioxide This section presents costs for the addition of chlorine dioxide to a conventional wastewater treatment plant using chlorine gas for disinfection. The plant is assumed to have an existing disinfection contact basin with a theoretical detention time of 30 minutes at peak hourly flow. The contact basin is assumed to be sufficient to serve as a chlorine dioxide contactor, and as such, no costs for contactor upgrades are included. In the conversion to chlorine dioxide, it is assumed that the existing chlorine gas will be used as a feed chemical for the generator and reacted with sodium chlorite to form chlorine dioxide, which will then be used as the disinfectant. The new facilities associated with chlorine dioxide include the chlorine dioxide generator, pressure regulation valves, flow meters, control valves, sensors, a storage tank (approximately 30-day supply) for the sodium chlorite stock (25% active), chemical feed pumping and piping, and a building to house the chemical storage and feed facilities.

Table 6-1 presents estimated capital costs for the addition of chlorine dioxide to plants that use chlorine gas. These costs are based upon a previous study by International Consultants and Malcolm-Pirnie (1998) that included capital cost estimates for the addition of chlorine dioxide (and other disinfection alternatives) to a conventional drinking water treatment facility for a wide range of design flow rates. Costs do not include any demolition or removal of existing facilities, and miscellaneous costs (e.g., for site work, subsurface conditions, contractor and engineering costs, etc.) and costs unrelated to chlorine dioxide were removed. The raw capital cost was then adjusted for the chlorine dioxide production rate (in lb/day), and normalized to the first quarter of 2007 CCI of 7870. The capital costs do not include the cost of any demolition or removal of existing facilities.

Table 6-1. Estimated Capital Costs for Addition of Chlorine Dioxide to a Water Treatment Plant with Chlorine Gas. [a]

Avg. Flow (MGD)	Chlorine Dioxide Dose (mg/L)	Chlorine Dioxide Generation (lb/day)	Chlorine Dioxide ($)	Residual Quenching ($)	Other Costs @ 20% ($)	Total Capital Cost ($)
1[b]	5	125	$1,700,000	$220,000	$380,000	$2,300,000
1[b]	10	250	$3,400,000	$280,000	$740,000	$4,400,000
1[b]	20	500	$6,900,000	$280,000	$1,400,000	$8,600,000
10[c]	5	1250	$15,000,000	$410,000	$3,100,000	$19,000,000
10[c]	10	2500	$30,000,000	$500,000	$6,100,000	$37,000,000
10[c]	20	5000	$60,000,000	$690,000	$12,000,000	$73,000,000
100[d]	5	12500	$150,000,000	$780,000	$30,000,000	$180,000,000
100[d]	10	25000	$290,000,000	$950,000	$58,000,000	$350,000,000
100[d]	20	50000	$590,000,000	$960,000	$120,000,000	$710,000,000

[a]All costs adjusted to first quarter 2007 (ENR CCI of 7870).
[b]Peak daily flow rate of 3 MGD.
[c]Peak daily flow rate of 30 MGD.
[d]Peak daily flow rate of 300 MGD.

The chlorine dioxide dosages in Table 6-1 are the same as those presented for chlorine (on a mass basis) in Table 3-7. Several studies comparing the germicidal efficiency of chlorine dioxide to chlorine were summarized by White (1992). The disinfection efficiency of chlorine dioxide depends upon the target organism. For a conventional wastewater treatment plant the organisms of greatest concern, with respect to permit compliance, are typically fecal coliform. White (1992) cites several studies that demonstrate that chlorine dioxide is more efficient (up to 1-log greater reduction in fecal coliform) than chlorine, within the first 5 minutes of contact time, for disinfection of secondary effluent. However, for contact times between 30 minutes and 1 hour, chlorine dioxide and chlorine have similar performance. Therefore, for the conditions conceptualized in this study, the doses of chlorine dioxide are the same as those of chlorine to provide similar disinfection performance.

In Table 6-1, the costs of residual quenching (removing the chlorine dioxide residual prior to discharge of the disinfected effluent) were estimated using the dechlorination costs presented by Darby et al. (1995). Although that study's costs were for gaseous chlorine, they can be used to estimate the costs for chlorine dioxide. Chlorine dioxide generally has a lower residual than free chlorine in disinfected secondary effluents, because the chlorine dioxide demand is generally higher than that of chlorine and chlorine dioxide dissipates rapidly in open basins exposed to sunlight (White, 1999). For the purposes of this study, the cost of removing the remaining chlorine dioxide residual was assumed to be half of the cost of dechlorination for an equivalent mass dose of chlorine and flow rate.

Changing from Chlorine Gas to Chlorine Dioxide without Chlorine Gas Facilities that do not want to use chlorine gas to generate chlorine dioxide can instead use a three-part chlorine dioxide generator technology. This generator reacts sodium chlorite, sodium hypochlorite, and hydrochloric acid, to produce chlorine dioxide. The capital cost for this scenario is similar to the costs presented above in Table 6-1, except that the additional cost of storage and feeding of sodium hypochlorite must be included.

The hypochlorite system consists of the following components:

- Chemical feed system consisting of a metering pump with motor speed control and flow meter
- Schedule 80 PVC pipe (with diaphragm valves) to transfer the hypochlorite from the chemical feed area to the application point
- At least two perforated pipe diffusers (so that one diffuser can be removed from service to clean scale from its orifices, especially if the hardness of the water is high)
- Two storage tanks, each with sufficient capacity to store 10 days worth of hypochorite,

Table 6-2 presents estimated capital costs for conceptual treatment plants with hypochlorite feed and storage facilities for three-part chlorine dioxide systems. The costs of the hypochlorite systems are based upon information provided by Kawamura (2000). These costs have been normalized to a 2007 CCI of 7870, and include 20% for other costs, such as engineering and administration during facility design and construction. The estimates do not include costs associated with any required demolition, removal of the existing gaseous chlorine system, and a building to house the chemical storage tanks.

Table 6-2. Estimated Capital Costs for Chlorine Dioxide using 3-Part Generators (with Hypochlorite Storage and Feed Facilities). [a]

Avg. Flow (MGD)	Chlorine Dioxide Dose (mg/L)	Chlorine Dioxide Generation (lb/day)	Chlorine Dioxide ($)	Residual Quenching ($)	Other Costs @ 20% ($)	Total Capital Cost ($)
1[b]	5	125	$1,720,000	$220,000	$388,000	$2,330,000
1[b]	10	250	$3,430,000	$280,000	$743,000	$4,460,000
1[b]	20	500	$6,950,000	$280,000	$1,450,000	$8,680,000
10[c]	5	1250	$15,100,000	$410,000	$3,100,000	$18,600,000
10[c]	10	2500	$30,200,000	$500,000	$6,130,000	$36,800,000
10[c]	20	5000	$60,300,000	$690,000	$12,200,000	$73,100,000
100[d]	5	12500	$151,000,000	$780,000	$30,200,000	$181,000,000
100[d]	10	25000	$291,000,000	$950,000	$58,300,000	$350,000,000
100[d]	20	50000	$591,000,000	$960,000	$118,000,000	$710,000,000

[a]All costs adjusted to first quarter 2007 (ENR CCI of 7870).
[b]Peak daily flow rate of 3 MGD.
[c]Peak daily flow rate of 30 MGD.
[d]Peak daily flow rate of 300 MGD

Adding Chlorine Dioxide at Plants Without Chlorine Infrastructure Most municipal applications incorporate existing chlorine feed facilities into the chlorine dioxide system, to save costs and to make use of the chemical storage and feed equipment already at the plant. For plants without chlorine infrastructure, the cost of a chlorine dioxide system would be substantially more than that presented above in Tables 6-1 or 6-2, because the cost of a contact basin would be required in addition to that of the chlorine dioxide-related equipment. Cost estimates under this scenario should be developed for the site-specific conditions and application.

6.2.6.2 Operation and Maintenance Costs

Operation and maintenance costs include electrical power, maintenance materials, labor costs, sample analysis, and chemicals. The predominant factor influencing the cost of chlorine dioxide disinfection is the cost of sodium chlorite, which can be ten times more expensive than chlorine on a unit mass basis. In addition, the maintenance required for the chlorine dioxide generator and appurtenant equipment can be substantial and is important for sustaining reactor performance and optimizing chlorine dioxide purity. Finally, analytical labor is required to regularly monitor the byproducts chlorite and chlorate, as well as residual free chlorine that can persist in the final chlorine dioxide solution.

Table 6-3 presents estimated O&M costs for chlorine dioxide disinfection. These costs are based upon estimates previously developed by International Consultants and Malcolm-Pirnie (1998). Those estimates were adjusted to the first quarter of 2007 Consumer Price Index of 610. Chemical costs are based upon a two-chemical system using sodium chlorite ($0.52/lb) and chlorine gas.

Table 6-3. Estimated O&M Costs for Chlorine Dioxide Disinfection.

Average Flow (MGD)	Chlorine Dioxide Dose (mg/L)	Chlorine Dioxide Generation (lb/day)	Total O&M[a] ($/yr)	Total O&M[a] ($/1000 gal)
1	5	40	$140,000	$0.38
1	10	80	$220,000	$0.60
1	20	170	$380,000	$1.04
10	5	420	$860,000	$0.24
10	10	830	$1,600,000	$0.45
10	20	1700	$2,700,000	$0.74
100	5	4200	$5,500,000	$0.15
100	10	8300	$10,000,000	$0.28
100	20	17000	$19,000,000	$0.53

[a]Includes equipment maintenance and analytical/technical labor.

6.2.7 Summary

Chlorine dioxide disinfection has been used for the disinfection of drinking water at full scale, but it is not a widespread practice and has not been used at major POTWs for wastewater disinfection. Although many POTWs seek alternatives to chlorine disinfection due to DBPs and safety/security concerns, use of chlorine dioxide does not necessarily alleviate these issues. Both technologies use hazardous materials that require special precautions. Chlorine dioxide eliminates THM formation, but instead generates chlorite, chlorate, and chlorinated organics as DBPs. In addition, chlorine dioxide systems are more complex and more expensive to operate than chlorination systems. Thus, the benefits gained by the use of chlorine dioxide relative to chlorine generally do not outweigh the disadvantages; these advantages and disadvantages are summarized below.

6.2.7.1 Advantages

Chlorine dioxide is equally or more effective than chlorine and chloramines for the inactivation of bacteria, viruses, and cysts. It is a stronger and more selective oxidant (e.g., does not react with ammonia, unlike chlorine), and its biocidal properties are less sensitive than chlorine to pH. In addition to disinfection, chlorine dioxide can be used to oxidize iron, manganese, sulfides, and phenolic compounds. Under proper generation conditions (i.e., no excess chlorine), THMs are not formed. Chlorine dioxide also provides residuals, which are needed in reuse applications and are useful for process control (U.S. EPA, 1999; WEF, 2006).

6.2.7.2 Disadvantages

Because chlorine dioxide is unstable and hazardous, it must be generated on-site and requires precautions for safe handling and storage. The chemical reagents used to form chlorine dioxide are also hazardous and require safety precautions, storage space, and special feed equipment. The generation, analytical requirements and process control for chlorine dioxide disinfection are complex. Consequently, costs are high for training, sampling, and laboratory testing for chlorite and chlorate. Sodium chlorite is also expensive, and as a result, the O&M costs for chlorine dioxide disinfection are relatively high. In addition, the process forms chlorite and chlorate as byproducts; regulatory limits may restrict the chlorine dioxide dose that can be applied. THMs may also be formed if excess chlorine is added during the chlorine dioxide generation process. Finally, chlorine dioxide may decompose in sunlight, and some systems may produce noxious odors (U.S. EPA, 1999; WEF, 2006).

6.3 Developing Technologies: Chemical Methods

6.3.1 Peracetic Acid

6.3.1.1 Background

Historical Use Peracetic acid (PAA), or peroxyacetic acid, can be used as a bactericide, virucide, fungicide, or sporicide agent (Rossi et al., 2005). Other uses include decoloring and disinfection in textile, pulp and paper industries; disinfection of ion exchangers and cooling towers; pathogen reduction in biosolids; sludge debulking; and reduction of biosolid odors (Koivunen and Heinonen-Tanski, 2005a). It has been applied to the food, beverage, medical and pharmaceutical industries, and its use has recently expanded to include wastewater treatment plants, particularly in Europe (Santoro et al., 2005). In this project's survey of practice no major POTW in the U.S. was using this technology for disinfection. However, a disinfection trial was performed in Ohio and recently, EPA has approved Proxitane® WW12 available from Solvay Chemicals (www.solvaychemicals.us) for use as a disinfectant to treat wastewater.

How PAA Disinfects The chemical formula of PAA is CH_3CO_3H, and the dissociation constant for the acid is approximately pK_a 8.2 (Sanchez-Ruiz et al., 1995). The undissociated acid is a stronger disinfectant than the dissociated anion (Liberti and Notarnicola, 1999).

PAA is generally supplied as a mixture of PAA, hydrogen peroxide, acetic acid, and water, with the following equilibrium (Gehr et al., 2002):

$$CH_3CO_3H + H_2O \Leftrightarrow CH_3CO_2H + H_2O_2 \qquad (6.3.1.1\text{-}1)$$

The hydrogen peroxide formed from PAA is also a disinfectant; however, literature reports that PAA is approximately 100 times stronger, so most of the disinfection from the solution mixture is expected to result from the PAA (Wagner et al., 2002).

PAA is inherently unstable and can be consumed by spontaneous decomposition, hydrolysis, and decomposition catalyzed by transition metals. At neutral pH ranges (5.5-8.2), spontaneous decomposition is the primary removal mechanism (Gehr et al., 2002):

$$2\ CH_3CO_3H \Rightarrow 2\ CH_3CO_2H + O_2 \qquad (6.3.1.1\text{-}2)$$

A 1% PAA solution has a hydrolysis half-life of six days, and pure PAA has a half-life of 100 minutes. However, commercially available solutions at 10-15% concentrations contain stabilizers and decompose much more slowly (Rossi et al., 2005).

For bacterial cells, PAA is thought to operate by releasing active oxygen or reactive hydroxyl radicals that attack the bacterial cell and disrupt the cell wall, membranes, some enzymes and/or DNA. For coliphage, PAA is suspected to damage nucleic acids or surface structures, e.g., protein coats or attachment sites for infection of host cells (Koivunen and Heinonen-Tanski, 2005a).

6.3.1.2 Disinfection Efficacy

Effect of Water Quality on Disinfection Disinfection efficacy is reported to depend strongly on water quality (Liberti and Notarnicola, 1999). Sanchez-Ruiz et al. (1995) found wide variations in disinfection efficacy among five raw wastewaters that they tested. Some of the variability was attributed to pH; log removals increased when the pH dropped below 7, as the species equilibrium shifted toward the more powerful acid disinfectant and away from the anion.

Suspended solids are likely to reduce the disinfection efficacy, but the data are conflicting. Baldry et al. (1995) reported that disinfection was insensitive to suspended solids levels. However, other sources indicate that high COD and suspended solids concentrations may scavenge PAA, thereby increasing PAA consumption and/or decreasing disinfection efficacy (Gehr et al., 2003; Koivunen and Heinonen-Tanski, 2005a). This is supported by data showing that disinfection efficiencies are much higher for biologically treated secondary effluents than raw wastewater (Sanchez-Ruiz et al., 1995). Gehr et al. (2002) reported that PAA demand was lower in secondary effluents than in primary effluents, which have higher levels of suspended solids, turbidity, and COD. Lazarova et al. (1998) also indicated that PAA requirements increased as total suspended solids (TSS) increased, up to 20 mg/L; increasing TSS from 10 to 40 mg/L did not further impact PAA requirements.

In terms of its ability to adapt to changes in flow rate, PAA is probably similar to chlorine. For combined sewer overflows, which have highly variable flow rates, the U.S. EPA lists PAA as one of five alternatives to chlorine (U.S. EPA, 1999).

Inactivation of Pathogens Disinfection efficiencies have been reported to be similar for hypochlorite and PAA at equal concentrations in tap water and secondary effluent (Koivunen and Heinonen-Tanski, 2005b). However, PAA was found to be more effective than hypochlorite in water with high organic content; this was attributed to the organics scavenging more hypochlorite than PAA (Koivunen and Heinonen-Tanski, 2005b).

Disinfection data for a variety of pathogens are given in Tables 6-4 to 6-7. The effectiveness of PAA against various pathogens generally decreases as follows: bacteria > viruses > bacterial spores > protozoan cysts (Liberti and Notarnicola, 1999). For example, F-RNA coliphage are generally more resistant to disinfection than enteric bacteria (Koivunen and Heinonen-Tanski, 2005a). In addition, PAA is more effective against *E. coli* and the coliphage MS2 than against poliovirus, echovirus, and coxsackievirus (Santoro et al., 2005). Studies have also shown that PAA disinfection removes limited or no clostridia (Baldry et al., 1995; Gehr et al., 2003), and is ineffective against the protozoa *Giardia* and *Cryptosporidium parvum* (Liberti and Notarnicola, 1999).

Table 6-4. Disinfection Efficacy of PAA Against Total Coliform.

Water Source	PAA Dose (mg/L)	Contact Time (min)	Log Reduction	Source
Primary effluent, pH < 8	10-30	15	1.9 - >6.5	(Sanchez-Ruiz et al., 1995)
Primary effluent, pH > 8	2-80	15	0.7-3.5	(Sanchez-Ruiz et al., 1995)
Primary effluent	5-15	4-27	0.4-3.9	(Koivunen and Heinonen-Tanski, 2005a)
Secondary effluent	0.5-4	8-38	0-4	(Veschetti et al., 2003)
Secondary and tertiary effluents	2-7	4-27	0.8-3.3	(Koivunen and Heinonen-Tanski, 2005a)
Tertiary effluent	0.9-2.1	17-240	1-3.5	(Baldry et al., 1995)
Wastewater	10	10	3	(Lazarova et al., 1998)

Table 6-5. Disinfection Efficacy of PAA Against Fecal Coliform.

Water Source	PAA Dose (mg/L)	Contact Time (min)	Log Reduction	Source
Primary effluent	1.5-6	60	0-4	(Gehr et al., 2003)
Secondary effluent	0.6 - >4	30-120	0.8-1.6	(Gehr et al., 2002)
Secondary effluent	0.5-4	8-38	0-5	(Veschetti et al., 2003)
Tertiary effluent	0.9-2.1	17-240	1.5-4	(Baldry et al., 1995)
Wastewater	10	10	3	(Lazarova et al., 1998)

Table 6-6. Disinfection Efficacy of PAA Against Other Bacteria and Protozoa.

Organism	Water Source	PAA Dose (mg/L)	Contact Time (min)	Log Reduction	Source
E. coli	Secondary effluent	0.5-4	8-38	0-4.5	(Veschetti et al., 2003)
	Peptone water (high organic content)	1-3	10	0.04-3	(Koivunen and Heinonen-Tanski, 2005b)
Fecal Streptococci	Secondary effluent	0.5-4	8-38	0-3.5	(Veschetti et al., 2003)
	Tertiary effluent	0.9-2.1	17-240	0.5-5	(Baldry et al., 1995)
	Wastewater	10	10	3	(Lazarova et al., 1998)
Enterococci	Primary effluent	4.5	60	1	(Gehr et al., 2003)
	Primary effluent	5-15	4-27	0.3-4.0	(Koivunen and Heinonen-Tanski, 2005a)
	Secondary and tertiary effluents	2-7	4-27	0.3-2.6	(Koivunen and Heinonen-Tanski, 2005a)
	Peptone water (high organic content)	0.5-3	10	0.1-3	(Koivunen and Heinonen-Tanski, 2005b)
Pseudomonas	Secondary effluent	0.5-4	8-38	0-5	(Veschetti et al., 2003)
	Secondary effluent	10	30	2-5	(Liberti and Notarnicola, 1999)
Salmonella	Secondary effluent	0.5-4	8-38	0-3	(Veschetti et al., 2003)
	Peptone water (high organic content)	1-3	10	0.05-2	(Koivunen and Heinonen-Tanski, 2005b)
Giardia lamblia	Secondary effluent	10	30	~0	(Liberti and Notarnicola, 1999)
Cryptosporidium parvum	Secondary effluent	10	30	0	(Liberti and Notarnicola, 1999)

Table 6-7. Disinfection Efficacy of PAA Against Viruses and Coliphage.

Organism	Water Source	PAA Dose (mg/L)	Contact Time (min)	Log Reduction	Source
Anti-E. coli bacteriophage	Secondary effluent	0.5-4	8-38	0-1.5	(Veschetti et al., 2003)
F-RNA coliphages	Primary effluent	5-15	4-27	0.2-0.8	(Koivunen and Heinonen-Tanski, 2005a)
	Secondary and tertiary effluents	2-7	4-27	0.2-0.9	(Koivunen and Heinonen-Tanski, 2005a)
MS2 coliphage	Primary effluent	1.5-3	60	<1	(Lazarova et al., 1998)
	Wastewater	10	0-120	0-3.5	(Lazarova et al., 1998)
	Peptone water (high organic content)	3-15	10	0.2-1.3	(Koivunen and Heinonen-Tanski, 2005b)
Poliovirus	Wastewater	10	0-120	0-2.5	(Lazarova et al., 1998)
$\phi\chi$174 bacteriophage	Wastewater	10	0-120	0-7.5	(Lazarova et al., 1998)

In pilot tests with a secondary effluent that was passed through a pressurized multimedia filter, extremely high doses (400 mg/L) of PAA with a 20 minute residence time were able to achieve 5-log removal of total coliform and meet a disinfection standard of 2 CFU/100 mL (Liberti and Notarnicola, 1999). A lower dose of 10 mg/L PAA with a 30 minute residence time could not achieve this standard. However, PAA was consistently able to meet the World Health Organization standard of 1,000 CFU/100 mL of fecal coliforms with secondary effluent and filtered secondary effluent (Liberti and Notarnicola, 1999). Koivunen and Heinonen-Tanski (2005a) suggest that PAA could be a competitive disinfectant for secondary effluents where the microbial limits are ~100 to 1,000 CFU/100 mL total coliforms or *E. coli*, and where limits for viruses and protozoa are not strict. Tertiary waters could achieve total coliform or *E. coli* concentrations of <10 to 100 CFU/100 mL; however, meeting low limits such as 2 CFU/100 mL would require high doses of PAA (Koivunen and Heinonen-Tanski, 2005a).

6.3.1.3 System Design and Implementation

PAA has been tested at wastewater treatment plants in Europe and Canada, primarily as a replacement for chlorination or to supplement UV disinfection in secondary treatment plants; however, there is little information on use or pilot studies within the U.S. As a result, there is currently some uncertainty over the regulatory requirements, public acceptance, and other potential constraints for the application of this technology within the U.S. Given the variability of the available data (see Section 6.3.1.2 for more details) and the apparent dependence on water quality, bench or pilot-scale studies would be necessary before implementation of this technology.

Literature indicates that the infrastructure for chlorine can also be used for PAA, so retrofitting hypochlorite systems for PAA is relatively easy, although storage and dosing systems may require modification of the hypochlorite equipment, as PAA may have more stringent storage and safety requirements (see details in Section 6.3.1.4). PAA solution can be fed directly into the wastewater to be disinfected, or can be added to carry water to achieve good mixing before being added to the wastewater. For facilities expanding or designing a new PAA system, the space and power requirements should be similar to those for hypochlorite. In designing the PAA system, it should be noted that the CT approach (product of the disinfectant concentration and contact time) may not apply to PAA disinfection (Rossi et al., 2005). The pretreatment requirements for PAA are not yet clear; however, data suggest that cleaner, better effluent water qualities are more economical and possibly more reliable for PAA treatment (see Section 6.3.1.2 for details).

6.3.1.4 Operation and Maintenance

PAA and hypochlorite are expected to have similar requirements for operation and maintenance. Some authors state that PAA systems are simple to install and operate (Koivunen and Heinonen-Tanski, 2005a); however, the U.S. EPA describes the operation as "complex" (U.S. EPA, 1999).

For maintenance of PAA facilities, the U.S. EPA gives the following guidelines. Feed lines, storage areas, leakage detection equipment, chemical injectors should be inspected periodically. The chemical housing facility should be cleaned and inspected periodically to ensure structural integrity, and the fire-extinguishing sprinkler systems must be inspected regularly. Spent chemical containers must be discarded appropriately, and spilled chemicals should be cleaned immediately. Personnel must wear special protective clothing and positive pressure self-contained breathing apparatus while working with PAA (U.S. EPA, 1999).

Although commercially available PAA solutions contain additives to enhance stability (Santoro et al., 2005), safe usage of PAA requires a number of safety precautions. The facility that houses PAA should have backup power at all times (U.S. EPA, 1999). PAA should be used only in areas free of ignition sources, and should be stored in a cool, dry, well-ventilated area in its original DOT-approved containers with hazard labels intact (U.S. EPA, 1999; National Research Council, 2000). The storage room should be kept separate from all other processes, as well as from acids, alkalis, organic materials, heavy metals, oxidizable compounds, and flammable substances (U.S. EPA, 1999; National Research Council, 2000). Contamination from any source, including metals and dust, should be avoided, because it could cause rapid decomposition and generate high pressure and large quantities of oxygen gas (U.S. EPA, 1993). PAA may only be stored in closed drums up to 30 gallons in volume, and may not be stacked. No more than 4,800 gallons of PAA mixture should be stored inside a facility at a time; greater quantities require multiple buildings. The storage building should be constructed of materials with maximum resistance to fire and explosion, and should include deflagration venting. The interior walls of the facility must be capable of withstanding 125 pounds per square foot of internal overpressurization and ceilings must be insulated (U.S. EPA, 1999).

6.3.1.5 Impacts to the Environment

Disinfection Byproducts PAA is generally reported to produce few or no disinfection byproducts or chemical residuals (U.S. EPA, 1999). Santoro et al. (2005) report that at doses and pH values commonly used for wastewater treatment, the principal end products are acetic acid, oxygen, water, and some relatively harmless carboxylic acids. Given the relatively benign end products, the U.S. EPA (1993) waived the requirements for avian and aquatic studies for the use of PAA as indoor microbiocides.

However, PAA has been reported to interact with amino acids, phenols and other aromatic compounds in sewage to form aldehydes, though not at toxic levels (Liberti and Notarnicola, 1999). Booth and Lester (1995) suggested that when organics or metals were present to promote the formation of radicals, PAA instigated the formation of free chlorine and bromine radicals, and hypobromous acid (but not hypochlorous acid) from chloride and bromide. As a result of these reactions, treatment of wastewater with 30 mg/L PAA resulted in the formation of 4-50 ng/L of chlorinated and brominated phenols, although the concentrations formed with a dose of 6 mg/L PAA were slight to negligible (Booth and Lester, 1995). Others have also found insignificant formation of halogenated organic compounds with treatment using PAA at doses up to 10 mg/L (Liberti and Notarnicola, 1999; Veschetti et al., 2003). More information is needed for the full range of PAA doses and wastewater qualities that might be applied in disinfection, as well as the impacts PAA might have if discharged to receiving waters (e.g., seawater, which naturally contains higher levels of bromide and chloride).

Changes in Water Quality There are conflicting reports on PAA residuals. The U.S. EPA (1999) reports no persistent residual, while others report PAA residual concentrations ~1-4 mg/L lower than the dosing concentration in secondary and tertiary effluents treated with 0.5-7 mg/L PAA (Veschetti et al., 2003; Koivunen and Heinonen-Tanski, 2005a), with significantly less residual (0-2.5 mg/L) in primary effluents treated with 5-15 mg/L PAA (Koivunen and Heinonen-Tanski, 2005a). There are generally no requirements for disinfectant removal (U.S. EPA, 1999; Koivunen and Heinonen-Tanski, 2005a).

There are also conflicting reports on the effect of PAA on other water quality parameters such as pH, and carbon load/oxygen demand. For example, some tests showed no change in pH (Baldry et al., 1995), while others showed decreases of 0.2-0.5 pH units in primary, secondary and tertiary effluents treated with 7-15 mg/L of PAA (Koivunen and Heinonen-Tanski, 2005a). The carbon load and oxygen demand of effluents are expected to increase with PAA treatment, because PAA decomposes to acetic acid. Lazarova et al. (1998) found that total organic carbon, TOC, increased by a factor of 1.5 after treating wastewater with 5 mg/L of PAA, while 10 mg/L of PAA tripled the TOC; biodegradable dissolved organic carbon increased by a factor of four.

Koivunen and Heinonen-Tanski (2005a) also report that PAA increases the carbon content of the water, but this increase may not be significant at the doses relevant for wastewater treatment. On the other hand, Baldry et al. (1995) reports that COD levels dropped from 52 mg/L to 45 mg/L and BOD concentrations dropped from 23 mg/L to 16 mg/L.

Toxicity in the Treated Effluent PAA has not been found to be carcinogenic, nor shown to cause reproductive or developmental toxicity in humans. Its acute toxicity is low, but it is extremely irritating to the skin, eyes, and respiratory tract, and exposure to 40% solutions can cause serious burns (National Research Council, 2000). Genotoxicity tests with bacteria and plants exposed to secondary effluent showed high inherent variability; treatment of the effluent with 2-4 mg/L PAA added little to no toxicity or mutagenicity (Monarca et al., 2000; Crebelli et al., 2005). As another indication that there is no residual toxicity after PAA disinfection, *E. coli* grew as well or better in PAA treated wastewater than in untreated wastewater (Lazarova et al., 1998).

Potential Risks As with all disinfection technologies, regrowth or recovery of inactivated cells is a potential risk. Based on literature, it appears that there is no significant regrowth or recovery in the first several hours after PAA disinfection, but that microorganism concentrations may increase over a period of days. Rossi et al. (2005) observed no regrowth or recovery in the first five hours after disinfection, and in fact observed continued disinfection if the PAA was not quenched. Sanchez-Ruiz et al. (1995) observed no increases in coliform concentrations in batch experiments where PAA-disinfected wastewater was diluted with seawater and exposed to laboratory illumination, but did see increases after a day when the batches were kept in the dark. Lazarova et al. (1998) also observed increases in fecal coliforms and heterotrophic bacteria two days after disinfecting wastewater with PAA and subsequently diluting it with seawater. More research is needed to determine whether this phenomenon is significant under field conditions.

The other major risks with PAA are related to its instability. PAA explodes when heated to 110°C, and pure PAA is very shock sensitive. The presence of PAA in a fire can accelerate the combustion of other materials; these fires can be extinguished with water, dry chemical, or halon extinguishers. PAA also reacts exothermically and potentially violently with substrates such as ethers, metal chloride solutions, olefins, and some alcohols and ketones. They may also form shock-sensitive peroxides, and runaway peroxide decomposition may occur with exposure to some metal ions, including iron, copper, cobalt, chromium, and manganese (National Research Council, 2000). Given the concerns about stability and explosiveness, the U.S. EPA (1999) stated that PAA requires on-site generation. In Italy, to address the concerns about transport, handling and storage, the commercial formulations for wastewater disinfection typically consist of 15% w/w PAA, 25% hydrogen peroxide, up to 35% acetic acid, with water making up the remainder (Santoro et al., 2005).

Additional Benefits PAA provides residual disinfectant, which is useful for controlling algae and biofilms or slime, and is beneficial for reuse applications. PAA may also reduce odors in wastewater, as it has been used for odor control of solids (Koivunen and Heinonen-Tanski, 2005a). There are few other data in the literature about other potential benefits (e.g., removal of microconstituents).

6.3.1.6 Costs

Thus far, the high price of PAA has limited its use for wastewater disinfection; however, recent studies have demonstrated that PAA could be a viable alternative, when compared to chlorination-dechlorination, UV, ozone or chlorine dioxide (Koivunen and Heinonen-Tanski, 2005a). For the primary effluent at one particular wastewater treatment plant, Gehr et al. (2003) indicated that PAA doses of up to 1.5-2 mg/L would be economically viable.

Capital Costs The capital cost associated with PAA disinfection is estimated to be on the order of 12-14% of the cost of constructing a facility with secondary treatment and filtration; this cost is on the same order as building the infrastructure for NaOCl with dechlorination or UV disinfection, and about half the cost of ozone infrastructure (Nurizzo et al., 2001). For facilities that currently use chlorine, the capital costs for converting to PAA are expected to be minimal because both chemicals can utilize the same infrastructure (Liberti and Notarnicola, 1999).

Operation and Maintenance Costs Table 6-8 gives estimated O&M costs to treat wastewater to various disinfection standards. The costs are significantly higher than for chlorine, UV and ozone; however, the O&M costs for PAA are comprised entirely of the chemical costs (including transport), which may decrease if PAA use and the associated manufacturing capacity increase.

Table 6-8. Relative O&M Costs of PAA for Disinfection of Filtered Secondary Effluent.

Standard	PAA Dose (mg/L)	O&M Cost	Relative Cost	Source
2 CFU/100 mL total coliforms	400	€2600/1000 m^3 ~$12,000/MG[a,b]	100X UV	(Liberti and Notarnicola, 1999)
1000 CFU/100 mL fecal coliforms	10	€65/1000 m^3 ~$3,000/ MG[a,b]	13X UV 2-4X ozone	(Liberti et al., 2000)
23-1000 MPN/100 mL total coliforms	4-14		2-4X UV or NaOCl 1.5X ozone	(Nurizzo et al., 2001)

[a]MG = million gallons
[b]Assumes the average 2006 conversion rate of $1.25/euro

6.3.1.7 Summary

Advantages Because PAA is a chemical disinfectant like chlorine, it can be flow-paced and adapts well to changes in flow rates. Because many treatment plants currently use chlorine and chlorine infrastructure can be adapted for PAA use, conversion to PAA would be relatively easy. In addition, the use of PAA eliminates most, if not all, disinfection byproducts and does not appear to create toxicity in effluents. It also provides some residual, which is useful for water reuse distribution systems.

Disadvantages PAA is unstable and requires precautions to ensure safe storage and use. Like most disinfectants, PAA does not disinfect all pathogens and indicator organisms equally well; it is effective against coliform bacteria, but is weaker against viruses, and has little effect on spores and protozoa. Disinfection performance is also affected by water quality, but the important parameters for determining the dose are not yet clear and there are few data on compliance history to establish the consistency of its performance. Based on the available data, extremely high doses of PAA would be required to reach very low coliform concentrations in the effluent. PAA is relatively expensive, which leads to high operations costs. However, the capital costs for construction are on par with chlorine and UV disinfection, and less expensive than ozone disinfection. If PAA use increases, the manufacturing capacity may increase, and the price may drop. Finally, PAA adds organic carbon to the effluent, and at high concentrations, may form aldehydes, chlorine and bromine radicals, and hypobromous acid.

Gaps in Knowledge More research is needed on the fate of PAA in effluents and receiving waters. For example, it is important to know whether the disinfection products observed by Booth and Lester (1995) appear under conditions that might be used at a wastewater treatment plant. Research is also needed to determine the consistency of performance with different water qualities, and to determine the relevant parameters that determine the required PAA dose. Finally, more research is needed to determine whether regrowth or recovery of damaged cells is significant under field conditions.

6.3.2 Ferrate

6.3.2.1 Background

Ferrate Chemistry Ferrate, or $FeO_4^{=}$, is a form of hexavalent iron (Fe (VI)) that is a powerful oxidant. The redox potential of ferrate is 2.2 volts (V) in acid and 0.7 V in base, compared with redox potentials of 1.4 V for chlorine gas, 1.5V for hypochlorous acid, and 0.8 V for hypochlorite ions (Safe Drinking Water Committee, 1980; Jiang and Lloyd, 2002). The protonated form, $HFeO_4^-$, is a stronger oxidant than the anionic $FeO_4^{=}$, and the acid dissociation constant (pKa) is 7.23 (Sharma et al., 2005).

Sharma et al. (2005) also designate FeO_4^{-3} (Fe(V)) as a ferrate. Fe (V) is three to five orders of magnitude more reactive than Fe (VI) for the oxidation of pollutants and amino acids. As with Fe (VI), reaction rates are faster with protonated Fe (V) in the form of $HFeO_4^{2-}$, than with the unprotonated form, FeO_4^{3-}; the pKa value for the $H_2FeO_4^-/HFeO_4^{=}$ dissociation is 7.5, and the pKa value for the $HFeO_4^{2-}/FeO_4^{3-}$ dissociation is 10.1 (Sharma, 2004; Sharma, 2005).

Ferrate can be produced through electrolysis with an iron anode, dry synthesis with iron oxide, or wet synthesis with iron hydroxide (Jiang and Lloyd, 2002):

$$Fe + 2\ KOH + 2\ H_2O \rightarrow K_2FeO_4 + 3\ H_2 \qquad (6.3.2.1\text{-}1)$$

$$Fe_2O_3 + 3\ Na_2O_2 \rightarrow 2\ Na_2FeO_4 + Na_2O \qquad (6.3.2.1\text{-}2)$$

$$2\ Fe(OH)_3 + 3\ NaOCl + 4\ NaOH \rightarrow 2\ Na_2FeO_4 + 3\ NaCl + 5\ H_2O \qquad (6.3.2.1\text{-}3)$$

Recent developments in technology have made wet synthesis more economical (Alig, 2007); however, it should be noted that this process produces sodium and chloride ions, thereby increasing total dissolved solids.

Stability of ferrate solutions is a concern, as ferrates can decompose or react with water or other species within the water (metals, organics, ammonia, etc.). The rate of decomposition depends on the initial ferrate concentration, pH, temperature, and the surface properties of the resulting Fe(III) solids. Dilute solutions of ferrate are generally more stable than concentrated solutions, and ferrate is most stable in strong base at pH 10-11 (Safe Drinking Water Committee, 1980). One study found that 1.4 to 2.9 mg/L solutions of ferrate decomposed within 10 minutes (based on disappearance of the characteristic violet color), while another study showed residuals after 20 minutes in 1.2 and 12 mg/L solutions (Fagan and Waite, 1983; Kazama, 1995).

Historical Use Potassium ferrate was first synthesized in the mid-1800s by Fremy (Safe Drinking Water Committee, 1980); Murmann and Robinson (1974) were the first to publish its disinfectant properties in water. Its use thus far has been limited by its chemical instability, difficulties in producing high purity ferrate, and cost, but there has been recent interest in it for green organic synthesis, "super iron" batteries, and water disinfection (Sharma et al., 2005). Based on this project's survey of practice, no major POTWs currently use this disinfection technology.

How Ferrate Disinfects Little has been published about the disinfection mechanism of ferrate, although studies have shown that ferrate can oxidize *E. coli* DNA polymerase I and degrade deoxyribonucleosides (Sharma et al., 2005). In addition, steric similarities to phosphate allow it to bind enzyme phosphoryl groups and inactivate phophatases, dehydrogenases, phosphorylase b, phosphoglucomutase and other proteins (Jessen et al., 2006). Studies by Kazama (1994, 1995) indicate that viral inactivation continues after the ferrate itself has decomposed, but addition of the reducing agent sodium thiosulfate stopped disinfection. These results suggest that ferrate decomposes to form an unidentified, long-lived oxidative intermediate, which can also act as a disinfectant.

6.3.2.2 Disinfection Efficacy

Table 6-9 provides data on the disinfection efficacy of ferrate against various organisms. Ferrate can be used to disinfect *E. coli*, enterococci, bacterial phages, endospores, and protozoan oocysts (Reimers et al., 2007). Murmann and Robinson (1974) found that the concentration of *Pseudomonas* bacteria in a buffered solution could be reduced using doses of 1.5 to 50 ppm, while bacterial concentrations in Missouri River water were reduced by 2 to 100 ppm of ferrate. Waite (1979) used a dose of 10 mg/L for 3.7-log inactivation of total coliform and 4-log inactivation of fecal coliform, while another study used a dose of 8 mg/L ferrate in secondary effluent for 3-log inactivation of total coliform, and 1.5-log inactivation of total bacteria (Jiang and Lloyd, 2002). In terms of relative resistance to ferrate, *Escherichia coli*, *Salmonella typhimurium*, and *Shigella flexneri* showed similar susceptibility to ferrate, but *Streptococcus faecalis*, *Bacillus cereus*, *Streptococcus bovis*, and *Staphylococcus aureus* were more resistant (Safe Drinking Water Committee, 1980).

The inactivation efficacy appears to be affected by ferrate dose, contact time, initial pathogen concentration, solids, pH, buffering capacity, water quality, and temperature; however, ferrate is affected less than chlorine by changes in organic content, pH, and temperature. Decreasing initial bacterial concentrations increased disinfection: a 10-fold dilution increased inactivation by two orders of magnitude. As with other disinfection methods, bacterial clumps limit disinfection efficacy; centrifuging and/or vortexing samples to break up clumps increased disinfection (Jessen et al., 2006). Sharma et al. (2005) indicated that suspended solids such as clay and organic particles do not affect inactivation, but another study reported that higher ferrate doses were required for bacterial inactivation in the presence of organic matter (Safe Drinking Water Committee, 1980).

Salinity (in artificial sea water, 10 to 36 parts per thousand) does not have a significant effect on inactivation. At lower temperatures, bacterial inactivation increased as temperature increased; however, the rate of ferrate decomposition also increased with rising temperature and became significant at higher temperatures. Inactivation of *E. coli* also increased as pH decreased from 8 to 6, despite a concurrent increase in the rate of ferrate decomposition, and inactivation decreased very slightly at pH 9.5. The bacteriophages f2 and Qβ were more sensitive than bacteria to pH. With a dose of 1.2 mg/L ferrate, the time required for 2-log inactivation of f2 was 4 minutes at pH 6 but 13 minutes at pH 8 (Safe Drinking Water Committee, 1980; Schink and Waite, 1980; Sharma et al., 2005; Jessen et al., 2006).

Ferrate has also been tested for biofilm control, where concentrations of 1.2 to 12 mg/L were sufficient to inhibit growth (but 0.12 mg/L was not); however, these concentrations could not remove an established biofilm of 60 μm thickness. The effectiveness of biofilm control did not depend on contact time (between 5 and 20 min) or frequency of exposure (once every two days to twice per day) (Fagan and Waite, 1983).

Table 6-9. Disinfection Efficacy of Ferrate Against Various Organisms.

Organism	Water	Dose (mg/L)	Time (min)	Temp. (°C)[1]	pH	Average Log_{10} Reduction	Source
Total bacteria	Secondary	8				3	(Jiang and Lloyd, 2002)
Total coliforms	Secondary	8				1.5	(Jiang and Lloyd, 2002)
E. coli	Artificial sea	5	5		~8	6.8	(Jessen et al., 2006)
	Unknown	6.0	2.5	20	7	2	(Safe Drinking Water
		1.2	5	20	7	2	(Safe Drinking Water
		6.0	7		8	3	(Jiang and Lloyd, 2002)
Enterococcus	Artificial sea water	5	5		~8	5.5	(Jessen et al., 2006)
Klebsiella pneumoniae	Artificial sea water	5	5		~8	7.3	(Jessen et al., 2006)
Vibrio cholerae	Artificial sea water	5	5		~8	4.4	(Jessen et al., 2006)
Strepotococcus faecalis	Unknown	60	15		8	2	(Safe Drinking Water Committee, 1980)
		12	5		8	2	(Safe Drinking Water Committee, 1980)
		6.0	4	20	7	2	(Safe Drinking Water Committee, 1980)
		1.2	365	20	7	2	(Safe Drinking Water Committee, 1980)
Sulfite-reducing *Clostridia*	River water	4.3	30			>5	(Sharma et al., 2005)
Aerobic spore formers	River water	4.3	30			>3	(Sharma et al., 2005)
Qβ	Phosphate buffer	1.4 – 2.9	5 – 10	25	6	6	(Kazama, 1995)
		2.9	10	25	7	6	(Kazama, 1995)
		1.4	10	25	7	3.5	(Kazama, 1995)
		2.9	10	25	8	4	(Kazama, 1995)
		1.4	10	25	8	1.5	(Kazama, 1995)
f2	Secondary effluent	1	0.8		5.9	2	(Schink and Waite, 1980)
		1	5.7		6.9	2	(Schink and Waite, 1980)
		1	22		7.8	2	(Schink and Waite, 1980)
		10	30		7.8	3	(Jiang and Lloyd, 2002)
	Unknown	6.0	1.5	20	6	2	(Safe Drinking Water Committee, 1980)
		1.2	4	20	6	2	(Safe Drinking Water Committee, 1980)
		1.2	13		8	2	(Safe Drinking Water Committee, 1980)

6.3.2.3 System Operation

There is currently very little information available on the operational requirements of ferrate disinfection. New technology for ferrate production can be integrated into existing hypochlorite infrastructure and may reduce costs (Alig, 2007); however, this method uses chlorine for production, thereby retaining the issues associated with chlorine storage and use.

6.3.2.4 Impacts to the Environment

Depending on the synthesis method, ferrate may increase the total dissolved solids concentration (TDS) in the water; this may be problematic for facilities with stringent TDS requirements. Little work has been done on possible disinfection byproducts from ferrate treatment; however, an Ames test indicated no mutagenicity in ferrate-treated waters (Jiang and Lloyd, 2002). Jessen et al. (2006) investigated regrowth of *E. coli* at ferrate doses of 5 and 10 mg/L with one-minute exposure. The 10 mg/L dose yielded complete inactivation (6-log) with no signs of regrowth after 48 hours; inactivation with the 5 mg/L dose was incomplete (4-log), and data indicated some metabolic activity, although bacterial levels continued to decrease between one minute and 48 hours after ferrate addition.

Ferrates provide many benefits in addition to disinfection. As powerful oxidants, ferrates have the ability to transform a number of other compounds. They rapidly oxidize sulfur compounds and can be used for odor removal and control (Murmann and Robinson, 1974). They can also lower the concentrations of As (III), Pb(II), Cu(II), Zn(II), Fe(II), Fe(III), Hg(II), Cd(II) and Cr(III) (Murmann and Robinson, 1974; Sharma et al., 2005).

When reduced, ferrate forms insoluble Fe(III), as an iron hydroxide or iron oxyhydroxide (Murmann and Robinson, 1974; Safe Drinking Water Committee, 1980).

$$FeO_4^{2-} + 4\,H_2O + 3\,e^- \rightarrow Fe(OH)_3 + 5\,OH^- \qquad (6.3.2.4\text{-}1)$$

$$2\,FeO_4^{2-} + 3\,H_2O \rightarrow 2\,FeO(OH) + 1.5\,O_2 + 4\,OH^- \qquad (6.3.2.4\text{-}2)$$

These solids can remove algae, phosphorus, contaminants (e.g., endocrine disruptors, NDMA, MTBE) and color from water, either through co-precipitation with other species (e.g., metals), or by acting as a flocculant to help settle suspended solids in water. Higher concentrations of ferrate (> 10 mg/L) result in faster precipitation of solids and therefore faster clarification of water; for example, a ferrate dose of 15 mg/L reduced COD from 40 to 7 mg/L. Ferrate appears to destabilize colloidal particles approximately 30 times faster than ferric or ferrous salts, and can be used at smaller doses than ferric chloride and aluminum sulfate, thus yielding lower sludge generation (Murmann and Robinson, 1974; Jiang and Lloyd, 2002; Sharma et al., 2005; Reimers et al., 2007).

6.3.2.5 Costs

Very little cost information is available for ferrate disinfection. Older literature sources generally indicated that ferrate salts are expensive and not commercially available (Safe Drinking Water Committee, 1980), but recent developments have decreased costs from ~$20/pound to ~$2/pound (Alig, 2007).

6.3.2.6 Summary

Ferrate is an oxidizing agent that shows promise for disinfection, removal of organics and metals, flocculation, and odor control. However, it decomposes in solution (i.e., loses activity), and the traditionally high cost of buying or generating ferrate has limited its use. Recent advances may reduce chemical costs, which will facilitate its use, but more research is needed on optimization of dose and contact time, disinfection efficacies against a wider range of organisms, effects of wastewater quality on disinfection efficacy, and potential for regrowth or disinfection byproducts in wastewater.

6.3.3 Bromine Compounds

This section describes several bromine compounds that have been used as biocides: elemental bromine (Br_2), bromine chloride (BrCl), and 1-bromo, 3-chloro, 5,5-dimethylhydantoin (BCDMH). Bromine compounds have been used for water disinfection for swimming pools, spas, and cooling towers. BCDMH has also been used for combined sewer overflow (CSO) disinfection under wet weather conditions.

6.3.3.1 Elemental Bromine

Pure elemental bromine is a heavy, red liquid. It is a strong oxidizing agent that reacts with surrounding materials if spilled. It is corrosive to steel and is therefore stored in glass jugs that must be specially crated to guard against shock damage and breakage during shipping and handling. Fumes from uncapped containers may be harmful to people and damage surroundings.

Hydrolysis of bromine produces hypobromous acid (HOBr), which is the species responsible for pathogen inactivation (see Section 6.3.3.2 for more detail). Hypobromous and hypochlorous acids are comparable disinfectants, but elemental bromine is more costly than gas chlorine for equivalent disinfection performance.

6.3.3.2 Bromine Chloride

Bromine chloride is a heavy, dark red, fuming agent formed from atomic chlorine and atomic bromine. It can be generated onsite from sodium or potassium bromide and hypochlorous ions. In general, the properties of BrCl fall between those of the individual halogens, chlorine and bromine. For example, it is less corrosive to steel than bromine, and can be stored in containers similar to those used for chlorine gas. It dissociates more quickly than chlorine in water, because it is twelve times more soluble (U.S. EPA, 1986; WEF, 2006). Bromine chloride hydrolyzes in water to form hypobromous acid (HOBr) and chloride as follows:

$$BrCl + H_2O \leftrightarrow HOBr + H^+ + Cl^- \qquad (6.3.3.2\text{-}1)$$

Hypobromous acid is the primary species responsible for pathogen inactivation; reactions are analogous to those with hypochlorous acid (HOCl). For both compounds, the acid is a more powerful disinfectant than the dissociated ion. HOBr has a higher pKa (8.7) than HOCl (pKa of 7.5), and consequently, bromine remains in the acid form and is effective over a larger pH range than chlorine. Bromine chloride also reacts quickly with ammonia to form bromamines, which have disinfection efficiencies comparable to free bromine residuals. Bromamines are more effective than chloramines for inactivation of bacteria and viruses, especially when organic and inorganic interferences are present (U.S. EPA, 1986; WEF, 2006).

Bromine species may also react with organic materials to form unwanted THMs. The relatively short half-life of its residuals may lower the formation potential of harmful disinfection byproducts in receiving waters. However, brominated organics formed during disinfection are often considered more harmful than the analogous chlorinated organics (WEF, 2006).

In practice, bromine chloride is dispensed through a feeder and diffused into the wastewater to be treated. Due to its rapid decomposition, good mixing is required to ensure immediate and thorough dispersion of the chemical. However, because of its high water solubility and rapid dissociation, contact tanks can be smaller than with chlorination. Processes are typically controlled using bromine residuals, which can be analyzed with methods directly analogous to those for chlorine residuals (U.S. EPA, 1986; WEF, 2006).

Bromine chloride should be stored in a building with proper ventilation because it is volatile. It is also extremely toxic and can cause burns to skin and other tissues. Therefore, the operators should completely cover their skin and wear rubber gloves and boots when entering a containment area (WEF, 2006). Bromine chloride is not widely available commercially and is more expensive than chlorine.

6.3.3.3 BCDMH

BCDMH (1-bromo, 3-chloro, 5,5-dimethylhydantoin), $C_5H_6BrClN_2O_2$, is a strong oxidizing agent, with each molecule containing two halogens (one bromine and one chlorine). In water, BCDMH hydrolyzes into dimethylhydantoin (DMH, $C_5H_8N_2O_2$), hypobromous acid, and hypochlorous acid as shown below:

$$BCDMH + 2\ H_2O \rightarrow DMH + HOBr + HOCl \qquad (6.3.3.3\text{-}1)$$

$$HOCl + Br^- \rightarrow HOBr + Cl^- \qquad (6.3.3.3\text{-}2)$$

BCDMH was developed in the late 1950s and first introduced to cooling water systems in 1980 (Boner et al., 2002). It is now commonly used for slime control in industrial, commercial, and household applications, and as a disinfectant for disinfecting water in pools, spas, and cooling towers. NSF International (originally the National Sanitation Foundation) certified its use as a disinfectant for public water supplies in 2000. Currently, only a handful of facilities in Japan have implemented BCDMH for disinfection of wastewater, although several have used BCDMH to disinfect combined sewer overflows (CSOs) since 2001 (Yamamoto et al., 2003; Fuchu et al., 2005).

Disinfection of CSOs is challenging, due to their intermittent nature and highly variable flow rates and water quality. Ammonia concentration, immediate chemical demand, pH, and temperature may vary widely and adversely affect performance and operations. Currently, CSO disinfection relies primarily on sodium hypochlorite (NaOCl), but BCDMH offers an alternative, with advantages such as a smaller footprint, better safety, and less storage space. For a given mass dose, BCDMH appears to be a more effective disinfectant than sodium hypochlorite, peracetic acid, or bromine chloride; it is also reacts quickly, with 90% of the log inactivation occurring within three minutes after addition (Boner et al., 2002; Moffa and Somerlot, 2007).

With regard to impacts on the environment, Yamamoto et al. (2003) detected no bromate when BCDMH was used; bromoform was the main THM species. In separate tests with doses between 3 and 12 mg/L and contact times of 3 to 5 minutes, THM levels did not exceed drinking water MCLs. However, HAAs exceeded MCLs by 40% after the highest BCDMH dose of 12 mg/L and the longest contact time of five minutes (Moffa et al., 2006). BCDMH addition produced less toxicity than sodium hypochlorite to fish, algae, and Daphnia species (Yamamoto et al., 2003). It showed some toxic effects on Daphnia, although the toxicity did not increase with increasing BCDMH doses; however, it appeared to have little toxic effect on fathead minnows (Moffa et al., 2006).

Facility requirements for CSO disinfection by BCDMH are similar to those for sodium hypochlorite: a contact chamber to treat the peak flow for a minimum of 5 minutes, high rate induction mixers, feed pumps and equipment, and miscellaneous piping, control, and safety items. The main differences are that BCDMH does not need chemical storage tanks, but does require a dissolution and injection unit because the chemical is typically supplied as a powder. The regulatory safety requirements for storing powdered BCDMH are not known at this time. Although experience is limited, BCDMH is expected to have fewer handling problems and lower exposure risk, but require more equipment maintenance relative to sodium hypochlorite systems (Moffa et al., 2006). Moffa and Somerlot (2007) estimated that the capital cost for BCDMH would be 11% higher than for NaOCl for CSO disinfection at one site, with the extra cost of the BCDMH dissolution and injection equipment being slightly higher than the savings from the smaller contact chamber. However, the capital costs are balanced by energy and chemical costs, which were 38% lower for BCDMH than for NaOCl (Moffa and Somerlot, 2007).

6.4 Developing Technologies: Physical Methods

6.4.1 Pasteurization

6.4.1.1 Background

Pasteurization is a technology that uses heat to denature proteins and kill or inactivate microorganisms. The key parameters in determining the effectiveness of the process are the temperature and exposure time. The technique is named for Louis Pasteur, who developed the process to reduce microbial degradation of wine by heating at a relatively low temperature (55°C/131°F), which inhibited bacterial growth but preserved the flavor of the wine. This process has subsequently been applied to a wide variety of food and beverage products, most notably to milk (Madigan et al., 1997). It has also been applied to sewage sludge (U.S. EPA, 1995), and has recently been investigated as a disinfection alternative for wastewater.

6.4.1.2 Disinfection Efficacy

Little to no information is currently available on the flexibility and reliability of the process with respect to wastewater quality. However, previous research with skim milk has shown that for a given level of deactivation, milk with higher fat content required longer exposure times (Bidawid et al., 2000); this suggests that water quality may affect performance.

Disinfection efficacies for various organisms in wastewater are given in Table 6-10. These experiments were conducted in Petri dishes, which were incubated in ovens for pasteurization. Table 6-11 shows removals as a function of time and temperature for flow-through pasteurization processes tested at the pilot scale.

Table 6-10. Pasteurization of Various Organisms in Raw Sewage.
(Moce-Llivina et al., 2003 with permission from American Society of Microbiology)

Organism	Average Log_{10} Reduction[a]
Fecal coliforms	>5.6 – 6.2
E. coli	6.0
Fecal *Streptococcus*	3.4
Spores of sulfate-reducing *Clostridia*	0.1
Culturable enterovirus	>1.7
Added culturable enteroviruses:	4.3-5.4
poliovirus1, coxsackievirus B4, coxsackievirus B5, Echovirus 6, EV1, EV2[b]	
Added bacteriophage MS2 and ΦX174[b]	2.1 - 2.8
F-specific RNA phages	2.1
Somatic coliphage	0.8
Phages infecting *B. fragilis* RYC2056	0.4
Added bacteriophage: MY2, SC12, SS13, B40-8[b]	0.3-0.5

[a]All pasteurizations were conducted for a contact time of 30 minutes, at a temperature of 60°C/140°F.
[b]In some experiments, bacteriophages and viruses were added to raw sewage before pasteurization. Organisms not marked as "added" were naturally-occurring.

Table 6-11. Pilot-Scale Pasteurization of Total Coliform with Varying Exposure.

Organism	Water	Temperature (°C)	Time (sec)	Average Log_{10} Reduction	Source
Total Coliform	Secondary effluent	73-85	3-6	>4.4 - >5.9[a]	(Knapp, 2006)
		71	5.6	4.6	(Knapp, 2006)
		68-71	3-4	2.8-4.4	(Knapp, 2006)
	Tertiary effluent	69-78	8-30	1.8-3.5	(Salveson et al., 2006)
		64-68	8-30	1.1-3.1	(Salveson et al., 2006)
		54-61	8-32	0.0-2.0	(Salveson et al., 2006)
MS2	Tertiary effluent	72-78	16-28	1.1-6.0	(Ryan Pasteurization & Power, 2006)
		64-70	8-13	0.0-1.0	(Ryan Pasteurization & Power, 2006)

[a]Effluent coliform values were generally below detection limits, so only a minimum reduction could be calculated.

6.4.1.3 System Design and Implementation

Pasteurization for wastewater treatment is currently being tested in pilot units, and has not been applied at full scale. As such, relevant data on design and implementation are limited, so any application of this technology should involve pilot tests to determine the disinfection efficacy for that specific water quality. Regulators will require proof of system reliability for disinfection; thus far, at least one pasteurization process for wastewater disinfection has been submitted to the California Department of Health Services for approval under the Title 22 regulations for disinfection of reclaimed water (Ryan, 2006). Because the application of pasteurization to wastewater is relatively new, public acceptance is uncertain; however, the lack of chemicals and the public's familiarity with milk pasteurization is likely to facilitate acceptance.

Despite the uncertainties, the technologies for pasteurization are fairly mature, so all of the components are expected to be normal stock equipment (Knapp, 2006). Based on pilot tests, several design criteria can be estimated. In terms of space requirements, pasteurization systems at facilities with <10 MGD are expected to require approximately the same or less space than a chlorine contact chamber with 90 minutes of contact time; the eventual process is envisioned as modular, which should facilitate expansion (Knapp, 2006; Ryan, 2006). As with most disinfection technologies, flow equalization is recommended (Ryan, 2006). For power, the turbine that drives the pasteurization requires an external supply of natural gas, which can be supplemented with methane produced from anaerobic digestion (Knapp, 2006; Ryan, 2006). This use of the anaerobic digester gas suggests that pasteurization may work well at POTWs that currently use digester gas for power generation; at these facilities, the turbines could generate power, provide the driving force for disinfection, and serve as a conduit for gas disposal. As with all gas turbines, compliance would be required with regulations on CO_2 emissions from the combustion of natural gas. More design information should become available as this technology is developed further.

6.4.1.4 Operation and Maintenance

There is currently little experience and information on the operation and maintenance needs for this process. Any system shutdown due to maintenance, repair, or emergencies (e.g., loss of power) will require an alternate, effective disinfection system. Some maintenance will likely be required to remove scaling on the heat exchangers, but there are insufficient data to estimate the frequency or intensity that might be needed. In general, effluents higher in calcium and alkalinity will require more attention due to scaling of the heat exchangers. Redundant pumping is necessary to ensure continuous flow of water and avoid heat damage. Standard gas turbine maintenance will also be necessary. Operator safety issues should be similar to those for a combined cycle power system (e.g., dealing with high temperatures, rotating equipment, etc.). As a side benefit, pasteurization provides power stability; because the gas turbines that provide the heat for pasteurization also provide electricity, facilities that use pasteurization will be less affected by fluctuations or outages on the power grid. Another potential benefit is the use of waste heat to heat the anaerobic digestors making this process more efficient.

6.4.1.5 Impacts to the Environment

There is currently no information available on disinfection byproducts, but few if any DBPs are expected, because no chemicals are added; as a result, there are no requirements for disinfectant removal. However, it is unknown whether any reactants or byproducts might be formed by the higher temperatures that occur during pasteurization.

Pasteurization raises the temperature of the effluent water, although the increase can be controlled to approximately 4°F or 2°C (Ryan, 2006), which would decrease dissolved oxygen concentrations by ~3-6% (Tchobanoglous et al., 2003). This may be a significant issue for POTWs discharging to sensitive water bodies such as those designated as cold water habitats, with temperature TMDLs, or for plants in California, where a statewide thermal discharge plan limits effluent temperatures, particularly when the effluent comprises a large portion of the receiving water. Most other environmental impacts are expected to be minimal, e.g., dissolved solids may decrease imperceptibly, as carbonates will deposit on the exchanger surfaces (Knapp, 2006).

Potential risks from accidental release or system failure are expected to be relatively low, because no chemicals are used. Under normal operating conditions, the temperature range for the water is not expected to cause problems with pyrogenic organics (Knapp, 2006). The risk of regrowth of microorganisms after pasteurization has not been investigated in wastewater, but it is a possibility, as it has been observed in sewage sludges (U.S. EPA, 1995). In sludges, it was attributed to the lack of competitive bacteria, which were eliminated during the pasteurization process (U.S. EPA, 1995).

6.4.1.6 Cost

Because this technology is new to wastewater treatment, little cost information is currently available; however, an economic analysis has indicated that this process is likely to be competitive with chlorination and UV disinfection (Ryan, 2006).

6.4.1.7 Summary

One of the key advantages of pasteurization is that it does not use any chemicals, which provides several benefits. It eliminates the need for disinfectant removal, minimizes the impacts on water quality (e.g., from disinfection byproducts), and reduces the operating costs and safety risks posed by chemicals. In addition, the gas turbines that generate the heat for pasteurization provide electricity for the POTW, thus making the facility less reliant on the external power distribution system. The pasteurization process also provides a productive outlet for the methane gas produced in anaerobic digestion and/or the waste heat from gas turbines.

However, this technology will be less cost effective for facilities without an existing heat source or turbine-generated electric power. In addition, facilities that discharge to sensitive waters may be concerned about the slight increase in effluent water temperature, and the associated decrease in the concentrations of dissolved oxygen. Finally, as with all emerging technologies, the lack of data makes this technology riskier to adopt currently.

The currently available data on disinfection efficacy have been collected either in batch laboratory studies, or for short time periods in-line with treatment processes. More data are needed to better understand the effects of contact time, temperature, and water quality on disinfection, and to optimize these parameters. Additional data on microbial regrowth and the resistance of various organisms (e.g., spores and thermophiles) to pasteurization will also help determine which organisms would serve as good indicators. More information is also needed on any potential byproducts or reactants that could be formed by the elevated temperatures during pasteurization, and the effects of the warmer effluent on receiving waters. Finally, more information is needed on operation and maintenance requirements and costs.

6.4.2 Ultrasonic Cavitation

6.4.2.1 Background

There are two types of cavitation: transient or inertial cavitation, and stable cavitation. Transient cavitation is defined as the rapid formation, expansion, and collapse of gas bubbles within a liquid with oscillating pressures. The oscillations can be caused by sound waves, lasers, hydrodynamic fluctuations, or electrical discharge through a spark. Bubbles form when the pressure of the liquid falls below the vapor saturation pressure, and implode when the pressure rises above it. The collapse of these bubbles is quite violent, with local temperatures up to thousands of degrees Celsius or Fahrenheit, and pressures up to hundreds of atmospheres or thousands of psi (Hua and Thompson, 2000; Gaines et al., 2006). The implosion can also form damaging micro-jets and chemical species, such as hydroxyl radicals and hydrogen peroxide (Gaines et al., 2006). Stable cavitation refers to the oscillation of bubbles within a pulsating liquid field. This oscillation causes the liquid immediately adjacent to the bubble interface to flow in a "microstream"; microstreaming has been shown to stress and disrupt cell membranes (Scherba et al., 1991) and create destructive shear forces within cells (Joyce et al., 2003).

Disinfection of microorganisms by cavitation is generally attributed to high temperatures, hydrodynamic forces, and chemical reactions (Gaines et al., 2006). High temperatures cause pasteurization: denaturing of proteins, which kills or inactivates cells. Hydrodynamic forces include shear forces from microstreaming and micro-jets, pressure gradients during bubble collapse, and stresses on cell walls and membranes from surface resonance induced by cavitation (Joyce et al., 2003). Once the cell wall or membrane is damaged, cells may fall apart or be susceptible to internal damage from radicals and other species produced during cavitation (Hua and Thompson, 2000). Ultrasound also decreases the heat resistance of bacterial spores and enhances the effectiveness of ozone disinfection for viruses and bacteria (Hua and Thompson, 2000). Scherba et al. state that the thermal contribution to disinfection is minimal for ultrasonic cavitation at 26 kHz and "various intensity levels" (Scherba et al., 1991). However, with a power input of 700 watts per liter (W/L) at 20 kHz, Madge and Jensen estimated that 52% of the disinfection they observed was due to heat (consistent with data from three other experiments), and 36% was due to mechanical stresses (Madge and Jensen, 2002). The remaining 12% was suggested to be synergistic effects from the combination of thermal and mechanical stresses, but may also have been due to chemical disinfection.

Ultrasound is the most commonly used source of cavitation for disinfection, although hydrodynamic cavitation has also been tested for disinfection of recycled water in an industrial cooling tower (Gaines et al., 2006). For ultrasonic disinfection, the most important operational parameters appear to be power, exposure time, and radiation frequency. Typical power values are in the range of 50 to 500 W/L, with reported values between 10 and 1500 W/L. Exposure times are generally in the range of seconds to hours. Ultrasound frequencies are defined as being greater than 20 kHz, the limit of human hearing. The term "power ultrasound" is sometimes used and refers to frequencies between 20 and 100 kHz, which appear to have the highest disinfection potential (Madge and Jensen, 2002), with the strongest hydrodynamic forces at low frequencies (Neis and Blume, 2003). At higher frequencies, disinfection efficacies decrease (Hua and Thompson, 2000; Joyce et al., 2003).

From a practical perspective, sonication offers the advantage of small units, which should fit relatively easily into existing spaces and which allow for easier expansion at wastewater treatment plants (Blume et al., 2002). Although disinfection tests thus far have been conducted primarily on industrial effluents (Gaines et al., 2006) or in controlled laboratory settings with ultrapure water or nutrient broth (Hua and Thompson, 2000; Joyce et al., 2003), recent work has tested wastewater as well (Madge and Jensen, 2002; Blume and Neis, 2005). Many researchers have concluded that the energy requirements for effective disinfection via cavitation would be economically prohibitive, but note that cavitation could be very effective in combination with other disinfectants (Mason et al., 2003a ; Blume and Neis, 2004; Blume and Neis, 2005); more details are given in Section 6.5.8. Additionally, some results indicate that larger scale continuous sonication reactors are more efficient than laboratory tests (Madge and Jensen, 2002; Neis and Blume, 2003); thus, disinfection efficacy and cost may improve when scaled up. Two such processes have been tested recently: one treated zooplankton in a flow of 2 MGD (300 m^3/hr) with an ultrasound dose of 3 W-min/L; the other treated a concentrated algae solution at a flow rate of 9 gallons per minute (2 m^3/hr) using a dose of 14 W-min/L (Mason et al., 2003b). Overall, however, very few projects have been constructed for continuous flow at scales that are practical for major POTWs, so currently no information exists on design, implementation, operations, maintenance, or cost.

6.4.2.2 Disinfection Efficacy

Disinfection is not significantly affected by the water quality parameters tested thus far: total suspended solids (20-600 mg/L), pH (6-8.5), alkalinity (150-300 mg/L), BOD, total organic carbon, and initial fecal coliform concentration (Madge and Jensen, 2002).

Disinfection does increase as the exposure time and supplied power increase (Scherba et al., 1991; Madge and Jensen, 2002; Dehghani, 2005; Mahvi et al., 2005). The time dependence is quite strong, and the process is often modeled as first order (Hua and Thompson, 2000; Neis and Blume, 2003). The quantitative dependence of disinfection on power is less robust, although disinfection generally improves as power increases, with frequency and exposure time held constant (Scherba et al., 1991; Hua and Thompson, 2000; Madge and Jensen, 2002; Blume and Neis, 2004; Dehghani, 2005; Mahvi et al., 2005). At low power, the sonication energy may be enough to fragment bacterial clumps but insufficient to inactivate them; this can cause an increase in the measured bacterial counts, because clumps containing multiple bacteria may appear as a single colony when analyzed before sonication, but would form many colonies once broken up (Joyce et al., 2003). At higher power, disinfection outweighs the fragmentation, and overall bacterial numbers decrease (Joyce et al., 2003). Madge and Jensen fit disinfection with a second order dependence on power (Madge and Jensen, 2002); this was a much stronger effect than was observed in other experiments. The researchers attribute the strong effect to 1) much larger temperature increases of up to 60°C, compared to <10°C increase in another experiment (Hua and Thompson, 2000; Madge and Jensen, 2002); and 2) a relatively high concentration of suspended solids. Waters with higher levels of solids can actually increase cavitation intensity because the solids can serve as nucleation-sites for bubbles to form (Madge and Jensen, 2002; Blume and Neis, 2005).

Ultrasound has been tested with a number of species of bacteria and bacterial spores, fungi, algae, viruses, and zooplankton (Scherba et al., 1991; Madge and Jensen, 2002; Mason et al., 2003b). It has also been shown to damage all tested bacteria and bacterial indicators, including total coliforms, fecal coliforms, *E. coli*, fecal *Streptococcus*, *Staphylococcus aureus*, *Bacillus subtilis*, and *Pseudomonas aeruginosa* (Scherba et al., 1991; Blume et al., 2002; Madge and Jensen, 2002; Mahvi et al., 2005). Exposure to 26 kHz ultrasound for 60 minutes slightly reduced the growth of the fungus *Trichophyton mentagrophytes*, and significantly reduced viral infectivity of the feline herpesvirus type 1, but not feline calcivirus; disinfection was independent of the initial viral concentration (Scherba et al., 1991).

Table 6-12 provides more data on disinfection efficacy. Gram-positive bacteria such as *Streptococcus*, *Staphylococcus* and *Bacillus* have thicker cell walls and a more tightly adherent layer of peptidoglycans than gram-negative bacteria such as coliform bacteria and *Pseudomonas* (Scherba et al., 1991). This thicker cell wall provides protection and resulted in lower removals of *Streptococcus* than *E. coli* for a given treatment (Neis and Blume, 2003). Scherba did not detect any significant differences between gram positive and negative bacteria, possibly due to the large amount of scatter in the data, which may have hidden any differences.

Table 6-12. Disinfection Efficacy of Sonication Against Various Microorganisms.

Organism	Water	Frequency (kHz)	Power (W/L)	Time (min)	Average Log_{10} Reduction	Source
Total Coliform	Unknown	42	85-235	1-30	0-1.1	(Mahvi et al., 2005)
Total Coliform	Unknown	42	85-235	40-90	1.7-3.2	(Mahvi et al., 2005)
E. coli	Unknown	42	115-350	1-30	0-1.1	(Dehghani, 2005)
E. coli	Unknown	42	115-350	40-90	1.7-3.2	(Dehghani, 2005)
E. coli	Distilled water	20	260-470	10-60	0.5-3	(Hua and Thompson, 2000)
E. coli	Secondary effluent	20	400	30-60	2.0-2.9	(Neis and Blume, 2003)
Fecal *Streptococcus*	Unknown	42	400	30-60	0.4-0.9	(Neis and Blume, 2003)

6.4.2.3 Impacts to the Environment

Cavitation increases the temperature of the water, but estimations of the magnitude vary (Table 6-13). Cavitation also strips CO_2 from the water, which increases the pH (Gaines et al., 2006); the pH change is independent of exposure time and increases as the power input rises, up to an increase of 1.5 pH units with 1,250W/L of power (Madge and Jensen, 2002). This increase in pH shifts the carbonate species equilibrium toward the ion and away from the acid, thereby increasing the concentration of carbonate ion. Scaling, or the precipitation of calcium carbonate, is an endothermic reaction, so the increased temperatures may increase precipitation of carbonate species on process equipment and decrease carbonate content of the effluent water (Gaines et al., 2006); however, no changes in alkalinity were observed by Madge and Jensen (2002). Sonication does not alter BOD levels (Madge and Jensen, 2002).

Table 6-13. Increases in Water Temperature Due to Sonication.

Temperature Change (ºC)	Power (W/L)	Sonication Time (min)	Source
<1.2-3.5 (0.00096ºC/second)	"Various intensities"	20-60	(Scherba et al., 1991)
<10	270-460	60	(Hua and Thompson, 2000)
60	700	20	(Madge and Jensen, 2002)

One of the main effects of sonication on water quality is the change in particle size distribution, as large particles or clumps of bacteria are broken up into smaller pieces (Table 6-14). Blume and Neis characterized a wastewater as having 63% of its solids >50 µm in size; after 20 seconds of sonication at 30 W/L, only 5% of the solids were > 50 µm in size (Blume and Neis, 2004). The net effect of cavitation is to decrease total and volatile suspended solids, with higher power inputs causing greater decreases in solids concentrations (Madge and Jensen, 2002). The total organic carbon content increases concurrently, and likely indicates the disintegration of solids to such a small size that they are considered dissolved, rather than suspended (Madge and Jensen, 2002).

Table 6-14. Changes in Particle Size Distribution Due to Sonication (from Blume and Neis, 2004).

Mean Particle Diameter (μm)	Power (W/L)	Sonication Time (sec)
68	N/A	0
35	170	10
20	310	10
11	30	20

As a side benefit, ultrasound has been tested and shown to remove a variety of organic compounds, such as aromatics and halogenated hydrocarbons, and inorganic contaminants, such as hydrogen sulfide (Weavers et al., 2002).

6.4.2.5 Summary

Ultrasonic cavitation has several advantages over other wastewater disinfection alternatives: it requires a relatively small footprint, uses no chemicals, and actually works better in waters with higher solids contents. However, it has not been tested extensively so there are many unknowns, and it appears that the energy requirements to achieve significant disinfection currently make it economically unfavorable, though this may change as the technology improves. At the moment, cavitation looks most promising in combination with more traditional disinfection techniques, such as chlorination or UV, because a small amount of ultrasound energy may drastically increase disinfection efficacy, and/or reduce energy, chlorine or UV requirements.

6.4.3 Electron Beam (E-beam) and Gamma Irradiation

6.4.3.1 Background

High energy electrons are generally produced by electron accelerators or gamma ray sources. Gamma irradiators typically use radioactive cobalt 60 as the energy source. In electron accelerators, electrons are produced by a cathode and accelerated by an electric field. The applied voltage of the electric field determines the speed and energy of the accelerated electrons; an applied voltage of 1.5 megavolts (MV) will accelerate electrons to an energy level of 1.5 mega-electron volts (MeV). The beam power is simply the product of the current and voltage, e.g., a 1.5 MeV accelerator operating at 50 mA produces 75 kW of beam power (Kurucz et al., 1991).

The absorbed radiation dose is measured in gray (Gy) units; one Gy is equivalent to the absorption of one joule of radiation energy by one kilogram of matter. The gray replaces the older radiation dose units of the rad (one rad is equivalent to 0.01 gray) and the roentgen equivalent physical (rep: one rep is equivalent to 0.93 mGy). Gamma sources generally deliver dose rates of 0.2 to 3 kGy/hour, while e-beams can deliver dose rates of 400 to 600 kGy/hour (Kurucz et al., 1991).

Historical Use Electron beam (e-beam) equipment has been in commercial use since the 1950's, and has received interest for water, wastewater and sludge treatment in the last few decades (Kurucz et al., 1991). Accelerators of 1-5 MeV, up to 40 or 50 mA, and 100 kW are readily available; higher energy and current machines are also available (Kurucz et al., 1991). Gamma irradiators have been used since the 1960's for the sterilization of medical products and consumer goods (Swinwood et al., 1994). These technologies have been proven effective for disinfection for some time, but the cost of treatment has thus far made the technology less attractive (Electric Power Research Institute, 1999).

How It Disinfects High-energy electrons interact with material (in this case, the water and its contaminants) to produce electrons of lower and lower energy. A single high-energy electron can initiate several thousand reactions as it dissipates energy. Eventually, a large number of low energy (<50 eV) electrons are generated, and they ionize the material to produce electrons and positive ions. These ions and electrons then react with water or its constituents; this process is known as indirect radiolysis (direct radiolysis occurs in low dielectric materials and is the chemical reaction caused when a positively ionized species recombines with its lost electron) (Kurucz et al., 1991).

E-beam irradiation of water creates a variety of oxidative and reductive species through indirect radiolysis (Cooper et al., 2006):

$$\begin{aligned} &0.7\,H_2O + \text{energy} \rightarrow 0.28\,OH\bullet + 0.06\,H\bullet \\ &+ 0.27e^-_{aq} + 0.05\,H_2 + 0.07\,H_2O_2 + 0.27\,H_3O^+ \end{aligned} \qquad (6.4.3.1\text{-}1)$$

Gamma radiation produces similar species: hydroxyl radicals (OH•), hydrogen atoms (H•), and hydrated electrons (e^-_{aq}). These reactive species can produce chemical lesions in nucleic acids, proteins and other molecules in cells and viruses. Second-order rate constants with proteins, enzymes, DNA, and RNA range from 107 to 1011 $M^{-1}\,s^{-1}$. Some literature indicates that the hydroxyl radical is primarily responsible for biological damage, but other data suggest that the hydrogen radical and aqueous electrons also contribute to inactivation. (Farooq et al., 1993)

E-beam and gamma irradiation technologies are unique relative to many other treatment processes in that they form both highly oxidizing (hydroxyl radical) and highly reducing (e-) species (U.S. EPA, 1993). This allows the processes to promote a wider range of reactions, with compounds that are amenable to either oxidative or reductive degradation (U.S. EPA, 1993). Other advantages of the e-beam technology include that it produces no additional waste streams and the byproducts are relatively non-toxic and at low levels (U.S. EPA, 1993; Larson, 1999).

6.4.3.2 Disinfection Efficacy

Disinfection requirements increase with the depth of the water being treated by e-beam or gamma irradiation; for 3-log inactivation, the required dose increased from 0.39 to 0.44 kGy as water depth rose from five to seven mm (Miyata et al., 1990). Although some studies suggest that disinfection is not impacted by pH between 3 and 11, suspended solids (up to 5%), COD, and irradiation temperature (Miyata et al., 1990; U.S. EPA, 1992), the Orange County Water District found that water quality does affect performance (Larson, 1999). Other studies support this finding: solvent removal in raw and secondary wastewaters required higher doses than potable water (U.S. EPA, 1992), and unchlorinated wastewater yielded less inactivation than chlorinated wastewater (Kurucz et al., 1991). However, Farooq et al. (1993) found little difference in inactivation between raw and secondary wastewaters, with slightly better performance in raw wastewater. Data on the effect of dissolved oxygen also conflicts, with Miyata et al. (1990) observing no effect, while others have found enhanced DNA degradation and biological inactivation in the presence of oxygen (Safe Drinking Water Committee, 1980; Farooq et al., 1993;).

Available information suggests that doses <1 kGy would be sufficient to disinfect potable water supplies; however, efficacies against recently discovered pathogens are uncertain, and data for wastewater are quite limited. Table 6-15 summarizes literature results for disinfection of wastewater using e-beam and gamma irradiation. Attempts to find radio sensitizers (chemicals to enhance disinfection) showed that acetic acid and hydrogen peroxide improved disinfection, but vitamin K, tertiary hydrobuylquinone and lactic acid did not. Unfortunately high chemical doses were required to reduce the e-beam dose: 100 mg/L of acetic acid, and almost 3,000 mg/L of hydrogen peroxide (Electric Power Research Institute, 1999).

Table 6-15. Disinfection Efficacy of E-Beam and Gamma Irradiation Against Various Pathogens.

Radiation Source	Dose (kGy)	Organisms	Water	Log_{10} Removal	Source
E-beam (2 MeV)	0.39-0.44	Total population	Wastewater	3	(Miyata et al., 1990)
E-beam (1.5 MeV)	5	Coliform, total bacteria, coliphage	Secondary effluent	2-4	(Farooq et al., 1993)
Gamma	5	Coliform, total bacteria, coliphage	Secondary effluent	2-6	(Farooq et al., 1993)
Gamma	0.05-0.14	*Bacillus subtilis, Enterobacter aerogenes, Escherischia coli, Salmonella* sp., *Salmonella paratyphi B, Salmonella typhi, Shigella flexneri, Shigella sonnei, Staphylococcus aureus, Streptococcus faecalis*	Unknown	>2	(Safe Drinking Water Committee, 1980)

In general, larger organisms appear to be more susceptible to inactivation by e-beam and gamma irradiation. For example, in sludge, viruses appear to be more resistant than bacteria or parasitic worms to e-beam irradiation. Similarly, in wastewater, coliphages were more resistant than coliforms or total bacteria to both e-beam and gamma irradiation. A dose of 5 kGy from e-beam or gamma irradiation removed 0.5 to 3 log units fewer coliphages than coliform or total bacteria from raw or secondary wastewater (Kurucz et al., 1991; Farooq et al., 1993).

Within bacteria, even a single species can have significantly different resistances between strains: for 2-log inactivation, *Salmonella typhimurium* (24) required four times the applied radiation dose of *Salmonella typhimurium* (LT2). In this study, the susceptibility of bacteria to e-beam inactivation, in order of increasing resistance, was as follows: *Salmonella typhimurium* (LT2), *Micrococcus* sp., *Escherichia coli* (K12), *Aspergillus niger* (spores), *Salmonella typhimurium* (24), *Streptococcus fecalis*, and *Clostridium perfringens* (spores). A gamma radiation study showed increasing resistance of organisms as follows: *Enterobacter aerogenes, Escherischia coli, Shigella flexneri, Salmonella typhi, Shigella sonnei, Salmonella* sp., *Staphylococcus aureus, Salmonella paratyphi B, Streptococcus faecalis, and Bacillus subtilis* (Safe Drinking Water Committee, 1980).

6.4.3.3 System Operation

Caution should be used in extrapolating laboratory batch-scale results to pilot or field-scale continuous-flow systems; literature indicates that higher doses (2 to 3 kGy) are required at the larger scales (Kurucz et al., 1991). At the pilot-scale Electron Beam Research Facility in Miami, Florida, the maximum penetration depth of 1.5 MeV electrons was approximately 7 mm in flowing water. The equipment applied an e-beam to a 48" wide falling film of water or sludge at a flow rate of 120 gallons per minute (0.2 MGD); excess radiation was used to ensure that all of the water was irradiated. As a result, some electrons passed through the water without being absorbed, and some of the energy was lost from the water. The efficiency of energy transfer was approximately 60 to 85%. The doses at 50 mA/75 kW were 6.5 to 8 kGy, and total power consumption including pumps, chillers, and other equipment was approximately 120 kW. Shielding (e.g., housing the equipment in a concrete structure) is an important safety requirement with e-beam and gamma technologies, to protect operators from x-rays that are produced incidentally (Safe Drinking Water Committee, 1980; Swinwood et al., 1994; Electric Power Research Institute, 1999)

6.4.3.4 Impacts to the Environment

Because e-beam and gamma irradiation technologies are relatively undeveloped, only a few studies have investigated reaction pathways, kinetics, and potential byproducts. Byproducts are certainly possible, as the radical species generated by irradiation are known to react with organics in the water. For example, one study suggested that halonitromethanes may form phosgene and peroxynitrite (Cooper et al., 2006); however, more research is needed on this.

One of the key benefits of radiation technologies is the ability to remove organics from water. A gamma irradiation study of effluents from the Riyadh Wastewater Treatment Plant showed reductions in the levels of BOD, COD, and TOC (Basfar and Abdel Rehim, 2002). Other susceptible compounds include aromatic hydrocarbons and halogenated methanes, ethanes, and ethenes (U.S. EPA, 1993). The Orange County Water District found that e-beam met or approached the maximum contaminant limits during simultaneous treatment of a variety of priority contaminants, successfully removed NDMA at low cost, and eliminated methyl tert-butyl ether with lower energy requirements than other oxidation technologies (Larson, 1999).

6.4.3.5 Costs

Because e-beam and gamma irradiation are experimental technologies, there is very limited information on costs. The pilot-scale Electron Beam Research Facility in Miami, Florida treated 120 gallons per minute (0.2 MGD) at doses of 5 to 9 kGy; the cost of treatment in 1999 dollars was approximately $1.00 per thousand gallons (Electric Power Research Institute, 1999). Table 6-16 shows an analysis done by Raychem Corporation (now part of Tyco Electronics) in 1995 that indicated that e-beam was cost-competitive with other oxidation technologies for organics removal with a flow of 600 gallon per minute (0.9 MGD) (de St. Remy, 1995). The Orange County Water District similarly found that e-beam installation and operation costs were cost-competitive with other technologies for the removal of organics (Larson, 1999). However, it should be noted that some of the competing technologies (e.g., UV treatment) have undergone significant development since these cost analyses were done, and that the cost for organics removal may differ from disinfection costs.

Table 6-16. Comparison of Costs (in 1995 Dollars) for Advanced Oxidation Technologies. (de St. Remy, 1995)

Cost	E-Beam	UV/H_2O_2	UV/Ozone/H_2O_2	Peroxidation
Capital	$2,550,000	$1,575,000	$1,510,000 - 1,695,000	$1,325,000
Annual O&M	$139,000	$1,571,000 - $1,915,000	$354,000 - $690,000	$1,146,000
Energy	$79,000	$572,000 – $735,000	$158,000 - $263,000	$773,000
Replacement Parts	$60,000	$147,000 - $189,000	$23,000	$90,000
Chemicals	-	$647,000 - $1,196,000	$68,000 - $510,000	$183,000
Life-Cycle Cost	$4,154,000	$19,732,000 - $23,704,000	$5,782,000 - $9,487,000	$14,569,000

6.4.3.6 Summary

Electron beam and gamma irradiation show some promise for wastewater treatment. They are able to treat waters of varying quality and even sludges, though the required doses may vary with water quality. Limited data indicate that they disinfect reasonably well, and can also degrade organic contaminants in water. However, much more information is needed on disinfection efficacy against a wider range of organisms, identification and quantification of the effects of water quality, investigation of the possibility of regrowth and disinfection byproducts, practical applicability (e.g., the restriction to a thin film of water, < 7 mm thick), and the economic viability of the process.

6.4.4 Photocatalysis/TiO_2

6.4.4.1 Background

Catalysts are materials that facilitate, but are not consumed by, a chemical reaction; photocatalysts are catalysts whose activity is promoted by light. Photocatalyzed reactions can occur either when a reactant sorbed to the catalyst absorbs radiation directly, or when the photocatalyst absorbs radiation and excites an electron to the conduction band. This creates an electron hole in the valence band of the photocatalyst, which can react with sorbed compounds to form radical species that can undergo further reactions.

For environmental applications, photocatalysts were first investigated for the treatment of refractory organics, and only more recently for disinfection (Watts et al., 1995; Rincon and Pulgarin, 2005). The catalyst most commonly used for disinfection is titanium dioxide (TiO_2), which is activated by radiation in the UV light range. Researchers have also developed TiO_2 catalysts that are activated by visible light, by doping the catalyst with small amounts of impurities (such as nitrogen, sulfur, etc.). Disinfection with TiO_2 was first demonstrated in 1985, and has since been shown to inactivate a variety of pathogens. The disinfection mechanism has been attributed to hydroxyl radicals, which damage cell membranes, cell walls, spore coats, and internal structures (Gogniat et al., 2006; Wu et al., 2006a; Wu et al., 2006b; Yasui and Kamiko, 2005); these radicals form from the reactions of hydroxide and water with the electron holes created when the photocatalyst absorbs radiation (Rincón et al., 2001).

In addition to the above disinfection mechanisms, the radiation applied to photocatalysts often has disinfectant properties, so the observed disinfection results from a combination of photocatalysis and radiation. Disinfection by the catalyst itself under dark conditions varies, from no significant disinfection over periods up to 120 minutes (Lee et al., 1997; Dillert et al., 1998; Rincón et al., 2001), to <1 log removal after 30 minutes (Yasui and Kamiko, 2005) and up to 2.5

log removal over 120 minutes in the presence of organics (Rincon and Pulgarin, 2004). In the presence of light, photocatalysts generally enhance disinfection (Lee et al., 1997; Rincon and Pulgarin, 2005; Yasui and Kamiko, 2005; Wu et al., 2006c), although no effect (Herrera Melian et al., 2000) or decreases in disinfection efficacy (Dillert et al., 1998) have been observed. The latter were attributed to catalyst particles shielding pathogens from the radiation, and competitive use of the radiation for reactions other than disinfection (Dillert et al., 1998).

The TiO_2 technology is currently being tested at the laboratory scale, with research aimed at increasing the rate of disinfection. As such, no process information (design, implementation, operations, maintenance, cost) is available. However, TiO_2 is known to improve water quality by removing refractory organics, total organic carbon, and chemical oxygen demand (Dillert et al., 1998; Herrera Melian et al., 2000). The applied radiation may also cause temperature increases in the effluent; for example, irradiation with an 800W UV lamp for three hours caused a 5-9°C increase in temperature (Herrera Melian et al., 2000).

6.4.4.2 Disinfection Efficacy

Table 6-17 summarizes the disinfection efficacy of photocatalysts against various microbes. Bacterial inactivation with TiO_2 is relatively independent of pH between 5 and 8 (Watts et al., 1995; Herrera Melian et al., 2000). Organic matter does affect disinfection; this can be attributed to competition between the organics and pathogens for the photocatalytically formed radicals (Herrera Melian et al., 2000; Rincon and Pulgarin, 2005). Data on the effect of dissolved oxygen conflict, with some reporting no effect (Herrera Melian et al., 2000), and others reporting that increased oxygen levels increased disinfection (Rincon and Pulgarin, 2005).

Photocatalytic disinfection has been tested in the laboratory with a number of organisms: bacteria, bacterial and fungal spores, algae, and coliphage (Lee et al., 1997; Rincón et al., 2001; Yasui and Kamiko, 2005; Wu et al., 2006c). Table 6-17 shows the disinfection efficacy of photocatalysts with some of these organisms. *E. coli* have been found to be more susceptible to disinfection with TiO_2 than *Streptococcus*, *Bacillus* species, *Enterococcus* species, or other coliform and gram-negative bacteria (Herrera Melian et al., 2000; Rincon and Pulgarin, 2005). *Bacillus* species were observed to form protective spores in response to the disinfecting conditions (Rincon and Pulgarin, 2005).

Investigations on regrowth of organisms after disinfection showed very little regrowth in the first 2-3 hours after treatment, but significant changes after 24 and 48 hours. For gram negative bacteria and the gram positive *Enterococcus*, photocatalysis slowed or eliminated regrowth, or promoted continued inactivation in the dark. Under the same conditions for a mixed microbial population in wastewater, bacterial concentrations rose under dark conditions after photocatalysis. Growth increased as disinfection times increased; this may indicate increased nutrients from photocatalyzed reactions of organics in wastewater, increased culturability of organisms after photocatalysis, or that treatment inactivated select bacteria, thereby eliminating competitors for bacteria that were not susceptible to photocatalysis. After 24 hours in the dark, bacterial concentrations increased above the pre-disinfection levels, implying true growth rather than simple recovery of damaged cells. The difference in regrowth behavior between the mixed population and the gram-negative bacteria/Enterococci indicates that photocatalysis affects various organisms quite differently (Herrera Melian et al., 2000; Rincon and Pulgarin, 2005).

Table 6-17. Disinfection Efficacy of Photocatalysts Against Various Pathogens.

Organism	Catalyst	Light	Water	Catalyst Conc (mg/L)	Time (min)	Average Log_{10} Reduction	Source
Unknown	TiO_2	UV	Filtered secondary effluent	5	360	1.5-2.5	(Dillert et al., 1998)
Total coliforms	TiO_2	UV	Primary effluent	0.25	150	3.5-5	(Herrera Melian et al., 2000)
	TiO_2	Sunlight	Primary effluent	0.25	150	2-3	(Herrera Melian et al., 2000)
Coliforms	TiO_2	Fluorescent	Secondary effluent	0.25	150	2	(Watts et al., 1995)
E. coli	TiO_2	Artificial solar	Milli-Q water	1	20	5	(Rincón et al., 2001)
	PdO/TiON[a]	Visible	Buffered water	0.6-1	30	>5	(Wu et al., 2006b)
Pseudomonas aeruginosa	TiON[a]	Visible	Buffered water	1	30-60	0.2-0.5	(Wu et al., 2006a)
	PdO/TiON[a]	Visible	Buffered water	0.6-1	30	>5	(Wu et al., 2006b)
Streptococcus faecallis	TiO_2	UV	Primary effluent	0.25	150	2	(Herrera Melian et al., 2000)
Staphylococcus aureus	PdO/ TiON[a]	Visible	Buffered water	0.6-1	30	>5	(Wu et al., 2006b)
Bacillus subtilis	TiO_2	UV	Demineralized water	1	60	5	(Rincon and Pulgarin, 2005)
	PdO/ TiON[a]	Visible	Buffered water	0.6-1	180	2	(Wu et al., 2006c)
Qβ coliphage	TiO_2	UV black	Milli-Q water or nutrient broth	0.1	20-60	2-4	(Lee et al., 1997)
	TiO_2	Suntan lamp	Milli-Q water or nutrient broth	Unknown	10-30	~6	(Yasui and Kamiko, 2005)
Poliovirus 1	TiO_2	Fluorescent	Secondary effluent	0.25	30	2	(Watts et al., 1995)

[a]PdO is palladium oxide. TiON refers to nitrogen-doped TiO_2.

6.4.4.3 Summary

Photocatalysis offers several benefits: elimination of chemical disinfectants, and the removal of recalcitrant compounds and other organics. It is not sensitive to pH changes between 5 and 8. However, like UV, photocatalysts are unlikely to be effective in low transmittance, turbid waters. In addition, most catalysts that are currently available are too slow to be of practical use for wastewater treatment, although improvements are rapidly being made. As with most emerging technologies, the biggest issue is likely to be the uncertainty over performance: it has been quite effective against *Enterococcus* and gram negative bacteria, including coliform and *E. coli*; however, it performed poorly against a mixed culture in wastewater. Similar discrepancies have been observed with regard to regrowth. Finally, deactivation of the catalyst (loss of activity) is possible, particularly in a complex matrix such as wastewater; this issue has not yet been investigated. More work is needed to clearly establish the applicability of this technology to wastewater treatment.

6.5 Multiple-Disinfectants Disinfection

6.5.1 Overview

In recent years, the sequential or simultaneous use of two or more disinfectants in wastewater treatment has been gaining attention. Using multiple disinfectants has the following potential benefits:

- Increased effectiveness: Disinfection technologies use different disinfection mechanisms and their germicidal efficacies for different microbial species vary significantly. As a result, a multiple-disinfectant system may inactivate a larger spectrum of pathogens than a single disinfectant. For example, a combined UV/chlorine system could inactivate *Cryptosporidium*, which is resistant to chlorine and chloramines but is susceptible to UV irradiation, as well as adenovirus, which is resistant to UV but is susceptible to low doses of free chlorine.
- Synergistic effects: The use of multiple disinfectants may have synergistic effects on pathogen inactivation and contaminant oxidation. For example, UV irradiation can enhance the hydroxyl radical production from ozone, and ozonation can increase the UVT of water to be disinfected. PAA or ultrasound can damage cells, thereby weakening them and decreasing the UV doses needed for disinfection.
- DBPs: Combining technologies can reduce the generation of specific DBPs. For example, a combined UV and chlorination system lowers chemical doses and shortens contact times; this reduces THMs and HAAs, relative to chlorination alone. The combined use of free and combined chlorine reduces the formation of THMs, HAAs, and NDMA, relative to using break-point chlorination or chloramination alone.
- Residual: For reuse applications, disinfectant residual is important to minimize biofouling in the distribution pipelines, to reduce slime in irrigation sprinklers and drip emitters, and to minimize algae and odors in reservoirs. Combining alternatives such as UV or ozone (which are strong disinfectants but provide little or no disinfectant residuals) with longer-lasting chemicals such as chloramines supplies the needed residual.
- Oxidation: In addition to pathogen inactivation, some disinfectants have the ability to oxidize emerging contaminants such as NDMA, pharmaceuticals, and personal care products. Adding these technologies into a multiple-disinfectant system would reduce the levels of emerging contaminants in the effluent.
- Operation: A multiple-disinfectant system offers operational flexibility, with the potential to use each disinfectant individually when appropriate. For example, the existing infrastructure for chlorination could be used to control biofouling of filter media and UV equipment, to disinfect wet-weather flow, or as a backup for the UV system.
- Cost: Multiple-disinfectant systems can be more cost effective. For example, when chlorination is also used, a UV system need not be over-designed for UV-resistant organisms or for wet-weather flows; this can be especially beneficial in cases where a chlorination system is already in place. However, one must ensure that the

materials in place such as plastics, elastomers, and gasket materials can with stand the effects of the additional disinfectant.

The potential disadvantages of a multiple-disinfectant include more complicated process control and operations as well as additional capital investment. However, a multiple-disinfectant system should be considered if the existing system is insufficient to meet the effluent microbial limit, specific pathogens that cannot be removed by a single technology become the target of regulatory limits, removal of emerging contaminants is required, or expansion of treatment capacity is needed.

The following sections discuss several combinations of multiple disinfectants, such as advanced oxidation processes (AOPs), which generate highly reactive hydroxyl free radicals that can oxidize various compounds in the water. The AOPs discussed in this section include ozone plus hydrogen peroxide, UV plus ozone, UV plus hydrogen peroxide, and UV plus peracetic acid. Subsequent sections discuss UV plus chlorine, the sequential use of free chlorine and chloramines, and the use of ultrasonic cavitation with other disinfectants. It should be noted that although these combinations have the potential benefits discussed above, few have been operated on a full-scale.

6.5.2 Ozone and Hydrogen Peroxide

Ozone is a disinfectant in and of itself, but also spontaneously decomposes to produce hydroxyl radicals, which can oxidize various compounds. Hydrogen peroxide can accelerate the rate of ozone decomposition, thereby elevating hydroxyl radical concentrations and increasing oxidation rates. The Metropolitan Water District of Southern California has conducted extensive research on this process, commonly called peroxone, to control organics and odor compounds in drinking water while providing sufficient levels of molecular ozone to guarantee CT values and primary disinfection. A report by the U.S. EPA (1999) states that peroxone is equally or more effective than ozone for pathogen inactivation, but a more recent study found that hydrogen peroxide decreased the disinfection efficacy of ozone in secondary wastewater effluent, i.e., ozone alone was a more effective disinfectant (Ferron et al., 2005). As with any applications of ozone, the bromate formation potential relative to the 10 μg/L MCL may limit its use at a particular site.

6.5.3 UV and Ozone

Landis et al. (2007) indicated that combining the strengths of ozone (oxidation, taste and odor reduction, virus inactivation, organics removal) and UV (inactivation of *Cryptosporidium*, *Giardia* and other pathogens) could result in a cost-effective strategy for water treatment. The combined use could also reduce the formation of halogenated DBPs, but may generate bromate, which has an MCL of 10 μg/L. This combination may also create synergies for disinfection, because ozonation can increase the UVT of wastewater, which would in turn reduce the required UV energy input. In one study, ozone doses of 1 to 5 mg/L followed by a UV dose of 40 mJ/cm^2 reduced fecal coliform to below the detection limit, while UV alone required a dose of at least 100 mJ/cm^2 for the same level of disinfection (Kinshella et al., 2007). Ried et al. (2004) demonstrated that UV/ozone can effectively disinfect wastewater and also reduce endocrine disruptors, pharmaceuticals, and personal care products. Benjamin et al. (2005) calculated that combined UV/ozone (at a dose of 1.5 mg/L) was more economical than UV alone based on a 20-year life cycle cost analysis.

6.5.4 UV and Hydrogen Peroxide

The combination of hydrogen peroxide and UV (UV/H_2O_2) is a commonly used advanced oxidation process (AOP) that has been proven effective for the destruction of emerging contaminants such as NDMA and several endocrine disrupting chemicals (Rosenfeldt and Linden, 2003). Another benefit of this combination is that peroxide in the effluent decomposes to oxygen on a 2:1 basis, thereby increasing the DO in the effluent. However, the required UV dose for these applications is approximately 10 times higher than for disinfection, and the target compounds may form other byproducts.

The order of the disinfectants may affect disinfection and oxidation efficacies. Koivunen and Heinonen-Tanski (2005) found that H_2O_2 addition followed by UV showed some antagonism for disinfection; in other words, the combined technologies were less effective than the sum of the individual components. When peroxide is added to a UV system, it forms hydroxyl free radicals almost instantaneously (Baxendale and Wilson, 1957). This formation reaction competes with disinfection and reduces both the effective UV dose and the amount of peroxide available for disinfection; however, it also increases the level of hydroxyl radicals available to oxidize organics in the effluent. As a result, peroxide addition between UV units may improve organics removal. For example, UV treatment of NDMA forms DMA (a precursor to NDMA); DMA is recalcitrant to UV irradiation but is susceptible to H_2O_2/UV. Thus, the combination of technologies can first transform NDMA to DMA via UV, and then remove DMA, thereby removing the potential for NDMA to reform. The Orange County Water District in California uses UV/H_2O_2 to treat reclaimed water for groundwater replenishment; the system is considered an appropriate technology for disinfection and provides an additional barrier to contaminants of concern (Royce et al., 2005).

6.5.5 UV and Peracetic Acid (PAA)

In a study evaluating reclaimed municipal wastewater as source of water and nutrients for plant nurseries, filtration and disinfection with PAA/UV effectively removed *Escherichia coli* and total coliform to below detection limits in all effluent samples from a pilot plant (Gori et al., 2004; Lubello, 2004). Another study showed that fecal coliforms were reduced to below detection limits in wastewater with 2 to 10 mg/L PAA followed by a UV dose of 40 mJ/cm^2; UV alone required a dose of at least 100 mJ/cm^2 to achieve the same level of disinfection. Following the highest PAA dose of 10 mg/L with a five-minute contact time, non-detect fecal coliform levels were achieved after a UV dose of only 10 mJ/cm^2 (Kinshella et al., 2007).

Koivunen and Heinonen-Tanski (2005) evaluated the synergistic effects of PAA/UV treatment, to determine whether the combined treatment was more effective than the sum of each technology alone. Combining PAA with UV increased disinfection efficacy with all tested enteric bacteria, but synergies with the coliphage MS2 were insignificant.

Caretti and Lubello (2003) conducted a five-month pilot plant study on four alternatives: PAA alone, UV alone, addition of PAA upstream of the UV device (PAA+ UV) and addition of PAA downstream of the UV device (UV+PAA). Under the investigated experimental conditions (2 to 8 mg/L PAA with 10 to 30 minutes of contact time and/or 100 to 300 mJ/cm^2 UV), the use of UV irradiation or PAA alone could not meet the microbiological limits. The UV + PAA treatment showed enhanced disinfection efficacy, and the PAA + UV treatment showed even greater improvement. In the latter case, the higher efficiency was attributed to the formation of free radicals due to the photolysis of PAA in presence of the UV radiation.

6.5.6 UV and Chlorine

Section 6.5.1 presents some of the potential advantages of combining UV and chlorine (UV/chlorine); this section provides more detailed information on the benefits. For example, the Vallejo Sanitation and Flood Control District in California uses low doses of hypochlorite to supplement UV disinfection and to prevent biological growth on the UV equipment. This practice was found to be more cost effective and reliable than UV alone (Tekippe, 1999).

Several studies have shown that UV/chlorine is much more effective than UV alone for the inactivation of adenovirus. In a laboratory study on drinking water disinfection, UV alone at a dose of 40 mJ/cm^2 caused only 1-log inactivation of adenovirus, but achieved 4-log adenovirus inactivation when coupled with free chlorine at a low CT value of ~0.15 mg-min/L (Durance et al., 2005). Monochloramine was less effective than free chlorine, with only 2.5-log adenovirus inactivation after a UV dose of 40 mJ/cm^2 followed by a chloramine CT of 350 mg-min/L. However, Ballester and Malley (2004) reported much better results with chloramines in laboratory experiments: 4-log adenovirus reduction with a UV dose of 40 mJ/cm^2 followed by a chloramine CT of 27.2 mg-min/L.

UV/chlorine can also reduce the disinfectant doses required for adequate treatment and provide synergistic effects. Kinshella et al. (2007) found that UV treatment alone required a dose of at least 100 mJ/cm^2 to achieve non-detect fecal coliform in a tertiary treated effluent. Pre-oxidation with 2 to 10 mg/L of chlorine reduced the required UV dose to 10 to 40 mJ/cm^2 to achieve the same effluent quality. Murphy et al. (2007) found synergistic effects: UV alone yielded 0.3-log reduction of suspended heterotrophic bacteria in drinking water, and chlorine alone yielded 0.4 log reduction, but the combination provided 1.2-log reduction. The results were even more striking for heterotrophic bacteria in biofilms, where bacterial counts increased 0.3-log units after UV treatment, decreased 0.7-log units after chlorine, and decreased by 1.7-log units after UV/chlorine.

The order of the disinfection technologies also impacts disinfection efficacy. Using secondary effluent samples from two wastewater treatment plants, Linden et al. (2004) found that UV doses of <40 mJ/cm^2 followed by chloramination achieved >4-log inactivation of *Clostridium perfringens* spores, somatic coliphages, and *Cryptosporidium parvum* oocysts. The reverse order (chloramination followed by UV doses of <40 mJ/cm^2) was less effective: it did not inactivate *Clostridium perfringens* spores, and achieved >3-log inactivation of somatic coliphage and *Cryptosporidium parvum* oocysts.

6.5.7 Free Chlorine and Chloramines

Although free chlorine is a stronger disinfectant than chloramines, many facilities choose to use chloramination for disinfection, in part due to concerns over byproducts such as THMs and HAAs. In addition, wastewater treatment plants that do not remove nitrogen have high ammonia concentrations, which make breakpoint chlorination impractical. In recent years, an increasing number of treatment facilities have implemented biological nitrogen removal (BNR). The BNR activated sludge process is usually designed and operated to produce an ammonia-free effluent, which allows the use of free chlorine for disinfection. However, many of these facilities continue to use chloramines for disinfection by adding a low dose of ammonia nitrogen before chlorination of the BNR effluent. While this reduces the formation of THMs and HAAs, the added ammonia may present a problem for plants that must meet stringent nitrogen discharge limits. In addition, research has shown that chloramines may react with compounds such as dimethylamine to form carcinogenic nitrosamines such as NDMA (Mitch and Sedlak, 2002).

The sequential addition of chlorine and chloramines may mitigate the disadvantages of each of these technologies. Laboratory experiments with secondary effluent showed that the sequential use of chlorine and chloramines produced less NDMA than chloramination alone; this was attributed to the reaction between free chlorine and NDMA precursors, which became unavailable for subsequent reactions with chloramines (Schreiber and Mitch, 2005).

The Sanitation Districts of Los Angeles County conducted laboratory and full plant-scale tests on the sequential application of free chlorine and chloramines at four water reclamation plants that implement BNR. They applied free chlorine upstream of media filters and chloramines (ammonia addition followed by chlorine) downstream of the filters. The pre-filter chlorine residuals provided initial disinfection and helped to control biofouling on the filter media. The subsequent addition of ammonia and additional chlorine stopped the reaction between free chlorine and THM precursors, and provided the CT necessary to meet California Title 22 disinfection CT requirements. Sequential chlorination was very effective in meeting the total coliform requirement of <2.2 MPN/100 mL for unrestricted reuse. Total THM concentrations in the final effluent were well within the drinking water standard of 80 μg/L, cyanide and cyanogen chloride concentrations were usually below the reporting limits or levels of concern, and no toxicity was detected in the disinfected effluent. NDMA concentrations were significantly lower than the levels observed when chloramination alone was used for disinfection (Huitric and Tang, 2007).

6.5.8 Ultrasonic Cavitation with Other Disinfectants

Many researchers indicate that the energy requirements for ultrasound as a stand-alone disinfectant are economically prohibitive for pathogen removals typically needed for wastewater treatment (Mason et al., 2003a; Blume and Neis, 2004; Blume and Neis, 2005). However, cavitation may synergistically improve disinfection efficacy when combined with other disinfectants because it 1) is very effective at breaking up bacterial clumps and particles that may shield pathogens from the disinfectants, 2) weakens cell membranes, making them more susceptible to chemical attack, and 3) may improve dispersion of chemical disinfectants in solution (Blume and Neis, 2005). For example, cavitation is reported to decrease the heat resistance of bacterial spores and enhances the effectiveness of ozone against viruses and bacteria (Hua and Thompson, 2000). Ultrasound also promotes chlorine degradation, thus allowing higher doses of chlorine to be used prior to sonication (Mahvi et al., 2005).

Several researchers have investigated the use of ultrasound with chlorine. In general, the combination of disinfectants was synergistic, i.e., the two treatments together removed more pathogens than the sum of each individually (Mason et al., 2003b). The timing of sonication and chlorination may affect disinfection. At 20 and 35 kHz, simultaneous disinfection removed more *E. coli* and total coliforms than sonication followed by chlorination (Mason et al., 2003a; Blume and Neis, 2005). However, pretreatment by sonication improved disinfection relative to chlorination alone in one case (Blume and Neis, 2005), while in another case, performance was worse with sonication pretreatment (Mason et al., 2003a); the cause of this discrepancy is unclear. At 850 kHz, sonication pretreatment improved disinfection by >1.5 log removal of *E. coli*, relative to chlorination alone; however, simultaneous sonication and chlorination actually decreased disinfection efficacy (Mason et al., 2003a). As a final note, simultaneous and sequential use of sonication were not found to have a significant impact on the concentrations of adsorbable organohalogens (AOX), relative to chlorine alone (Blume and Neis, 2005).

Research has also been conducted on the use of UV with ultrasound. UV radiation does not penetrate particles well, so the use of ultrasound to break up the particles can significantly improve disinfection. Treatment of water with five to ten seconds of 20 kHz ultrasound prior to UV exposure resulted in an additional 1-2 log removal of *E. coli* and an additional 1-1.5 log removal of fecal streptococci over UV exposure alone (Blume and Neis, 2004). Increasing the power input from 170 W/L to 310 W/L increased the removal by <0.5 log (Blume and Neis, 2004). An analysis of the energy requirements showed that the combined disinfection was able to achieve the same level of disinfection as UV alone using only 43% of the energy, and provided significantly better disinfection using only 60% of the energy of UV alone (Blume and Neis, 2004). These energy savings, combined with the relatively small reactor size (low capital cost) for ultrasound (Blume et al., 2002), could provide a method for cheaper wastewater disinfection.

6.6 Summary

A number of technologies have been developed in response to difficulties encountered with the use of chlorine for wastewater disinfection. Although chlorine dioxide is established for drinking water treatment, it is not used for wastewater, largely because the few benefits that it provides relative to chlorine disinfection of wastewater do not justify the additional system complexity and expense. The other six individual technologies discussed in this chapter (PAA, ferrate, pasteurization, ultrasonic cavitation, e-beam, and photocatalysis) are less established, but share the benefit of producing fewer known DBPs than chlorination. In addition, ultrasound, e-beam, photocatalysis, and ferrate have been shown to remove organics and other contaminants from water. Only PAA provides a disinfectant residual; this is often required to reduce microbial growth in distribution systems during water reuse but may be problematic if the effluent is discharged (though PAA has shown no toxicity in tests).

Data on disinfection efficacy with these technologies are generally limited; PAA and pasteurization appear to be similar to chlorine with good inactivation of bacteria, weaker inactivation or variable performance against viruses, and poor performance against cysts. The chemical disinfectants (PAA and ferrate) are probably better suited to wide variations in flow rate than the other alternatives, but also share some of the safety concerns of chlorine. PAA and sonication have been estimated to cost more than chlorine, and there is little published information on the costs of disinfection using pasteurization, e-beam, photocatalysis and ferrate. However, these cost estimates are likely to change as the technologies develop further, so recent literature should be consulted for current information. In addition, many factors beyond cost influence the ultimate choice of disinfection technology (see Section 8.3 for details), so the benefits and disadvantages of each technology must be weighed.

Because each technology has strengths and weaknesses, a recent trend has been to combine technologies to realize the benefits of each while minimizing the disadvantages. For example, ultrasonic cavitation may improve disinfection with each of the other technologies by breaking up particles and reducing their negative effects on inactivation of pathogens. The use of free chlorine and chloramines may reduce the overall concentrations of disinfection byproducts, and implementation would be relatively low-cost for facilities that currently use chlorine. The increasing use of UV disinfection has prompted research on the combined use of UV with chlorine, ozone, peracetic acid, and ultrasonic cavitation; each of these alternatives shows promise for improving the disinfection of wastewater, reducing disinfection byproducts, and/or decreasing the cost of disinfection.

CHAPTER 7.0

EXISTING DISINFECTION PRACTICE

The existing disinfection practice for major POTWs utilizes chlorine, UV, and ozone. This chapter discusses the use of each of these technologies as determined from this project's survey of disinfection; the methods used to gather the data are described in Chapter 2.0. The chapter focuses on UV, because there is already a wealth of information regarding chlorine disinfection, as documented in several outstanding references (e.g., Water Environment Federation (1996), Connell (2002), and Water Environment Federation (2006)) and in past surveys conducted by the Water Environment Federation (WEF) and WEF Disinfection Committee. The few POTWs using ozone for disinfection were also surveyed; however, due to the small number of facilities, information is limited.

7.1 Chlorine

The majority of major POTWs use chlorine as the primary means of disinfection, but the form of chlorine being used by each facility cannot be determined from existing databases. As determined during the UV survey (Section 7.3.2), safety was a major driving force away from chlorine. Therefore, the following subsections address safety by examining the reported releases of gas chlorine between January 1990 and June 2006, as described in Chapter 2.0.

Because the water and wastewater industries share similarities in owners, staff, and regulatory oversight, the chlorine gas safety records of both industries are included. This information serves as a general indication of the hazard level when working with gaseous chlorine, and provides information for utilities that own and operate both water and wastewater facilities.

7.1.1 Chlorine Gas Releases from Water and Wastewater Facilities

Table 7-1 summarizes the number of chlorine gas releases >10 pounds at all water and wastewater facilities (not only the major ones), by the type of facility and storage. The total number of POTWs in this country is estimated at more than 15,000, and most of them use some form of chlorine. Thus, releases of chlorine gas over this 16.5-year period occurred at only a very small percentage of the POTWs using gas chlorine. However, the risks should be taken seriously, because these releases can be a potent threat and have the potential to cause significant damage and loss of life.

Although it is beyond the scope of this project to estimate the number of water treatment plants, professional experience of the authors suggests that the U.S. has many more water treatment facilities than wastewater treatment plants, so one might expect more releases from the water industry. Water industry releases have been dominated by gas cylinders, whereas wastewater treatment industry releases have occurred roughly equally often from gas cylinders and 1-ton vessels.

Table 7-1. Chlorine Gas Releases by Facility and Storage Type.

Chlorine Gas Storage	Facility Type		
	Wastewater	Water	No ID
150 pound gas cylinder	14	44	3
1 ton vessels	13	11	0
Railcars	1	1	0
Not identified	3	18	0
Total number by facility type	31	74	3
Percent by facility type	28.7%	68.5%	2.8%

Table 7-2 summarizes the reported causes of chlorine gas leaks. The data fall into three general categories. The most common cause of releases by far (~50%) involves the chlorine gas storage vessel (cylinder, fusible link or plug, regulator or valve, and evaporator). The next most common causes are the feed line and associated piping for the gas (~16%), and operator error (~15%). These causes were similarly distributed between 150-pound cylinders (generally used in smaller facilities) and one-ton cylinders (generally used by larger utilities).

Table 7-2. Chlorine Gas Releases by Identified Cause and Facility Type.

Identified Causes	Type of Facility			Total	% of All Releases	Percent by Type of Storage			
	Water	POTW	No ID			150-lb	One-ton	Rail Car	No ID
Cylinder leak	6	3	1	10	9.3%	100.0%			
Fusible link or plug failure	2	3	0	5	4.6%	60.0%	20.0%		20.0%
Regulator or Valve	30	7	1	38	35.2%	65.8%	21.1%		13.2%
Evaporator	1	1		2	1.9%		100.0%		
Feed/piping failure	14	3	0	17	15.7%	29.4%	23.5%	11.8%	35.3%
Operator error	8	8	0	16	14.8%	50.0%	37.5%		12.5%
Fire	2		1	3	2.8%	33.3%			66.7%
Equipment failure		2		2	1.9%	100.0%			
Not identified	11	4		15	13.9%	46.7%	20.0%		33.3%

Table 7-3 summarizes the distribution of the states that have had releases. Between January 1990 and June 2006, 38% of states have experienced a chlorine release at a wastewater treatment facility, and 52% of states have experienced a release from the drinking water industry.

Table 7-3. Chlorine Gas Releases by State and Facility Type between January 1990 and June 2006.

State	POTW	Water	No ID	Total	State	POTW	Water	No ID	Total
Alabama	3	4		7	Minnesota			1	1
Arizona		4		4	Missouri		2		2
Arkansas	1	4		5	New Jersey		2		2
California		5	1	6	New Mexico	1			1
Connecticut	1	1		2	New York		4		4
Florida	5	5		10	North Carolina	1	1		2
Georgia	1			1	Ohio		6		6
Illinois		2		2	Oklahoma	1			1
Indiana	1	3		4	Pennsylvania	2	4	1	7
Iowa		2		2	South Carolina	3	1		4
Kansas	1	3		4	Tennessee		2		2
Kentucky	2	2		4	Texas	2	3		5
Louisiana	1			1	Utah		2		2
Maine	1			1	Virginia	1	5		6
Maryland		4		4	Vermont	1	1		2
Massachusetts	2	1		3	Wisconsin		1		1

Figure 7-1 summarizes the number of releases per year between January 1990 and June 2006 by the type of facility. The water industry usually had at least one release per year, and the numbers of releases has been trending downward since 1997. A similar trend was expected with POTWs, which have been shifting away from chlorine gas to hypochlorite and UV disinfection; however, the opposite was found. The wastewater treatment industry averaged 1.9 releases per year over the 16.5 years covered by these incident reports.

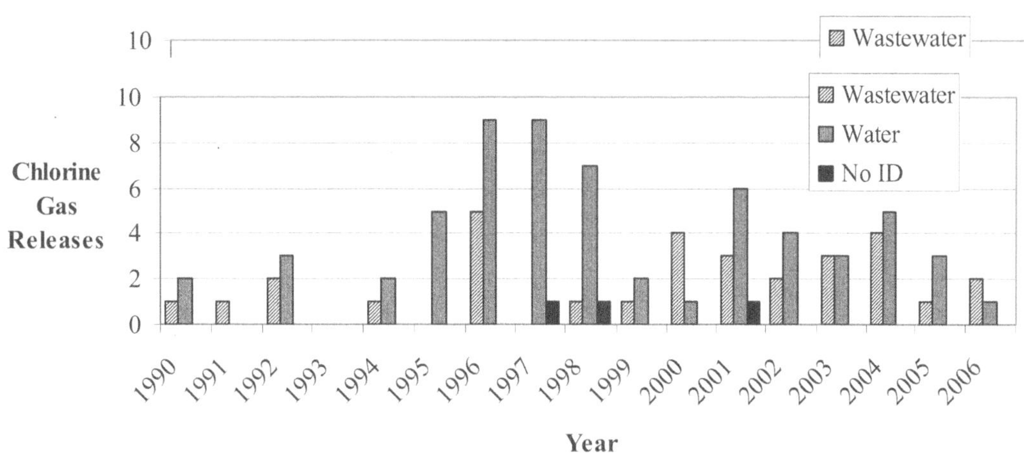

Figure 7-1. Gas Chlorine Releases at POTWs and Drinking Water Treatment Plants from January 1990 to June 2006.

7.1.2 Findings of Chlorine Release Data

The chlorine release information indicates the following:

- Reportable chlorine releases from sewage treatment plants are relatively rare, considering the large number of treatments plants and the change-out frequency of chlorine gas storage units. Eighteen states have had no reportable releases of chlorine gas from water or wastewater facilities in the last 17 years.

- Given the relatively low frequency of reportable chlorine releases, the change from gas chlorine at POTWs has been driven less by historical performance than by other factors, such as perceived risk; a broader view of the hazard of gas chlorine to the community such as its transport through the community to the point of use; regulatory compliance; the desire of the general public for fewer acutely toxic chemicals in their neighborhoods; and reducing the potential terrorist targets.

7.2 UV Database Results

This section presents information from the U.S. EPA and vendor databases that were used for the survey of practice. The methodology used to obtain the data is described in Section 2.3. Because this method likely underestimates the number of facilities using UV, the discussion in this section focuses on trends in data, rather than absolute numbers.

7.2.1 Historical Installation Rate of UV Systems

Although UV was first installed for the disinfection of drinking water in France in the early 1900s, the first UV system for municipal wastewater disinfection wasn't installed until the mid-1970s (Whitby, 2002). In 1986, UV was used by an estimated 50 POTWs in the U.S., mostly smaller plants with flows <1 MGD (WEF, 1996). Data from the current project show that the largest POTW in the U.S. using UV disinfection now has an average dry weather design flow rate of 250 MGD. Based on data from the 2000-2004 time frame, this project estimated that 930 major POTWs use UV; given the rate at which UV systems are being sold (see next paragraph), the error on this value is ~15%.

Figure 7-2 summarizes the number of systems sold to major POTWs per year from 1989 to 2005 (based on vendor databases). Data for 2005 were incomplete, and vendor records before 1989 were also incomplete due to mergers, loss of data, and lack of record keeping; these data are therefore not included. Figure 7-2 indicates that the number of systems sold each year increased from 1989 to 2004. For the last 10 years (1996 to 2005), an average of 68 POTWs per year converted to UV disinfection. This figure implies that 73% (680) of the 930 UV systems have been brought on-line since 1996, and that conversion by major POTWs to this technology is a relatively recent development.

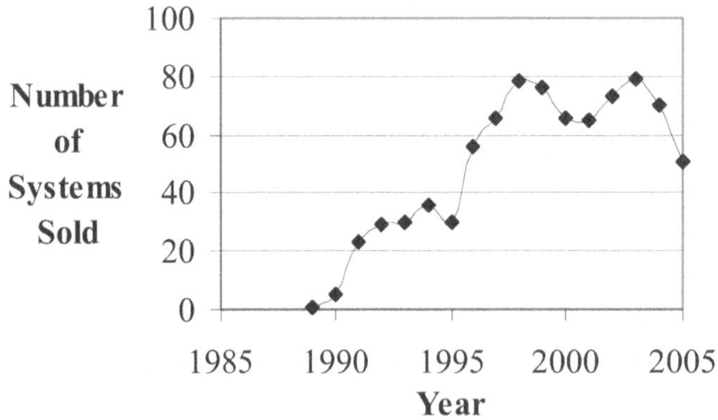

Figure 7-2. Number of UV Installation Sold by Year.

7.2.2 Geographical Distribution of UV Use

Table 7-4 summarizes the number of major POTWs and those using UV, categorized by state. This inventory is representative for the 2000 to 2004 time frame; however, as Figure 7-2 suggests, these numbers are rapidly increasing, with as many as 70 or more UV systems added each year. The error on the total number of UV plants (930) represented in Table 7-5 is estimated to be ~15%.

Table 7-4. Major POTWs Using UV Disinfection by State.

State	Number of POTWs	UV	Percent of Facilities with UV	State	Number of POTWs	UV	Percent of Facilities with UV
Alabama	121	31	26%	Montana	24	5	21%
Alaska	17	4	24%	Nebraska	27	10	37%
Arizona	52	20	39%	Nevada	18	2	11%
Arkansas	78	12	15%	New Hampshire	25	10	40%
California	268	30	11%	New Jersey	106	21	20%
Colorado	81	28	35%	New Mexico	27	9	33%
Connecticut	63	17	27%	New York	196	17	9%
Delaware	8	0	0%	North Carolina	123	29	24%
District of Columbia	1	0	0%	North Dakota	11	1	9%
Florida	201	19	10%	Ohio	212	70	33%
Georgia	144	31	22%	Oklahoma	61	6	10%
Hawaii	13	5	39%	Oregon	60	18	30%
Idaho	37	10	27%	Pennsylvania	258	42	16%
Illinois	187	29	16%	Rhode Island	17	3	18%
Indiana	112	43	38%	South Carolina	102	16	16%
Iowa	90	6	7%	South Dakota	17	5	29%
Kansas	58	47	81%	Tennessee	110	12	11%
Kentucky	76	23	30%	Texas	483	72	15%
Louisiana	99	20	20%	Utah	28	6	21%
Maine	36	1	3%	Vermont	18	2	11%
Maryland	51	18	35%	Virginia	95	22	23%
Massachusetts	75	8	11%	Washington	77	39	51%
Michigan	128	17	13%	West Virginia	48	18	38%
Minnesota	47	8	17%	Wisconsin	92	32	35%
Mississippi	61	13	21%	Wyoming	16	7	44%
Missouri	95	16	17%				

UV use is widespread; every state except Delaware (and Washington, D.C.) has at least one major POTW with UV disinfection. Kansas has the highest level of UV adoption, with more than 80% of major POTWs using UV technology. Based on UV adoption levels, the states were placed into one of three groups (Figure 7-3): below the median (<15%), near the median (15-30%), and above the median (>30%). There is some clustering of states with above median adoption of UV: the Pacific Northwest (Washington and Oregon), Colorado and the adjoining states (Nebraska, Kansas, New Mexico, Arizona, and Wyoming); and a cluster of states near the Ohio River Valley (Maryland, West Virginia, Ohio, Indiana, Kentucky). Hawaii, Wisconsin, and New Hampshire make up the other above the median adoption states.

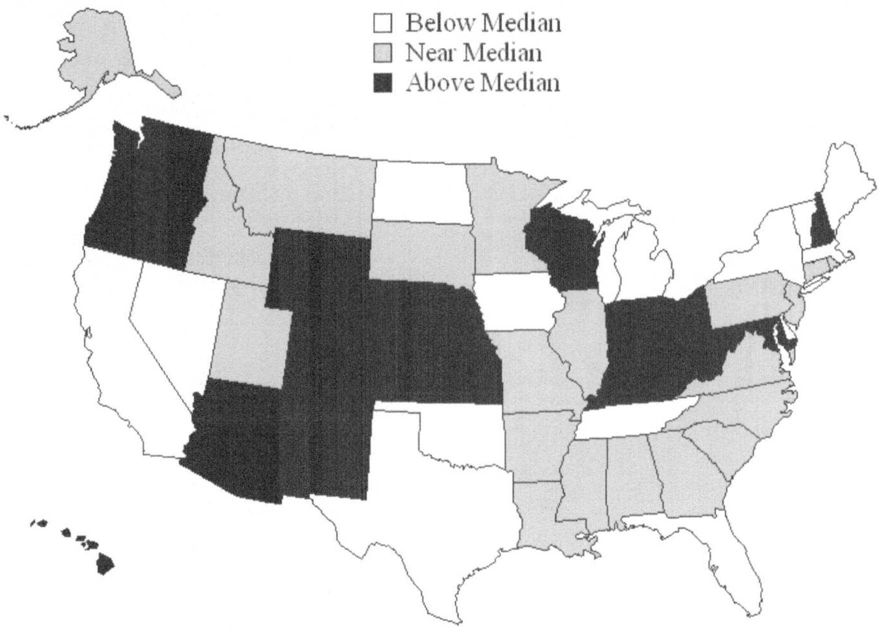

Figure 7-3. Adoption of UV by State.

7.2.3 Size of POTWs Using UV

Table 7-5 summarizes the number of major POTWs and those that use UV, categorized by flow rate. The majority of POTWs that use UV are relatively small, with ~75% of all UV facilities having an average dry weather design flow rate of between one and five MGD. Approximately 20% of all POTWs using UV are mid-size treatment plants with flow greater than five but less than 20 MGD. Only 4% of all major POTWs that use UV have flow >20 MGD. For the first three flow rate categories, the level of UV adoption is similar at ~20%; however, only ~13% of the largest plants use UV. This is consistent with the historical trend that the earliest adopters of UV were small plants, as mentioned in Section 7.2.1. One possible reason for the slow adoption rate of larger plants may be the considerable investment required to upgrade the disinfection process at these large facilities, which makes them slower to upgrade and more conservative in their decisions (i.e., less likely to choose new technologies). Large POTWs also tend to be located in areas with high populations, so another possible explanation is that many large plants may be located in urban centers on the coasts; these plants may discharge to the ocean, have less stringent discharge requirements, and therefore be less motivated to change to UV disinfection.

Table 7-5. Summary of POTWs using UV Disinfection by Flow Rate Category.

Flow Rate Category (MGD)	Number of Major POTWs	Percent of All POTWs	POTWs using UV	Percent of All UV Plants	UV Adoption Rate
1-2.5	2,168	48.7%	445	47.3%	20.5%
>2.5-5	1,094	24.6%	247	26.8%	22.6%
>5-20	884	19.9%	198	21.6%	22.4%
>20	304	6.8%	40	4.4%	13.2%
Total	4,450	100.0%	930	100.00%	20.9%

7.2.4 Type of Lamp

Table 7-6 summarizes the types of UV lamps used by POTWs, categorized by size. LPLO lamps are used in approximately 44% of all systems, while LPHO and MP lamps each comprise ~28% of all systems.

Table 7-6. Type of UV Lamp Systems by POTW Design Flow Rate.

Lamp Type	1-2.5 (MGD)	>2.5-5 (MGD)	>5-20 (MGD)	>20 (MGD)	Total POTWs	Percent of All POTWs
LPLO	235	97	66	7	405	43.5%
LPHO	113	77	55	10	255	27.4%
MP	91	69	75	23	258	27.7%
Unknown	6	2	4	0	12	1.3%
Total	445	245	200	40	930	100.0%

Figure 7-4 presents normalized data for those systems where the type of lamp was known. LPHO usage is relatively constant (approximately 25-30%) across all flow rates. The data used for the normalization is presented in Table 7-6. As flow increases, LPLO usage declines from ~55 to 20%, and MP usage increases from ~20 to 55%. The LPLO trend is consistent with larger plants being slower adopters, and therefore less likely to use the older LPLO lamp type. Another factor in both the LPLO and MP trends may be lamp energy output; MP lamps have the highest output and therefore require fewer lamps and a smaller footprint than the other lamp types. Large facilities, which require correspondingly large UV systems, are likely to be located in urban areas, with little space to expand. Consequently, they may be particularly interested in reducing the space requirements, complexity, and capital cost of their UV systems. In addition, as noted by survey respondents (Section 7.3.6.2), systems with more lamps require more maintenance, so facilities with larger and more complex UV systems may prefer to reduce maintenance time by using systems with fewer lamps.

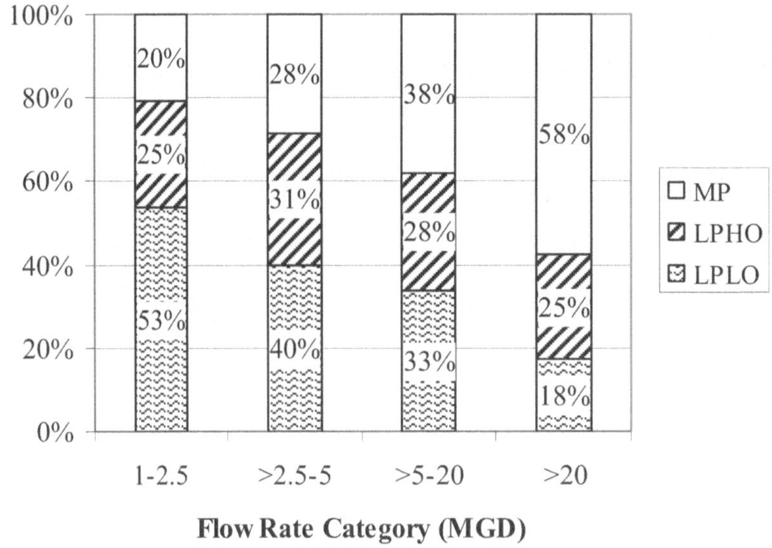

Figure 7-4. Normalized Distribution of Lamp Type.
(Percent of POTWs with Each Lamp Type Indicated Within Bars)

7.2.5 Reactor Configuration and Lamp Orientation

Table 7-7 summarizes the vendor information for UV systems by reactor configuration (open channel or pressure) and lamp orientation within the reactor. More than 95% of UV systems are open channel, and ~80% of UV systems use open channels with the lamps horizontal and parallel to flow.

Table 7-7. UV Reactor Configuration and Lamp Orientation.

Reactor Configuration	Lamp Orientation to Flow	Number of POTWs	Percent of Total	Median Flow Rate (MGD)	Min Flow Rate (MGD)	Max Flow Rate (MGD)
Open Channel	Horizontal and parallel	746	80.2%	3.0	1	250
Open Channel	Vertical and perpendicular	129	13.9%	2.5	1	37
Pressure	Horizontal and perpendicular	42	4.5%	1.5	1	12
Unknown		13	1.4%	2.9	1	9

7.2.6 Upstream Wastewater Treatment

The U.S. EPA databases contain information on unit process information for about half (49%) of the 930 POTWs that use UV for disinfection. This information is summarized in Table 7-8. According to the available data, more than 99% of the systems have some form of secondary treatment ahead of the UV disinfection process; one plant uses only advanced primary treatment. Approximately 79% of the systems use some form of suspended growth treatment (activated sludge, nutrient removing activated sludge, activated sludge plus a coagulant, or sequencing batch reactor). Slightly more than 20% use some type of fixed film secondary process (trickling filter or rotating biological contactors).

Table 7-8. Summary of Treatment Processes Used Ahead of UV Disinfection.

Type of Treatment	Number	Percent
Advanced Primary	1	0.2%
Suspended Growth Secondary		
Activated Sludge	196	43.7%
Activated Sludge with Nutrient Removal	119	26.5%
Activated Sludge + Ferric or Alum Addition	35	7.8%
Sequencing Batch Reactor	4	0.9%
Fixed Film Secondary		
Rotating Biological Contactors	24	5.3%
Trickling Filter	70	15.6%
Total	449	100.0%

7.3 UV Survey Results

This section of the chapter describes the results from the 103 major POTWs using UV that responded to the survey questionnaire. The first portion summarizes the profile of the facilities that were surveyed. The second portion provides the results of the survey.

7.3.1 Profile of Surveyed POTW Using UV

The survey target was to obtain responses from approximately 100 facilities; 103 facilities were covered. Figure 7-5 shows the geographical survey distribution; responses were returned from 38 of the 50 states. The goal for geographical distribution was to balance proportional representation and equal representation. For example, California has 6% of all major POTWs in the U.S., has 3% of all major POTWs with UV disinfection, and represents 2% of all states. Therefore, the goal was to have surveys from California represent 2-6% of all collected surveys. "Near target" means that the percentage of surveyed facilities was within one percentage point of equal or proportional representation (1-7% for California). Six states (California, Colorado, Connecticut, Georgia, Virginia, and Wyoming) were above target, and six states (Arkansas, Indiana, Louisiana, Missouri, Pennsylvania, and Tennessee) were below target.

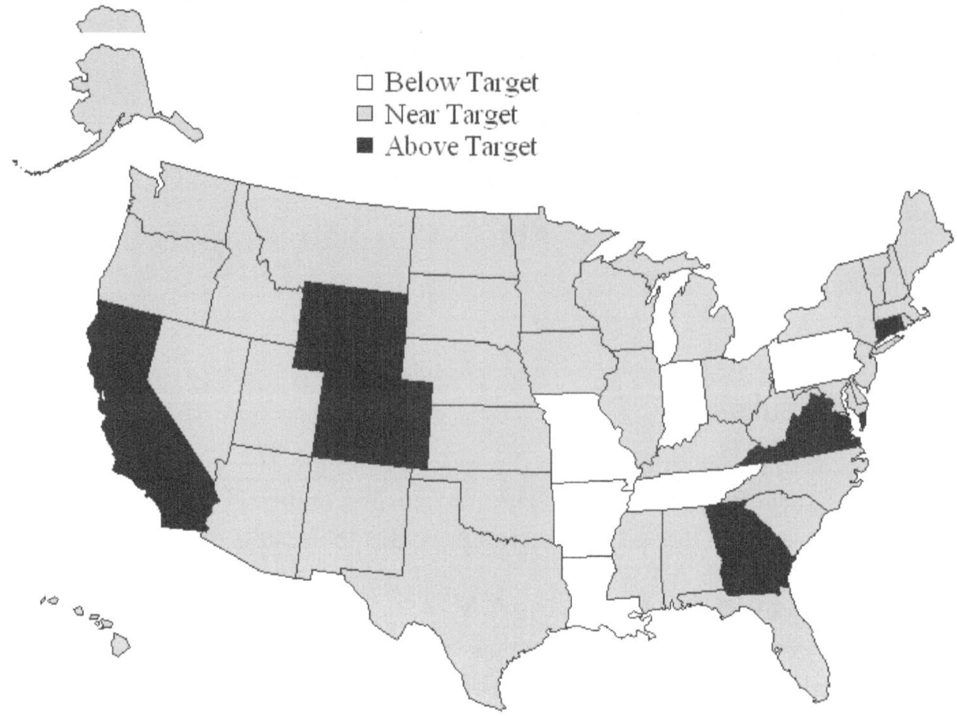

Figure 7-5. Representation of Surveyed POTWs with UV Systems.

Table 7-9 summarizes the survey distribution by plant size/flow rate for the 99 plants that provided flow rate information. As with geographical distribution, the size distribution was intended to strike a balance between proportional representation and equal representation across size categories. Survey responses were near or above targets for all categories except the smallest size. The low response for these plants may be due to the lack of resources, staff availability and knowledge, and technical support at these facilities.

Table 7-9. Summary of POTWs for Limited Project UV Survey.

Flow Rate (MGD)	# of UV POTWs Surveyed	Surveyed Percentage	Equal Representation	Proportional Representation of Major POTWs	Proportional Representation of Major UV POTWs	Status
1-2.5	17	17%	25%	49%	48%	Below target
>2.5-5	25	25%	25%	25%	27%	Near target
>5-20	39	39%	25%	20%	21%	Above target
>20	18	18%	25%	7%	4%	Near target
Total	99	100%	100%	100%	100%	

Table 7-10 provides a profile of the surveyed POTWs, with respect to lamp type. As with the other distributions, the goal was to balance proportional representation (given in Table 7-6) and equal representation (33% each). LPLO facilities were slightly underrepresented, while LPHO and MP facilities were slightly overrepresented. Each lamp type was represented by at least three vendors. All of the lamp types had installations that were relatively recent, i.e., within the last five years, and the LPLO and MP facilities had some older systems than the LPHO.

Table 7-10. UV Survey: Type of UV Lamp Systems.

Lamp Type	1-2.5 (MGD)	>2.5-5 (MGD)	>5-20 (MGD)	>20 (MGD)	No Flow Rate Reported	Total – Surveyed POTWs	Percent of Surveyed POTWs	No. of Vendors	Earliest Year Installed	Latest Year Installed
LPLO	3	9	10	2	2	26	26%	5	1988	2001
LPHO	6	12	15	4	1	38	37%	3	1996	2006
MP	8	4	13	12	1	38	37%	4	1982	2006
Total	17	25	38	18	4	102	100%			

7.3.2 Reasons for Installing UV

Table 7-11 summarizes the responses of 101 facilities that indicated at least one reason for installing a UV system. The top three reasons for making the change were 1) increased safety, 2) elimination of gas chlorine, and 3) elimination of chlorination/dechlorination. Over half of the respondents indicated that these were among the reasons for installing UV disinfection. Elimination of permit requirements and ease of operations were each cited as reasons in ~40% of the cases. About 25% indicated that a more stringent discharge or receiving water permit requirements was part of the justification for changing to UV disinfection.

Table 7-11. Summary of Responses for Changing to UV Disinfection.

Reason for Installing UV	Percent "Yes" out of 101 Facilities
Safer work environment	72%
Elimination of gas chlorine	65%
Elimination of chlorination/dechlorination	57%
Elimination of permit requirements such as Risk Management Plans	43%
Ease of operation	39%
More stringent discharge requirements	28%
More stringent receiving water requirements	22%
Less O&M labor	16%
Because the water is to be reused/reclaimed	15%
Elimination of gas chlorine upgrades to meet building code changes	12%
Lower operating costs	7%

In a follow up question, 59% of respondents indicated that UV eliminated chlorine use at the facility. For those POTWs that continued to use chlorine, potential applications include, but are not limited to shock chlorination of the UV system to control algae and other microbes, backup disinfection, and control of odors or algae, sludge bulking, and fouling of clarifiers or basins.

7.3.3 Design Related Information

Approximately three quarters (73%) of the surveyed facilities used UV systems with automatic wipers. The responses to the design-related portion of the survey also indicated that regulations are important in the implementation of a UV system. In ~70% of the cases, the UV system was required to meet a guideline or mandated requirement.

7.3.3.1 Design Flow

Table 7-12 summarizes information for the three lamp types, based on POTW average dry weather design flow and UV system design flow. The design flow rates generally increased with lamp output energy; POTWs using MP lamps clearly had the largest design flows for the UV system and overall POTW. As expected from Table 4-1, the number of lamps used per MGD of design flow decreased as lamp output energy increased.

Table 7-12. Design Flow Rates of Surveyed POTWs and Their UV Systems.

	POTW Design Flow Rate (MGD)			UV System Design Flow Rate (MGD)			No. of Lamps/MGD of Design Flow (UV System)		
Statistic	LPLO	LPHO	MP	LPLO	LPHO	MP	LPLO	LPHO	MP
Median	5.5	5.5	10	6	10	38	54	17	3
Min	1.0	1.2	1.2	1	2	1	6	3	2
Max	53	45	103	115	68	360	202	144	121
No. of Responses	24	37	37	23	25	32	23	25	31

7.3.3.2 Upstream Processes

Table 7-13 summarizes the information for the 100 facilities that described the processes upstream of the UV system. All have at least secondary treatment processes, and 74% of those are activated sludge processes. In addition, almost half (47%) use tertiary treatment. Because many facilities use multiple treatment processes, the number of the individual processes listed in Table 7-13 is greater than the total number of plants that responded.

Table 7-13. Summary of Treatment Processes Used Ahead of UV Disinfection.

Secondary Treatment	Number	Tertiary Treatment	Number
Activated Sludge - Conventional	32	Alum Addition	11
Activated Sludge - Anaerobic/Anoxic/Oxic	14	Ferric Chloride Addition	2
Activated Sludge - Extended Aeration	14	Clarification Using Tube Settlers	1
Activated Sludge - Other	21	Mixed Media Filter	12
Biological Nutrient Removal	58	Rapid Sand Filter	9
Other Suspended Growth Process	6	Slow Sand Filter	6
Oxidation Ditch	13	Sand Filtration/Intermittent	2
Rotating Biological Contactor (RBC)	3	Sand Filtration/Recirculating	4
Trickling Filter	10	Other Filtration	3
		Granular Activated Carbon	2
		Reverse Osmosis	1

7.3.3.2 UVT of Water to be Disinfected

Table 7-14 summarizes data on the average operating UV transmittance (UVT) and the design UVT (minimum UVT) for 63 valid survey responses. More than 50% of the systems operated between 60 and 80% UVT. More than 60% of the systems were designed to handle a UVT between 50 and 70%.

Table 7-14. Summary of Average and Design UVT for 63 Responses.

Range of UVT	Average Operational UVT	Design UVT for System
100-90%	3%	0%
90-80%	17%	3%
80-70%	30%	8%
70-60%	25%	28%
60-50%	19%	36%
<50%	2%	13%
Don't Know	3%	11%

7.3.3.3 Need for Bench or Pilot Study

Table 7-15 summarizes the information concerning the need for a bench or pilot study and whether one was performed as part of the installation. In 48 cases, bench and/or pilot studies were performed as part of the design process. The majority (79%) of these facilities felt that the study was needed to assist in the design of the UV system.

Table 7-15. Summary of Responses for the Need for a Bench or Pilot Study.

Tests Performed	No. of Responses	No. Recommending Tests	Percent
Bench only	20	13	65%
Pilot only	9	7	78%
Bench and Pilot	19	18	95%

7.3.3.4 Resources for a Successful Project

The survey requested information on the relative importance of five potential resources that may be used in the implementation of a successful project. These findings are summarized in Table 7-16. The consultant's skill was ranked as one of the top two factors ~75% of the time. Vendor quality and visiting/touring other facilities were the next most important and were ranked in the top two factors more than 40% of the time. Of less importance were the skill and knowledge of the facility's staff and the availability of technical resources.

Table 7-16. Summary of Ranking of Elements for a Successful Project.

		Rank, 1 being Most Important				
Items to Rank	Total Responses	1	2	3	4	5
Consultant's skill and knowledge	90	46%	27%	9%	13%	6%
Vendor	89	24%	19%	22%	11%	24%
Visiting and touring other facilities	88	22%	19%	14%	15%	31%
Own staff skill and knowledge	90	12%	20%	33%	23%	11%
Available technical resources (books, articles, etc.)	89	3%	10%	26%	30%	30%

7.3.3.5 Vendor Selection

Table 7-17 summarizes the vendor characteristics that have the highest value to the 97 owners of UV systems who responded. System design and technical knowledge were the top two desirable vendor characteristics, and were ranked as one of the top two factors ~60% of the time.

Responsiveness and system features were next, and size of the company and the sales staff were ranked least important.

Table 7-17. Summary of Desirable Vendor Characterisitics.

Vendor Characteristic	Ranking, 1 Being Most Desired					
	1	2	3	4	5	6
System Design	39%	20%	27%	11%	2%	1%
Technical Knowledge	30%	31%	25%	8%	5%	1%
Responsiveness	23%	21%	14%	35%	6%	1%
System Features	13%	22%	27%	30%	5%	3%
Size of Company	3%	2%	1%	11%	36%	47%
Sales Staff	1%	6%	3%	2%	48%	39%

7.3.4 Transition to UV Disinfection

Facilities were asked about the ease of converting to UV disinfection, including the complexity of the UV system, the ease of transition, and the training and maintenance times required. Chi-squared tests and analysis of variance (ANOVA) were performed to determine whether these parameters varied significantly with facility size (based on average dry weather design flow rate), lamp type, or the level of treatment. Details of the statistical analyses are provided in Appendix B. A significance level $(P) \leq 0.05$ indicates that the correlation or difference in mean values is likely to be significant. However, some caution should be used in applying these results. Because the analyses only indicate the probability of significance, they may miss correlations that exist or may find significance in data when no correlations exist in reality.

7.3.4.1 Ease of Converting to UV

Table 7-18 summarizes the responses concerning the complexity of the UV system and the transition from the previous disinfection system. Although more than 60% indicated that the UV system was more complex than the system it had replaced, more than 85% of the responses indicated that the transition to the UV system was relatively smooth.

Table 7-18. Summary of Transition and Complexity of UV Systems.

Element to Rate	Responses	More	Same	Less
Complexity of UV system compared to the one it is replacing	91	62%	11%	27%
	Responses	**Difficult**	**Not too bad**	**Easier than Expected**
Difficulty of the transition to UV	91	14%	41%	45%

Table 7-19 lists the results of the statistical analysis for these survey parameters; Appendix B provides a breakdown of the survey responses by flow category, lamp type, and treatment level. The difficulty of transition correlated significantly with the lamp type; plants using MP lamps were less likely to find the transition easy (Figure 7-6). The ease of transition did not correlate significantly with the other treatment system characteristics, and the relative complexity of the UV system did not correlate significantly with any of the treatment system characteristics. However it should be noted that other correlations may exist, but be undetected due to factors such as the variability in the data and/or the sample size. Although the correlations were not statistically significant, the data indicate that larger plants were less likely to find the transition easy, and MP users were more likely to feel that UV was more complex than their previous system (see Appendix B.2 for data).

Table 7-19. Significance Levels (P Values) for the Correlation Between Complexity/Transition and Treatment System Characteristics.

Parameter	UV Complexity	Transition
Average Dry Weather Design Flow	0.11	0.14
Lamp Type	0.08	0.04
Treatment Level	0.97	0.91

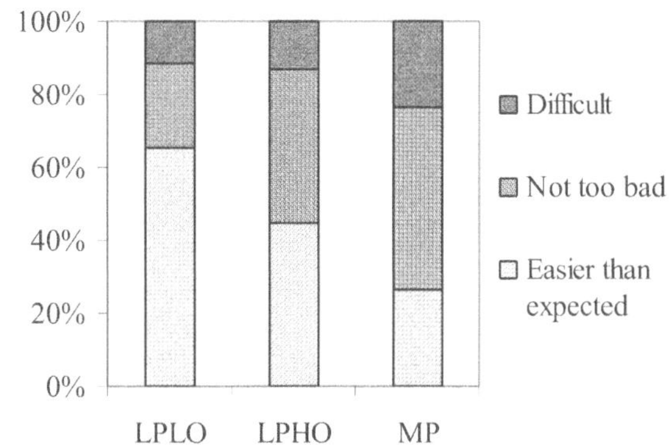

Figure 7-6. Relative Difficulty of Transition, by Lamp Type.

7.3.4.2 Training

Table 7-20 summarizes the responses concerning the classroom and hands on training times required for operation of UV systems. The range of reported training times is quite large and spans more than an order of magnitude, so data should be used with caution.

Table 7-20. Summary of Training Times.

Statistic	Classroom Training	Hands On Training
Median Hours	6	10
Minimum Hours	1	1
Maximum Hours	40	80
No. of Responses	71	87

Table 7-21 lists the ANOVA results for these survey parameters; Appendix B provides a breakdown of the survey responses by flow category, lamp type, and treatment level. The analysis indicates that the training times show no significant variation with POTW design flow rate, lamp type, or treatment level. However it should be noted that other correlations may exist, but be undetected due to factors such as the variability in the data and/or the sample size. For example, the data indicate that training times generally increased with flow rate (see Appendix B.2 for data), although the correlations were not statistically significant.

Table 7-21. Significance Levels (P Values) for Training Times and Treatment System Characteristics.

Parameter	Classroom Training	Hands On Training
Average Dry Weather Design Flow	0.09	0.45
Lamp Type	0.40	0.66
Treatment Level	0.48	0.14

7.3.5 Operational Survey Results

Facilities were asked about the maintenance times required, average lamp life, and the annual cost of lamp components. Appendix B details the statistical analyses that were performed to determine whether these parameters varied significantly with facility size (average dry weather design flow rate), lamp type, or the level of treatment. Significance levels of $P \leq 0.05$ are considered significant.

7.3.5.1 Maintenance Labor

With 69 survey responses, the median maintenance time required for operation of UV systems was 4 hours/week. The minimum reported maintenance time was 0.2 hours/week, and the maximum was 120 hours/week. Table 7-22 lists the ANOVA results for maintenance times; Appendix B provides a breakdown of the survey responses by flow category, lamp type, and treatment level. Maintenance times varied significantly with flow rate and increased with the POTW design flow (Figure 7-7). This is expected, because larger facilities have larger UV systems with more components to maintain and replace.

Table 7-22. Significance Levels (P Values) for Maintenance Times and Treatment System Characteristics.

Parameter	Maintenance Time
Average Dry Weather Design Flow	0.02
Lamp Type	0.79
Treatment Level	0.64

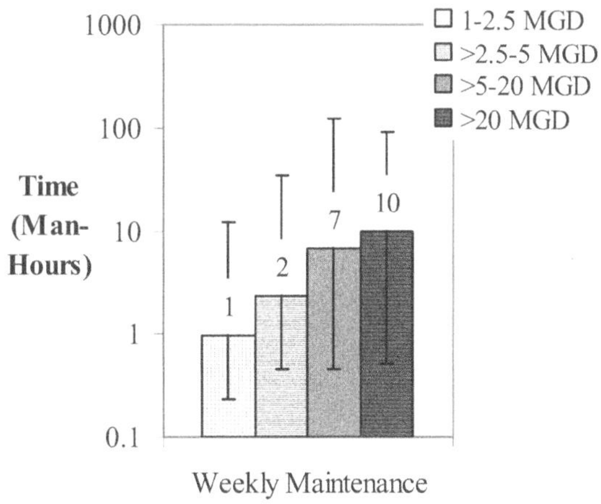

Figure 7-7. Summary of Median Weekly Maintenance Times by Size. (Data bars and listed numbers represent median values, error bars represent minimum and maximum reported values)

7.3.5.2 Lamp Life and Cost Data

With 84 survey responses, the median lamp life was approximately 9,000 hours. The minimum reported lamp life was 1,200 hours, and the maximum was 25,000 hours. Table 7-23 lists the ANOVA results for lamp life; Appendix B provides a breakdown of the survey responses by flow category, lamp type, and treatment level. Lamp life varies significantly with the lamp type. As expected from Table 4-1, MP lamps had the shortest life, approximately half the lifetime of the LPLO and LPHO lamps (Figure 7-8).

Table 7-23. Significance Levels (P Values) for Lamp Life and Treatment System Characteristics.

Parameter	Lamp Life
Average Dry Weather Design Flow	0.43
Lamp Type	0.004
Treatment Level	0.98

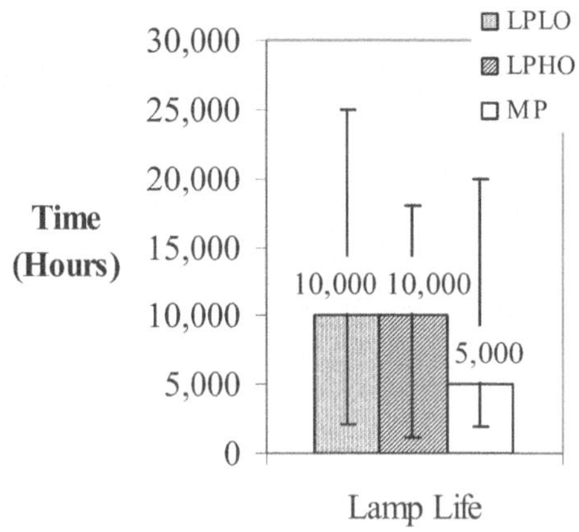

Figure 7-8. Summary of Lamp Life by Lamp Type.
(Data bars and listed numbers represent median values, error bars represent minimum and maximum reported values)

Table 7-24 presents the annual cost to replace lamp components for the UV systems of the surveyed POTWs; Appendix B provides a breakdown of the survey responses by flow category, lamp type, and treatment level. The median annual cost of lamp replacement is several times higher than the median costs of sleeve and ballast replacement.

Table 7-24. Summary of Annual Lamp Component Costs.

Statistic	Lamps	Sleeves	Ballasts
Median Cost	$11,000	$2,000	$3,000
Minimum Cost	$600	$130	$110
Maximum Cost	$250,000	$70,000	$200,000
No. of Responses	80	63	60

Table 7-25 lists the ANOVA results for lamp component costs. Lamp, sleeve, and ballast costs varied significantly with the facility size and lamp type. Costs generally increased with POTW design flow and increasing lamp energy output (Figure 7-9). This makes sense, because larger plants have larger UV systems with more components to replace, and the higher output MP lamps have shorter lives and need more frequent replacement.

Table 7-25. Significance Levels (P Values) for Lamp Component Costs and Treatment System Characteristics.

Parameter	Lamps	Sleeves	Ballasts
Average Dry Weather Design Flow	2.6E-07	0.0002	0.004
Lamp Type	0.001	0.004	0.002
Treatment Level	0.96	0.52	0.85

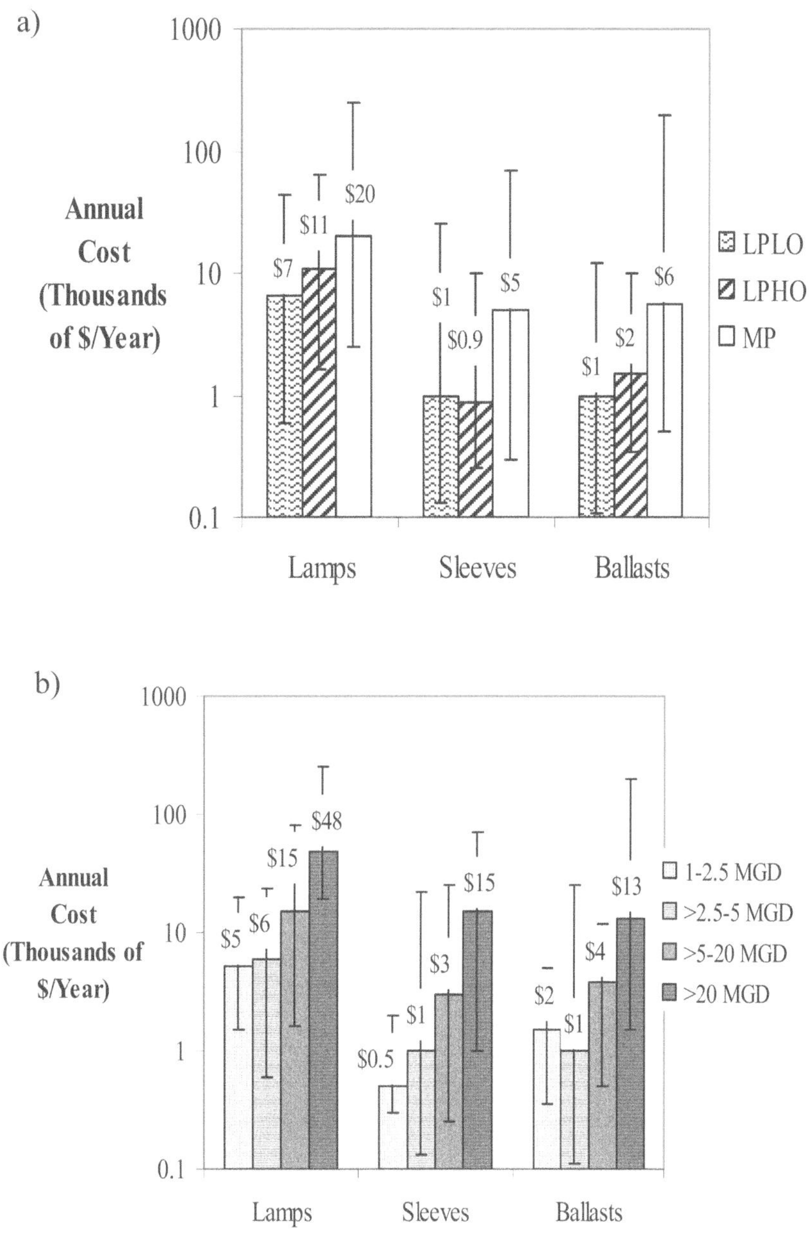

Figure 7-9. Summary of Lamp Component Costs by (a) Size and (b) Lamp Type.
(Data bars and listed numbers represent median values, error bars represent minimum and maximum reported values)

7.3.6 Discharge Requirements

The following information focuses on the discharge requirements, the type of receiving water, and its designated use.

7.3.6.1 BOD, TSS, and Microbial Levels

Most facilities (67 out of 88 responses, or 76%) are regulated based on fecal coliforms, while 10% are regulated on total coliforms, and the rest are regulated on other organisms such as

E. coli or fecal streptococci. Because of the low number of responses for total coliforms and other organisms, all of the analyses of microbial quality were performed using the data on fecal coliforms only. Table 7-26 shows survey results on effluent water quality parameters; Appendix B provides a breakdown of the survey responses by flow category, lamp type, and treatment level. As with the other data, the range of reported levels is quite large and spans several orders of magnitude, so results should be used with caution.

Table 7-26. Summary of Effluent Quality Parameters.

	BOD (mg/L)		TSS (mg/L)		Fecal Coliform (MPN/100 mL)
Statistic	Permitted	Typical	Permitted	Typical	Permitted
Median	23	3	30	4	100
Minimum	4	0.5	4	0.01	0.5
Maximum	150	205	175	120	2000
No. of Responses	95	87	95	85	67

7.3.6.2 Type of Receiving Water

Table 7-27 summarizes the types of receiving water for the UV treated effluent. In more than 80% of the cases, the receiving water is a river or stream. Ocean and estuary discharges comprise 10%. The "Other" category, at 4%, typically referred to reclamation uses instead of discharge to a body of water. Discharge to a lake accounted for only 2% of the responses.

Table 7-27. Summary of Receiving Water for UV Treated Effluent.

Category	Number
Receiving Water	
River	46
Stream	38
Lake	2
Ocean	6
Estuary	4
Other	4
No. of Responses	100

7.3.6.3 Designated Use of Receiving Water

Table 7-28 summarizes the types of receiving water for the effluent of the surveyed UV plants. There were 70 responses for this survey question. In some cases, multiple use was indicated so the total percentage exceeds 100%. The most frequent designated use for the receiving water was recreation, followed by aquatic life and irrigation/agriculture use. Drinking water was the designated use in ~9% of the cases. The "Other" category includes "none", electrical power generation, and flow into a sound or ocean.

Table 7-28. Designated Use of Receiving Water of UV Treated Effluent.

Category	Number / Percent
No. of Responses	70 / 100%
Designated Use	
Recreation	44 / 63%
Aquatic Life	11 / 16%
Irrigation/Agriculture	10 / 14%
Drinking	6 / 9%
Stream Augmentation	4 / 6%
Other	4 / 6%

7.3.7 Advice from Facilities

The survey included space for comments and four short-answer questions:

- What surprised you the most when you changed to UV?
- What has surprised you the most while operating your UV system?
- What advice would you give to someone considering switching to UV disinfection?
- How would you improve your UV system?

These questions were designed to allow the facilities to discuss issues that were important to them. The responses fell into four main categories: design issues, issues that affect both design and O&M, O&M issues, and installation issues. Many of their recommendations echo advice in the NWRI Guidelines. The following sections cover each of these categories in greater detail; each comment described here was mentioned by at least two POTWs, while the most common recommendations were voiced by 10-15 facilities. Appendix C provides a full list of the answers.

7.3.7.1 Design Issues

The most prevalent advice on design issues was that POTWs do thorough research, before choosing UV for disinfection. Many people who recommended this suggested visiting or talking with similar facilities that operate the UV system of interest, and have been operating UV for some time. These similar facilities can provide reasonable estimates on the time and cost requirements of UV, as well as potential problems that might be encountered during design, installation, or operation. Another recommendation from survey respondents was to ensure that the UV system capacity was sufficient for high flow conditions, such as during heavy rainfall. Finally, electrical surge protection and emergency power backup are important. Fluctuations in power can affect lamp output, and therefore disinfection efficacies, and backup power is needed to ensure continuous disinfection in the event of a power failure.

7.3.7.2 Design/O&M Issues

The most common recommendations for design/O&M issues were to use automatic wipers, which significantly reduce maintenance time, and to cover the UV system (e.g., to build a roof over it, or install it inside a building). Covering the UV system has several advantages. First, it makes maintenance much easier in adverse weather conditions, such as extreme cold, snow, or rain. One respondent was unable to work on electrical problems during storms because the system was uncovered, and another had to replace all of the electronics on the UV system after lightning struck. Covering the system also reduces algal growth, which would otherwise increase maintenance requirements and/or decrease disinfection efficacy.

Other recommendations include minimizing the number of lamps to reduce maintenance time, and carefully designing the hydraulics, to avoid flooding and properly control headloss and water depth above the lamps. People also suggested using multiple channels; even those with multiple lamp banks noted that there are times when multiple channels are helpful, e.g., to clean a channel or replace an effluent control valve.

There were also some comments about electronics and control systems. A number of respondents had issues with ballasts and circuit boards getting wet, and several had problems with overheating ballast cabinets. Overheating shortens the life of the electronics and requires more frequent replacement of the ballasts, which translates to more maintenance time and higher costs. Several people also complained that the control systems were overly complex; this comment was made for several different UV systems and vendors. People strongly recommended integrating the UV control system with the plant SCADA system. Finally, some respondents had difficulty controlling the UV dose with the UV intensity/UVT analyzers.

7.3.7.3 O&M Issues

For O&M, the most common issue was algal growth (discussed in Section 7.3.6.2) and animals (fish and insects, most likely attracted by the radiation) in and around the UV equipment. A number of people also mentioned that influent water quality affected disinfection and/or maintenance. Several facilities used iron in upstream processes; iron is well known for fouling UV lamp sleeves. Most of these facilities were able to control the problem through rigorous cleaning protocols, in addition to the use of automatic wipers. Other plants had problems with particles reducing disinfection efficacy; they recommended the use of filters upstream of the UV units. Finally, industrial effluents can have a significant impact on disinfection. One surveyed plant received an effluent that was high in organics and reduced the UVT below 50%. This problem could not be solved and the plant has since been converted to hypochlorite disinfection.

The survey also generated a number of comments about spare parts. Several people noted that most parts are proprietary and single-source; as a result, they can be expensive, and are not always available quickly. Therefore, POTWs should keep plenty of spare parts on-site. Some respondents also found that lamps do not last the nominal lifetime, particularly during start-up; facilities should expect higher costs for parts replacement during this time.

7.3.7.4 Installation Issues

Respondents noted that good vendor support is essential, particularly during start-up. Several recommended having the vendor validate the system. Finally people suggested testing the ballasts before installation, as this can make start-up more efficient.

7.4 Ozone

The use of ozone in both the wastewater and water treatment industries in the United States received significant federal funding from the Clean Water Act beginning about 30 years ago (Drago and Thompson, 2005; Thompson and Drago, 2005). This section summarizes the use of ozone and the evolution of the facilities associated with this technology.

7.4.1 Historical Ozone Installation Patterns by Wastewater Industry

Table 7-29 summarizes the historical installation of ozone at POTWs during a 30-year period (1975-2004), and predicted ozone use through the year 2010. In general, the wastewater industry was an early adapter of the first generation of this technology, but quickly moved away from it, due to the high cost and complexity of the systems, which made them unreliable and difficult to operate.

Most (43 of 50) of the POTWs that added ozone did so during the first 10 years of the 30-year period shown in Table 7-29. The median treatment capacity of these ozone systems was 2 MGD. Only four POTWs added an ozone system during the second 10 years (1985-1994), and two POTWs added ozone during the most recent 10 years (1995-2004). The median capacity of these six new POTWs (1985-2004) was 9 MGD.

Table 7-29. Ozone Use at POTWs.

Period (Years)	Installation at All POTWs		
	Number	Median Capacity for Period (MGD)	Median Ozone Capacity for the Period (lb/day)
1975 – 1984	43	2	400
1985 – 1994	4	21	1,900
1995 – 2004	2	14	1,800
2005 - 2010	2	60	2,000

Robson and Rice (1991) summarized the number of POTWs using ozone circa 1989 in the 1986 U.S. EPA "Design Manual, Municipal Wastewater Disinfection" (Table 7-30). The two U.S. EPA databases (CWNS-2000 and Envirofacts) indicated 32 major POTWs with ozone, ozone and chlorine, or ozone and UV equipment. This number was consistent with the information in Table 7-30 (i.e., the sum of the first two rows) indicating convergence between these two papers and the U.S. EPA databases.

Table 7-30. Summary of Status of Ozone Use by POTWs in 1989. (Robson and Rice, 1991)

Category	Number of POTWs	Percentage
Major POTW operating in 1989	17	39.5%
Major POTW, equipped but not operating in 1989	15	34.9%
Not a major POTW and equipped in 1989	11	25.6%
Total POTWs using Ozone as of 1986-1989	43	100.0%

7.4.2 Expansion of Ozone Facilities and Ozone Use in 2006

As indicated in Table 7-30, by 1989, almost half of the major POTWs with ozone equipment had switched to an alternative disinfection technology. A cross check of the identified facilities was performed using vendor information and a phone survey. The information indicated that all 11 POTWs with dry weather capacity < 1 MGD had replaced ozone with another disinfection technology. Of the 43 facilities using ozone in 1989, only five major POTWs were still using ozone in 2006, a drop of almost 90%. The current disinfection survey of practice for ozone indicated that seven POTWs currently use ozone for disinfection.

7.4.3 Projected Number of Major POTWs Using Ozone – 2010

Four POTWs are in the process of resuming the use of ozone or converting to ozone. Two of these had previously stopped using ozone, but are planning to convert back to ozone from chlorine in 2007. The other two POTWs will be converted from chlorine to ozone by 2010. One POTW indicated that it plans to switch from ozone to UV disinfection in 2010. Based on this information, the total number of POTWs using ozone for disinfection in 2010 is expected to be ten.

The combined capacity of POTWs that currently use ozone is 170 MGD. In 2010, the combined capacity of POTWs using ozonation is expected to be at least 500 MGD, based on the capacity of the POTWs that are adding or resuming use of ozone, less the capacity of the POTW that plans to stop using ozone.

7.4.4 Profile of Major POTWs Using Ozone

Table 7-31 provides a summary of the range of the sizes of the major POTWs in the 50 U.S. states and Washington, D.C. in comparison with the POTWs using ozone. Table 7-32 summarizes the secondary treatment processes used by the POTWs currently using ozone and the future plants (expected to operate by 2010) that are in design or the planning stages.

Table 7-31. Size and Median Comparison of All Major POTWs and POTWs Using Ozone.

POTW Category	Total # of POTWs	Median (MGD)	Range (MGD)
All Major POTWs	4,450	2.7	0.95-1200
Major POTWs with Ozone Disinfection	7	10	3-34

Table 7-32. Summary of Secondary Treatment at POTWs using Ozone.

Secondary Treatment	Current	Future
Oxidation Ditch	1	1
Air Activated Sludge	4	5
Pure Oxygen Activated Sludge	2	4

The major POTWs using ozone were surveyed and asked about average ozone dosage, maximum ozone dosage, minimum contact time (which occurs at maximum flow rates), and average contact time. Results are summarized in Table 7-33.

Table 7-33. Ozone Dosages and Contact Times at Major POTWs.

Statistic	Average Ozone Dosage (mg/L)	Maximum Ozone Dosage (mg/L)	Minimum Contact Time, at Peak Flow (min)	Contact Time at Average Flow (min)
Median of Reported Values	3.5	6.5	12	30
Minimum of Reported Values	3	3	7	15
Maximum of Reported Values	18	27	15	36
No. of Responses	4	4	4	3

The typical POTW using ozone has the following characteristics:

- Generally larger than the typical POTW (a median of 10 MGD for the ozone plants as compared to 2.7 MGD, nationally);
- Use some form of suspended growth biological secondary treatment (oxidation ditch, air activated sludge, or pure oxygen activated sludge);
- Of the four POTWs reporting microbial discharge limits, the regulated microbe was predominantly fecal coliform (FC), with ranges from none detected to 1,000 FC/100 mL;
- Five POTWs discharge to a river or creek; and
- Two POTWs indirectly supplement their drinking water supply and have a tertiary treatment train that includes flocculation, coagulation and settling with lime, filtration, and ozonation as multiple microbial barriers

CHAPTER 8.0

COMPARISON OF DISINFECTION ALTERNATIVES

8.1 Introduction

The initial goal of wastewater disinfection was the inactivation of pathogens to protect public health. However, wastewater disinfection has become much more complicated with the discovery of potentially harmful disinfection byproducts (DBPs) and the toxicity of chemical disinfectants to the environment. In order to select the most appropriate disinfection alternative, POTWs must now consider and balance multiple goals, including disinfection efficacy, DBP formation, costs, process reliability, and many others.

The purpose of this chapter is to summarize and compare the various viable wastewater disinfection alternatives presented in the previous chapters. Section 8.2 summarizes these alternatives. Section 8.3 compares the mature alternatives with respect to the criteria used to select wastewater disinfection alternatives (this also includes the factors that drive the changes in disinfection technology). Section 8.4 provides some general guidelines for choosing appropriate candidates for disinfection, as well as alternatives that are commonly selected during wastewater disinfection upgrades. Finally, Section 8.5 presents examples from the literature of the decision processes used by two districts in selecting wastewater treatment alternatives.

8.2 Technology Alternatives

Table 8-1 groups the wastewater technologies discussed in this report into the following four main categories, with subcategories as appropriate (e.g., sources of chlorine/chloramines and types of UV lamps):

- Mature technologies that are well understood and have been applied at full-scale. Facilities considering upgrades to their existing disinfection practices are likely to favor these technologies, because the advantages and disadvantages are well known, design information and operational experiences are readily available, and regulatory approval is generally easier.
- Technologies that have been tested and may be practicable, but are not widely used for wastewater disinfection at full-scale in the U.S.
- Combinations of technologies. This approach is likely to gain greater acceptance in the future, as POTWs upgrade existing disinfection practices to achieve multiple objectives or to provide multiple barriers for disinfection. Table 8-1 lists the most viable alternatives. As with the individual technologies, each alternative has multiple combinations based on the different forms of chlorine delivery, UV lamp types, and ozone generation methods.
- Emerging technologies that are still under development. More studies are needed for these technologies to be considered for field-scale applications.

Table 8-1. Wastewater Disinfection Alternatives.

Category	Wastewater Disinfection Alternatives
Mature technologies	▪ Chloramination Chlorine gas Sodium hypochlorite Calcium hypochlorite On-site generation (0.8%) On-site generation (12%) ▪ Breakpoint chlorination Chlorine gas Sodium hypochlorite Calcium hypochlorite On-site generation (0.8%) On-site generation (12%) ▪ UV irradiation Low-pressure low-output (LPLO) lamps Low-pressure high-output (LPHO) lamps Medium pressure (MP) lamps ▪ Ozonation Ozone generated from air Ozone generated from oxygen
Practicable technologies, but rarely used	▪ Chlorine dioxide ▪ Peracetic acid
Combination of technologies	▪ Breakpoint chlorination and chloramination[1] ▪ UV and chloramination[2] ▪ UV and breakpoint chlorination[2] ▪ UV and ozonation[3] ▪ UV and peracetic acid[4]
Emerging/innovative technologies	▪ Pasteurization ▪ Catalysis ▪ Brominated chemicals ▪ Ferrate ▪ Pulse UV ▪ Many others

[1]Five options are possible based on the different forms of chlorine: gas, sodium hypochlorite, calcium hypochlorite, onsite generation of 0.8% sodium hypochlorite, and onsite generation of 12% sodium hypochlorite.
[2]A total of 15 alternatives are possible based on three types of UV lamps (low-pressure low-output, low-pressure high-output, and medium pressure) and five forms of chlorine.
[3]A total of 6 alternatives are possible based on three types of UV lamps and two types of ozone generation methods (air or oxygen).
[4]Three alternatives are possible based on three types of UV lamps.

The rest of this chapter focuses on the disinfection alternatives that POTWs are most likely to consider: the mature technologies (chloramination, breakpoint chlorination, UV, and ozonation) and combinations of these technologies.

8.3 Comparison of Mature Technologies by Driving Factors/Decision Criteria

Selecting a wastewater disinfection alternative is a complicated process that must consider numerous factors; each treatment facility may have different motivations, constraints, priorities, and site-specific needs. This section discusses some of the key driving forces behind the trend of changing disinfection methods, as determined from the literature review and survey of existing practices. These driving factors generally become part of the larger set of decision criteria used in selecting a disinfection technology; they are grouped into the following categories:

- Changes in permit requirements for disinfection efficacy
- Changes in permit requirements for effluent water quality
- Plant expansions or upgrades
- Safety concerns
- Design issues
- Operation and maintenance (O&M) issues
- Costs
- Other factors

The effects of these factors are discussed briefly and compared across the mature disinfection alternatives in the following sections; details can be found in Chapters 3.0, 4.0, and 5.0. The first three factors are evaluated by type of residual for chlorine, UV, and ozone. The next four factors are evaluated by type of chlorination facilities (gas, liquid, or on-site generation), UV, and ozone. These comparisons are very general in nature; site-specific conditions should be considered before choosing a disinfection alternative.

Although not discussed explicitly in the following sections, combined disinfection (sequential or simultaneous use of two or more disinfectants, discussed in detail in Section 6.5) in wastewater treatment may also be considered whenever an existing process is changed or upgraded. The existing technology may provide certain benefits, and upgrading by adding a different technology can provide an additional level of protection against a variety of organisms, yield synergistic disinfection, reduce disinfection byproducts or other contaminants of concern, and/or provide disinfectant residuals in the effluent.

8.3.1 Changes in Permit Requirements for Disinfection Efficacy

Changes in the disinfection requirements of a plant's permits may also prompt changes in disinfection practices. Permits generally specify the maximum allowable concentrations of indicator microorganisms (e.g., total coliform or fecal coliform bacteria) or pathogens (e.g., enteric virus). As discussed in Section 7.3.5.2, these limits vary widely, by as much as several orders of magnitude. Permit requirements may also be seasonal, and may depend on whether the effluent is discharged to natural water bodies (e.g., rivers or oceans), or reused (e.g., for groundwater replenishment, irrigation, etc.). As permits are periodically renewed, the new permits may impose more stringent requirements such as lower microbial concentrations and/or the addition of new microorganisms to the list. The existing disinfection system may be insufficient to meet the new limits and may need upgrading or replacement.

Table 8-2 compares the mature alternatives with regard to disinfection efficiency and regrowth potential. Because the disinfection mechanisms are specific to each technology and target microorganism, disinfection efficacies vary. In general, all four mature alternatives effectively inactivate coliform, though chloramines are a weaker disinfectant and require a higher dose and/or longer contact time to achieve the same level of inactivation. *Giardia* and *Cryptosporidium* are resistant to free chlorine and chloramines, but are very susceptible to UV irradiation. The reverse is true for adenovirus, which is resistant to UV but very susceptible to low doses of free chlorine.

Table 8-2. Comparison of Mature Wastewater Disinfection Alternatives – Disinfection Efficacy.

Factors	Free Chlorine	Combined Chlorine	UV	Ozone
Disinfection Efficiency				
Total coliform	Good	Fair	Good	Good
Fecal coliform	Good	Fair	Good	Good
E. coli	Good	Fair	Good	Good
Poliovirus	Good	Poor	Good	Good
Adenovirus	Good	Fair	Poor	Good
MS2 coliphage	Good	Poor	Fair	Good
Enterovirus	Good	Poor	Good	Good
Cryptosporidium	Fair/Poor	Poor	Good	Fair
Giardia lamblia	Fair/Poor	Poor	Good	Fair
Regrowth after disinfection	Likely	Likely	Likely	Likely

Regrowth of microorganisms can occur in receiving water bodies or in the distribution systems for reclaimed water, if the microorganisms are not adequately inactivated during the disinfection process. Regrowth poses a greater concern for UV disinfection because pathogens are more easily shielded during disinfection, and regrowth has often been observed under both light and dark conditions.

8.3.2 Effluent Water Quality

Additional permit requirements may affect disinfection practices, because wastewater disinfection can positively or negatively impact the effluent through generation of disinfection byproducts (DBPs) such as trihalomethanes, haloactic acids, NDMA, and cyanide; alteration of water quality parameters such as TDS, chloride, and ammonia; increases in the toxicity of the final effluent; and incidental removal of compounds of concern. Table 8-3 summarizes these effects for each of the mature disinfection alternatives relative to each other.

Table 8-3. Comparison of Mature Wastewater Disinfection Alternatives – Effluent Water Quality.

Factors	Free Chlorine	Combined Chlorine	UV	Ozone
Formation Potential of Disinfection Byproducts				
THMs	High	Low	No	Low
HAAs	High	Low	No	Low
NDMA	Low	High	Destruction	Destruction
Cyanide	Low	Medium	No	No
Cyanogen chloride	Low	Medium	No	No
Bromate	No	No	No	Yes
Chlorite	In NaOCl soln	In NaOCl soln	No	No
Chlorate	In NaOCl soln	In NaOCl soln	No	No
Nitrite	Destruction	Destruction	Yes (MP lamps)	Destruction
Aldehydes	No	No	Low	High
Change of Other Water Quality Parameters				
Residual impacted by change in NH_3	Yes	Yes	No	No
TDS increase	Yes	Yes	No	No
Chloride increase	Yes	Yes	No	No
Temperature increase	No	No	Yes	No
Change of pH	Yes	Yes	No	No
Toxicity				
Persistence of disinfectant residuals	Yes	Yes	No	No
Potential for increasing effluent toxicity	High	Low	Little to none	Low
Accidental release of mercury	No	No	Yes	No
Incidental Benefits				
Removal of compounds of concern (hormones, pharmaceuticals and personal care products, endocrine disrupting compounds)	Some	Poor	Poor	Good
Removal of NDMA or its precursors	No	No	Some	Some
Increased DO in the effluent	No	No	No	Yes

As a physical disinfectant, UV produces the fewest known DBPs; for the other technologies, the main DBP concerns are THMs and HAAs with free chlorine, NDMA with combined chlorine, and bromate with ozone. Both chlorine-based technologies provide residuals, which can be helpful to prevent biogrowth in reuse applications, but also add toxicity to the effluent and may require removal before discharge to water bodies. Each of the technologies causes different changes to water quality, and some provide incidental benefits such as removal of contaminants or addition of oxygen to the effluent. These effects, along with concerns from the public and current or anticipated regulatory limits may favor or restrict the use of some technologies (e.g., on-site chlorine generation increases TDS concentrations and may not be viable for facilities with strict limits, strict ammonia limits may prevent the use of chloramination with dechlorination, etc.).

8.3.3 Plant Expansions or Upgrades

Expansions or upgrades of upstream treatment processes may alter water quality and often warrant an evaluation of existing disinfection practices. For example, the addition of membrane units into a process train improves effluent water quality by reducing particles that can shield pathogens from disinfectants. They may also increase UVT and lower demand for chlorine, chloramines, and ozone; this makes disinfection with these alternatives more efficient and may permit the use of previously impractical alternatives. Similarly, POTWs are increasingly required to include nutrient removal, which results in a better quality effluent that may increase the number of alternatives that can be used for disinfection. The removal of nitrogen from wastewater also allows the use of free chlorine, rather than combined chlorine for disinfection. Table 8-4 summarizes the sensitivity of the mature disinfection technologies to various influent water quality parameters. Suspended solids, dissolved organics and inorganics, low UVT, and high scaling potential can all reduce the disinfectant dose that reaches the target organisms. Temperature and pH affect reaction chemistry, and therefore the disinfection efficacy.

Table 8-4. Comparison of Mature Wastewater Disinfection Alternatives – Sensitivity to Influent Water Quality.

Factors	Free Chlorine	Combined Chlorine	UV	Ozone
Ammonia	High	High	Low	Low
Nitrite	High	High	Low	High
UVT	Low	Low	High	Low
Suspended solids/Particulates	Medium	Medium	High	Medium
Dissolved organics	Medium	Medium	Medium	Medium
Compounds with scaling potential	Low	Low	High	Low
pH	High	Medium	Low	Medium
Temperature	Medium	Medium	Medium	Medium

Improvements in disinfection alternatives may also prompt upgrades of disinfection systems. For example, UV lamp technologies have significantly improved over the last twenty years; older systems may require extensive maintenance, use a large footprint, or simply near the end of their design life. Therefore, POTWs equipped with earlier generations of UV lamps may be motivated to upgrade to a new UV system or change to other disinfection alternatives. Similarly, older ozone equipment that was used in wastewater disinfection was less reliable and had higher operating costs than current ozone systems; as a result, more than 90% of the original facilities abandoned ozone disinfection. However, newer ozone equipment is more efficient and reliable; as a result, two facilities recently chose ozonation for disinfection, as well as for the additional benefits of removal of color and emerging compounds of concern.

8.3.4 Safety Issues

As shown in the survey of UV facilities, safety issues are a major factor behind the shift in disinfection practices, especially for the many plants across the country that use chlorine gas. Table 8-5 summarizes some of these issues.

Table 8-5. Comparison of Mature Wastewater Disinfection Alternatives – Safety

Factors	Gas Chlorine	Hypochorite Solution	On-Site Generation	UV	Ozone
Safety					
Risk to employees from accidental releases	Yes	Yes	Yes	No	Yes
Potential hazard to neighbors	Yes	Yes	No	No	No
Potential hazards during transportation	Yes	Yes	No	No	No
Potential for adverse occurrences from malicious activities	Yes	Yes	Yes	No	No
Public acceptance	Low	Medium	High	High	Medium
Permitting					
On-site storage of hazardous materials	Yes	Yes	No	No	No
Preparation of risk management plan	Yes	Yes	No	No	No
Approval from fire department	Yes	Yes	Yes	No	No

8.3.5 Design Issues

Major factors to consider in designing wastewater disinfection systems include the facility requirements, complexity of design, and others. Table 8-6 summarizes the impacts of each of these factors.

Table 8-6. Comparison of Mature Wastewater Disinfection Alternatives – Design and Permitting.

Factors	Gas Chlorine	Hypochorite Solution	On-Site Generation	UV	Ozone
Complexity of design					
Status of equipment development	Mature	Mature	Emerging	Evolving	Evolving
Life cycle of equipment/technology	Long	Long	Short	Short	Short
Number of existing installations	Small	Large	Small	Medium	Small
Availability of multiple equipment vendors/suppliers	High	High	Low	Low	Low
Existing design guidelines available	Yes	Yes	Yes	Yes	Yes
Requirement for Bench/Pilot scale testing	Low	Low	Low	High	High
Complexity of monitoring and control system	Low	Low	Medium	High	High
Compatibility with existing upstream processes	Good	Good	Good	Good	Good
Redundancy requirements	Low	Low	Medium	High	High
Adaptability to water quality fluctuations	Good	Good	Good	Fair	Fair
Adaptability to flow rate fluctuations	Good	Good	Good	Fair	Fair
Regulatory acceptance					
Time required to gain regulatory acceptance	Long	Short	Short	Long	Long
Existing guidance for regulatory approval	Yes	Yes	Yes	Yes	No
Facility requirements					
Required footprint/space for process/equipment	Medium	Large	Large	Small	Medium
Chemical storage	Yes	Yes	Yes	No	Yes
Suitability for remote site	Yes	Yes	Yes	No	No
Need for consistent chemical supply	High	High	Medium	Low	Low
Need for consistent power supply	Low	Low	High	High	High
Other Design Considerations					
Usability for other process – sludge bulking	Yes	Yes	Yes	No	No
Usability for other process – filter cleaning	Yes	Yes	Yes	No	No
Usability for other process – odor control	Yes	Yes	Yes	No	Yes
Suitability for future expansion	Fair	Fair	Fair	Good	Fair
Disinfectant residual in effluent (ability to maintain for reuse application or need to remove prior to discharge)	Yes	Yes	Yes	No	No

The difficulty of design is an important factor. Designing a chloramination system is relatively simple, because the technology has been used widely for a long time. Consequently, the technology is mature, design guidelines are readily available, bench/pilot tests are generally unnecessary, and the time required for regulatory acceptance is relatively short. In addition, as chemical technologies, both free and combined chlorine can adjust easily to changes in flow rate or water quality, and the residuals can be used to monitor and control the process. However, few POTWs use free chlorine, so the time for regulatory approval is expected to be longer. UV and ozone are less adaptable to changes in flow rate and water quality, and therefore require redundant capacity and/or backup disinfection systems. With short-lived or no disinfectant residuals, the process monitoring and control systems for free chlorine, UV and ozone are more complicated to design. Finally, the technologies for ozone and UV are still evolving, so bench/pilot testing is recommended and the time for regulatory approval is longer.

From a practical standpoint, the requirements that a system places on a POTW must also be considered. For example, chloramination systems have relatively large footprints while UV systems are relatively small; as a result, UV systems are generally easier to expand. Both UV and ozone exert significant energy demands and may require facilities to alter or add to their power infrastructure. Finally, some disinfectants may provide additional benefits to a treatment plant, because they can be used at other locations within the facility. For example, both ozone and chlorine disinfectants can be used for odor control, and chlorine compounds can also be used for sludge bulking and filter cleaning.

8.3.6 Operation and Maintenance Issues

Table 8-7 compares the mature disinfection alternatives with regard to operation and maintenance. In order to maintain safe and effective disinfection, all of the mature technologies require well-trained, experienced operators. This is particularly true for UV and ozone systems, which have more complex systems and process controls; these systems tend to require more maintenance time, due to more frequent servicing of monitoring devices and replacement of parts. Spare parts for technologies like UV and ozone (that use non-standardized equipment) may not be available locally and should be kept in stock. UV also requires extra maintenance time for cleaning of the tanks and lamp sleeves; cleaning of the quartz sleeve can be very laborious for facilities with poor influent water quality and/or without automatic cleaning systems.

Table 8-7. Comparison of Mature Wastewater Disinfection Alternatives – Operation and Maintenance.

Factors	Gas Chlorine	Hypochorite Solution	On-site Generation	UV	Ozone
Operation					
Need for skilled labor	Medium	Medium	Medium	Medium	High
Need for operator training	High	High	High	High	High
Suitability for automated process control	Good	Good	Goo d	Good	Good
Complexity of process control	Medium	Medium	High	High	High
Risk to operator safety	High	High	High	Medium	High
Potential degradation of feed chemical(s)	Yes	Yes	No	No	No
Concerns with supply of feed chemical(s)	Yes	Yes	No	No	No
Concerns with interruption of power supply	Low	Low	High	High	High
Potential for scaling and corrosion of equipment	Yes	Yes	Yes	Yes	Yes
Maintenance					
Service frequency of monitoring devices	Low	Low	Medium	High	Medium
Cleaning frequency of the equipment/tank	Low	Low	High	High	Low
Replacement frequency of parts	Low	Low	Medium	High	Medium
Availability of local parts	Good	Good	Poor	Poor	Poor
Time required for maintenance	Low	Low	Medium	High	High
Disposal of waste	No	No	Yes	Yes	No

8.3.7 Costs

Table 8-8 compares the mature disinfection alternatives with regard to capital and operation and maintenance (O&M) costs. The capital cost may include site preparation, modification of existing process, equipment, process control instrument, ancillary facilities, design, installation, and permitting. Ozone is generally considered to have relatively high capital costs, although the actual capital cost for a given application is very site-specific and depends on many factors such as flow rate, water quality, and effluent discharge limit. For example, dechlorination, if needed, considerably increases the capital cost of chlorination. For O&M costs, labor costs for ozonation and UV disinfection may be higher than for chlorination. Chemicals dominate the O&M costs for chlorination, while energy and lamps/parts replacement dominate O&M costs for UV. The percentage of O&M costs for energy for ozone and on-site generation are also quite high.

Table 8-8. Comparison of Mature Wastewater Disinfection Alternatives – Costs.

Factors	Gas Chlorine + Dechlorination	Hypochorite Solution + Dechlorination	On-site Generation + Dechlorination	UV	Ozone
Capital cost without chlorine contact basin	Medium	Low	Medium	Medium	High
O&M cost					
Energy	Low	Low	High	High	High
Chemicals	High	High	Low	Low	Low
Parts replacement	Low	Low	Medium	High	Medium
Labor	Low	Low	High	High	High
Waste disposal	Low	Low	Low	Medium	Low

8.4 Selecting and Upgrading Disinfection Alternatives

This section presents a generalized framework to help POTWs identify disinfection alternatives to consider when selecting a new or upgrading technology at their facilities.

8.4.1 Selecting Disinfection Alternatives

Based on common driving forces and decision factors, Tables 8-9, 8-10, and 8-11 present some broad guidelines for the use of the mature disinfection technologies. These tables indicate general conditions when each disinfection alternative is recommended, when it should be considered, and when it should be avoided. These tables do not differentiate how the disinfectant is delivered (i.e., different forms of chlorine applied for disinfection, different types of UV lamps, or whether ozone is generated using pure oxygen or air).

In addition, these tables do not account for site-specific factors that may affect the decision process. As discussed in Section 8.3, multiple criteria must be considered when choosing the disinfection method, and these may conflict with one another, i.e., tradeoffs must be made between the advantages and disadvantages of each alternative. For example, ozone is more complicated and costly to construct and operate than the other technologies; consequently, utilities are unlikely to select it unless incidental benefits such as contaminant removal or increased dissolved oxygen levels are a high priority. Determination of the most appropriate alternative for each plant is site-specific and depends on the relative weight given to each decision factor for that facility. Descriptions of the decision processes for two facilities are given in Section 8.5.

Table 8-9. Disinfection with Chlorination/Chloramination.

Chlorination/chloramination recommended if:	Consider chlorination/ chloramination if:	Chlorination/chloramination not favorable if:
▪ Upstream processes include only primary or secondary treatment ▪ Suspended solids are relatively high ▪ Water quality and flow rate vary widely ▪ A disinfection residual is needed in the effluent	▪ Disinfection byproducts such as bromate are a concern ▪ Disinfectant residual is needed (e.g., for certain reuse applications)	▪ Disinfection byproducts such as THMs, HAAs, cyanide or NDMA are a concern ▪ There are strict effluent requirements on TDS, ammonia, chlorine residuals ▪ Regulations discourage the use of chlorine-base disinfectants ▪ Chemical handling, safe and reliable transport, or accidental releases are a major concern ▪ Space is limited ▪ Disinfection requirements include high levels of inactivation of *Cryptosporidium, Giardia,* or other protozoa

Table 8-10. Disinfection with UV.

UV recommended if:	Consider UV if:	UV not favorable if:
▪ Chemical addition must be eliminated (note that a requirement of zero-residual in the effluent still allows other options) ▪ Disinfection byproducts such as THMs, HAAs, NDMA Cyanide, or bromate are a concern	▪ Water to be disinfected is treated by tertiary treatment processes ▪ UV transmittance is high ▪ Chemical use (e.g., for safety, cost, or regulatory reasons) is to be minimized ▪ Space is limited ▪ Disinfectant residual is undesirable ▪ Regrowth is not a major issue	▪ Water to be disinfected is not treated by tertiary treatment processes, when pathogen discharge limit is low ▪ Water contains compounds that are likely to cause fouling of the quartz sleeve ▪ UV transmittance is low ▪ Maintenance requirements must be minimal (e.g., at remote locations) and there are complicating factors such as power fluctuations, rapid algae growth, etc. ▪ There are extreme changes in water flow rate and quality ▪ There are frequent power outages ▪ Adenovirus is a concern ▪ Disinfectant residual is needed (e.g., for certain reuse applications)

Table 8-11. Disinfection with Ozone.

Ozone recommended if:	Consider ozone if:	Ozone not favorable if:
▪ Incidental removal of chemicals in the effluent, such as odor-causing compounds or compounds of emerging concern (COEC) is important ▪ It is desirable to increase dissolved oxygen concentration in the discharge	▪ Water to be disinfected is treated by tertiary treatment processes ▪ Disinfectant residual is not required ▪ Disinfection byproducts such as THMs, cyanide, HAAs or NDMA are a major concern ▪ Disinfection byproducts such as bromate and aldehydes are not a major concern ▪ Facility uses pure oxygen activated sludge	▪ Water is not treated by tertiary treatment processes ▪ Maintenance, training, or operational requirements must be minimal ▪ Capital costs are a major issue ▪ There are extreme changes in water flow rate and quality, though ozone can handle some variation, within design specifications ▪ Bromate is a concern ▪ Disinfection requirements include high levels of inactivation of *Cryptosporidium, Giardia,* or other protozoa ▪ Disinfectant residual must be present (e.g., for certain reuse applications) ▪ There are frequent power outages ▪ Electrical energy costs are high

8.4.2 Upgrading Disinfection Technologies

The majority of POTWs in the U.S. currently use one of the following disinfection alternatives for disinfection: chlorination (chlorine gas, hypochlorite, or onsite generation of hypochlorite), UV, or ozone. Each of these alternatives may be upgraded because of site-specific issues and concerns. Table 8-12 lists the alternatives available for possible upgrade from the existing disinfection methods and presents some conditions for consideration during the selection process. The most likely upgrade scenarios are bolded. These plausible scenarios include changing from chlorine gas to hypochlorite if byproducts formed from chlorination are not a major concern; and changing to UV disinfection if byproducts are a major concern. This statement is probably overly simplistic; a more comprehensive evaluation by the consulting engineer of the factors discussed in Section 8.3 is recommended.

Table 8-12. Possible Upgrade of Wastewater Disinfection Technologies.

Existing Technology \ Upgraded Tech.	Hypochlorite	Onsite Generation of Chlorine	UV	Ozone	Combined Technologies
Chlorine gas	Common if DBPs are not a major concern	Possible	Likely if DBPs are a major concern	Possible if COEC removal is important	Not likely
Hypochlorite	--	Possible	Likely if DBPs are a major concern	Possible if COEC removal is important	Possible with UV
Onsite Generation of Chlorine	Not likely	--	Likely if DBPs are a major concern	Possible if COEC removal is important	Possible with UV
UV	Possible if water quality (e.g., UVT) is poor for UV disinfection, and DBPs are not a concern	Possible if water quality (e.g., UVT) is poor for UV disinfection, and DBPs are not a concern	Possible to improved UV lamp technologies	Possible if COEC removal is important	Possible with hypochlorite or ozone
Ozone	Possible	Possible	Possible if disinfection efficacy, system complexity, and bromate DBPs are a concern	Possible to improved ozone generation technologies	Possible with UV

At present, use of more than one disinfection technology at a POTW is not common, unless multiple objectives (e.g., disinfection efficacy and algae control) need to be accomplished. However, as discussed in Section 6.5, this practice has many potential merits, including more effective disinfection for a wide range of pathogens/indicator microorganisms, and possible synergistic effects. Because most existing facilities are already equipped with one disinfection system, upgrading to sequential or combined disinfection alternatives may simply mean adding a second disinfection system to the existing one. For example, a facility that practices chlorine disinfection may add UV disinfection due to concerns over byproduct formation. The facility may choose to keep the chlorination system in place but lower the dosage to reduce byproduct formation, inactivate adenovirus (which UV does not treat effectively), control biological growth on filter media, provide residual to the effluent, or reduce redundancy requirements for the UV system during wet weather or high flow events. Similarly, an existing ozonation system can benefit a new UV system by increasing UV transmittance, which allows more efficient design and operation of the UV system. It is anticipated that more and more POTWs will consider the use of sequential or combined disinfection to increase reliability of the system and to improve disinfection performance when they consider upgrading existing disinfection practices. However, there are tradeoffs between the potential benefits listed above and the disadvantages of having greater operating and maintenance requirements for two disinfection systems instead of one.

More information is needed to establish the relative benefits and shortcomings of the use of more than one technology for wastewater disinfection.

8.4.3 Summary of Advantages and Disadvantages of Mature Technologies

Below are tables that summarize the advantages and disadvantages of the three mature technologies reviewed by this report.

Table 8-13. Chlorination and Chloramination: Advantages and Disadvantages.

Advantages	Disadvantages
▪ Well-established technology with well-known inactivation mechanisms ▪ Relatively easy to apply and often less expensive than other alternatives, regardless of system size ▪ Chlorine chemicals are relatively inexpensive and available in gas, liquid, and solid form ▪ Useful in other areas of the plant, e.g., to reduce biofouling in tertiary filters, to control odors, for sludge bulking ▪ Provides a residual, which is useful for operational monitoring and control, as well as preventing growth in distribution systems during reuse. ▪ Easily adjusted for fluctuations in flow rate and water quality. ▪ Free chlorine is an effective biocide and virucide, may remove some emerging compounds of concern, and forms little or no NDMA. ▪ Monochloramines are more stable and longer lasting than free chlorine, and have much less potential to form THMs, HAAs, cyanide, and cyanogen chloride than free chlorine.	▪ Not very effective for protozoa ▪ Less effective at high pH ▪ Produces disinfection byproducts that are harmful to human health or the environment ▪ Increases total dissolved solids and chloride levels in the effluent ▪ May need to dechlorinate effluent before final discharge ▪ Chlorine gas is hazardous and corrosive, and requires special leak containment and scrubber facilities. Although hypochlorite avoids some of these disadvantages, higher concentrations of sodium hypochlorite are unstable and will produce chlorate as a byproduct ▪ Free chlorine can form halogenated disinfection products (e.g., THMs and HAAs), as well as oxygenated byproducts that can enhance biological regrowth in water reuse distribution lines if a chlorine residual is not maintained. Breakpoint chlorination is only viable at low ammonia concentrations ▪ Chloramines are weaker than other disinfectants, require a long contact time, and can generate NDMA, cyanide and cyanogen chloride. Facilities with nitrification or nitrification/denitrification may need to add ammonia, and the added ammonia in the final effluent may enhance biofilm growth in the transmission lines for water reuse

Table 8-14. UV: Advantages and Disadvantages.

Advantages	Disadvantages
▪ Effective inactivation of most viruses, spores, and cysts ▪ Physical process rather than a chemical disinfectant, which eliminates the need to generate, handle, transport, or store toxic/hazardous or corrosive chemicals ▪ No residual effect that can be harmful to human or aquatic life ▪ User-friendly and safer for operators ▪ Shorter contact time when compared to other disinfectants ▪ Smaller footprint requirement than other methods ▪ Little formation of disinfection byproducts ▪ No effects on TDS levels ▪ Incidental removal of emerging contaminants such as NDMA	▪ Typical disinfection dosages may not effectively inactivate some viruses (e.g., adenovirus), spores, and cysts ▪ Repair and regrowth possible under dark or light conditions ▪ Ineffective for wastewater with low UV transmittance (UVT), e.g., due to turbidity, suspended solids, or some industrial waste discharges ▪ On-line monitoring of delivered dose is challenging ▪ Matching UV dose and energy usage with flow is complex ▪ Restarting system to full energy and effectiveness require multiple channels, bypass, and/or storage ▪ O&M costs may be high due to energy needs, frequent lamp replacement, cleaning to control sleeve fouling, etc ▪ No disinfection residual remains in the effluent (biofouling, odors, slimes may form in the distribution pipeline and appurtenances for water reuse) ▪ Mercury waste at end of lamp life ▪ Potential mercury releases from catastrophic lamp failures ▪ May need alternative system to treat very large peak wet weather flows

Table 8-15. Ozone: Advantages and Disadvantages.

Advantages	Disadvantages
▪ Effective disinfectant for inactivating bacteria, viruses, parasites, and parasite cysts (better than chlorine) ▪ Short contact time (approximately 10 to 30 minutes) ▪ Rapid decomposition, so no harmful residuals that must be removed after ozonation ▪ No regrowth of microorganisms except for those shielded by particulates ▪ Few safety problems associated with shipping and handling because ozone is generated on-site ▪ Increases the dissolved oxygen concentration of the effluent ▪ Usually reduces effluent COD concentrations ▪ Helps with sulfide (odor), iron, and manganese removals ▪ Can oxidize some COECs, such as pharmaceuticals and endocrine disruptors ▪ No THMs or HAAs ▪ Smaller footprint than chlorine	▪ Requires complicated equipment and efficient contact systems to provide high transfer efficiency ▪ May require LOX transport to site and on-site storage ▪ Relatively complex to operate and maintain; more suitable for medium- to large-sized plants ▪ Ozone is very reactive and corrosive to metals, plastics, and elastomers, thus requiring corrosive resistant material such as stainless steel. ▪ Off-gases from ozone contactor must be destroyed to prevent worker exposure because ozone is extremely irritating and possibly toxic ▪ May not effectively inactivate some viruses, spores, and cysts at doses used for coliform reduction ▪ No long-term residuals, so chlorine may still be needed if disinfection residual is required ▪ Generates some disinfection byproducts, some of which (e.g., bromate and aldehydes) are potentially harmful to human health ▪ Suspended solids and organic matter reduce efficacy, so only secondary or tertiary effluents are feasible ▪ High capital and O&M costs (energy intensive)

8.4.4 Summary of Advantages and Disadvantages of Other Technologies

The following tables review the advantages and disadvantages of the other technologies for wastewater disinfection. In general, these technologies share the following disadvantages, which have not been included in each table to avoid redundancy:

- No full-scale capital and O&M cost data for the range of sizes (1-100 MGD) of major U.S. POTWs;
- Poor understanding of the effects of water quality on disinfection efficacy;
- Insufficient knowledge on disinfection byproducts;
- Little information on the potential for regrowth of organisms after disinfection;
- No long-term operational experience at full-scale;
- No history of reliability or compliance;
- Lack of regulatory requirements and guidelines.

Table 8-16. Chlorine Dioxide: Advantages and Disadvantages.

Advantages	Disadvantages
▪ Equally or more effective than chlorine and chloramines for the inactivation of bacteria, viruses, and cysts ▪ Biocidal properties are less sensitive than chlorine to pH ▪ Stronger and more selective oxidant than chlorine and chloramines (e.g., does not react with ammonia, unlike chlorine) ▪ Can also be used to oxidize iron, manganese, sulfides, and phenolic compounds ▪ Does not form THMs under proper generation conditions (i.e., no excess chlorine) ▪ Provides residuals, which are needed in reuse applications and are useful for process control ▪ Relatively low capital costs ▪ Large water treatment and pulp and paper experience	▪ Unstable and hazardous: must be generated on-site, and requires precautions for safe handling and storage ▪ Hazardous chemical reagents used to form chlorine dioxide require safety precautions, storage space, and special feed equipment ▪ Generation, analytical requirements, and process control are complex ▪ Forms byproducts chlorite and chlorate; regulations may limit the chlorine dioxide dose that can be applied ▪ May also form THMs if excess chlorine is added during chlorine dioxide generation ▪ Increases chloride and TDS content of the water ▪ Relatively high O&M costs, due to costs of 1) training, sampling, and laboratory testing for chlorite and chlorate; 2) the sodium chlorite used to form chlorine dioxide ▪ Chlorine dioxide may decompose in sunlight ▪ Some systems may produce nuisance odors

Table 8-17. Peracetic Acid: Advantages and Disadvantages.

Advantages	Disadvantages
▪ Similar disinfection efficacy to chlorine at equal concentrations ▪ Adapts well to changes in flow rates ▪ Conversion relatively easy for treatment plants that currently use chlorine ▪ Eliminates most, if not all, disinfection byproducts ▪ Does not appear to create toxicity in effluents ▪ Provides some residual, which is useful for water reuse distribution systems (controls algae, biofilms, slime) and may reduce odors ▪ Capital costs for construction are on par with chlorine and UV disinfection, and less expensive than ozone disinfection. ▪ EPA has approved chemical for wastewater disinfection ▪ Large European POTW experience (114 MGD)	▪ Requires precautions to ensure safe storage and use and may trigger hazardous materials permit requirements. ▪ Relatively weak against viruses, and has little effect on spores and protozoa ▪ Probably impractical for low effluent coliform limits, because extremely high doses of PAA would be required ▪ Chemical is expensive, which leads to high O&M costs; however, if use increases, manufacturing capacity may increase, and price may drop. ▪ Adds organic carbon (BOD) to the effluent, and at high concentrations, may form aldehydes, chlorine and bromine radicals, and hypobromous acid

Table 8-18. Ferrate: Advantages and Disadvantages.

Advantages	Disadvantages
▪ Strong oxidizing agent ▪ Promising for disinfection, removal of organics and metals, flocculation, and odor control ▪ Less sensitive than chlorine to changes in organic content, pH, and temperature of water	▪ Unstable and decomposes relatively quickly ▪ May increase TDS in water, depending on the ferrate synthesis method ▪ Historically expensive, although new technologies may reduce prices

Table 8-19. BCDMH: Advantages and Disadvantages.

Advantages	Disadvantages
▪ Appears to be a more effective disinfectant than hypochlorite, peracetic acid, or bromine chloride ▪ Rapid disinfection reaction ▪ Very good for use with variable flow rates (e.g., combined sewer overflows) ▪ Smaller footprint, better safety, and less storage space than hypochlorite ▪ Low levels of THMs (e.g., bromoform) at tested doses; no bromate detected thus far ▪ Uses infrastructure similar to hypochlorite ▪ Expected chemical and energy costs lower than hypochlorite	▪ May generate HAAs exceeding regulatory limits (MCLs) at high dosages ▪ Expected to require more equipment maintenance than hypochlorite ▪ Higher capital costs than chlorine

Table 8-20. Pasteurization: Advantages and Disadvantages.

Advantages	Disadvantages
▪ No chemicals and accosiated safety risks: no disinfectant removal needed, fewer impacts on water quality (e.g., from disinfection byproducts), lower operating costs due to absence of safety risks posed by chemicals. ▪ Less dependence on the external power distribution system, because gas turbines that generate the heat for pasteurization provide electricity for the POTW ▪ Provides a productive use for the methane gas produced in anaerobic digestion and/or the waste heat from gas turbines	▪ Less cost effective for facilities without an existing heat source or turbine ▪ Increase in effluent water temperature (as low as 4°F/2°C), with the associated decrease in the concentrations of dissolved oxygen ▪ Cold water habitat receiving water limits difficult to meet in summer ▪ May scale heat transfer equipment and require antiscalents and/or high maintenance cost for cleaning

Table 8-21. Ultrasonic Cavitation: Advantages and Disadvantages.

Advantages	Disadvantages
▪ No chemicals: no disinfectant removal needed, fewer impacts on water quality (e.g., from disinfection byproducts), and lower safety risks posed by chemicals. ▪ Small footprint ▪ Works better in waters with higher solids contents ▪ Can remove a variety of organic compounds ▪ Breaks up particles/clumps, and may therefore work well in combination with other technologies like UV or chlorine	▪ Energy-intensive for good disinfection ▪ Expensive ▪ Difficult to control

Table 8-22. Electron-beam and Gamma Irradiation: Advantages and Disadvantages.

Advantages	Disadvantages
▪ No added chemicals: no disinfectant removal needed, fewer impacts on water quality (e.g., from disinfection byproducts), and lower safety risks posed by chemicals. ▪ Produces both oxidizing (hydroxyl) and reducing (electron) species, thereby promoting a wider range of reactions than other alternatives ▪ Able to treat waters of varying quality, and sludges ▪ Can disinfect reasonably well ▪ Can degrade organic contaminants	▪ May be impractical for high flow rates, if limited to treating thin sheets of flowing water. ▪ Radioactive licensing and safety concerns for gamma irradiation ▪ Danger of irradiation to operators

Table 8-23. Photocatalysis: Advantages and Disadvantages.

Advantages	Disadvantages
▪ No added chemicals: no disinfectant removal needed, fewer impacts on water quality (e.g., from disinfection byproducts), and lower safety risks posed by chemicals. ▪ Not sensitive to pH changes between 5 and 8 ▪ Can degrade organic contaminants	▪ Lack of basic long term information (deactivation of the catalyst, practical applicability at high flow) ▪ Probably ineffective in low transmittance waters ▪ Most currently available catalysts are too slow to be applicable, though technology is improving.

Table 8-24. Combined Disinfection Alternatives: General Advantages and Disadvantages.

Advantages	Disadvantages
▪ Provides more operational flexibility (can choose to operate one or multiple technologies, as appropriate) ▪ May inactivate a larger spectrum of organisms ▪ May create synergistic effects, e.g., the combination of disinfectants may be more effective than the sum of each alone ▪ May reduce generation of disinfection byproducts ▪ May provide disinfectant residual for water reuse applications ▪ May remove other contaminants, such as trace organics ▪ May be more cost-effective overall (capital and O&M)	▪ More expensive to construct ▪ More complex to operate

8.5 Case Studies

This section presents two examples from the literature, in which extensive efforts were made by wastewater treatment plants to assess alternative technologies for effluent disinfection. These illustrate the processes used in selecting alternatives, and the criteria that were considered in reaching a decision for these specific facilities. In both cases, the utilities and their consultants first identified possible alternatives, and then narrowed the possibilities to the most likely candidates (e.g., by eliminating technologies that are not well-established). They then determined the criteria that were most important for the facility or facilities, and assigned weighting factors to those criteria. Each disinfection alternative was given a score for each criterion; this score was multiplied by the weighting factor to obtain a weighted score. In the first case, from Chicago, IL, the alternatives with the highest scores were then further investigated to estimate costs. In the second case, from Roseville, CA, the estimated costs were part of the weighted criteria, and the alternative with the highest score was recommended for pilot testing.

8.5.1 Chicago, Illinois

In 2004, the Metropolitan Water Reclamation District of Greater Chicago (MWRDGC) and its consultants began a study to evaluate potential disinfection options for three of the water reclamation plants that the MWRDGC operates. These plants have average design flow ranging from 333 to 1,200 million gallons per day (MGD). The objectives of the study were to: provide a final recommendation for one or more disinfection technologies that best fit the MWRDGC's three water reclamation plants; and prepare capital and O&M cost estimates. Details on the study are described in Zenz et al. (2007).

The MWRDGC began by assembling a task force of recognized experts in the field of wastewater disinfection. The task force began by identifying a wide range of disinfection alternatives, including both mature and cutting-edge technologies:

- Chlorination only: calcium hypochlorite, sodium hypochlorite solution, on-site generation of sodium hypochlorite, or chlorine gas (four options)
- Ozone: generated from air or oxygen (two options)
- UV radiation: low or high intensity lamps (two options)
- Chlorination-dechlorination: using one of the four chlorine source listed above and either sodium bisulfite or sulfur dioxide for dechlorination (eight combinations)
- Chlorine dioxide
- Bromine compounds
- Sequential disinfection processes
- Membrane processes

This initial list was evaluated based upon the experience of the task force, manuals of practice, and scientific literature and textbooks regarding pathogen destruction efficiency, disinfection byproducts and impacts on aquatic life and human health, full-scale operating experiences from existing systems, and site-specific considerations.

The task force then narrowed the field. Gas chlorination and gas dechlorination were eliminated due to MWRDGC's concerns with potential hazards to humans and the environment caused by accidents or terrorism either at the plant or during transportation. The task force expected that chlorine alone would be unable to meet future water quality standards for chlorine residuals, and eliminated that alternative. Chlorine dioxide, bromine compounds, sequential disinfection processes, and membrane processes were also eliminated, due to the lack of experience at large-scale wastewater treatments. The remaining technologies were the following:

- Ozone: generated from air or oxygen (two options)
- UV radiation: low or high intensity lamps (two options)
- Calcium hypochlorite with sodium bisulfite
- Sodium hypochlorite solution with sodium bisulfite
- On-site generation of sodium hypochlorite with sodium bisulfite

These technologies were further evaluated using a matrix of evaluation criteria that were chosen by consensus between the MWRDGC and the task force. The criteria did not include estimated capital or operations/maintenance costs, because the task force and the MWRDGC decided that the alternatives should be assessed without regard to the numerical costs. The experts in the task force assigned a weighting factor to each criterion. Table 8-25 lists the criteria and associated weighting factors.

Table 8-25. Evaluation Criteria, Weighting Factors, and Alternative Technology Scores for MWRDGC.

Criteria	Weighting Factors	Ozone Air	Ozone Oxygen	UV LPLO	UV MP	$Ca(OCl)_2$ + $NaHSO_3$	NaOCl+ $NaHSO_3$	Onsite NaOCl + $NaHSO_3$
Desirable performance requirements								
Safety								
Low risk from accidental releases of chemicals/waste products	3	2	2	2	3	2	1	2
Low hazard potential to neighbors	3	2	2	3	3	3	2	2
Low level of onsite chemical storage	3	3	3	3	3	1	1	2
Low potential for malicious adverse occurrences	3	3	2	3	2	2	2	3
Low hazard potential for transportation spills and releases	3	3	3	2	2	1	1	2
Operational flexibility								
Can be readily adjusted based on variation in flow	5	2	2	1	1	3	3	2
Can be readily adjustment based on variation in the water quality	5	2	2	1	1	3	3	2
Demand can be easily measured	5	2	2	3	3	3	3	3
Operational reliability								
Predictability of performance	8	2	2	2	2	2	2	2
Reliability of operations (low down time)	7	2	2	3	2	2	3	2
Disinfection byproducts								
Low potential for byproduct formation	4	2	2	3	3	1	1	1
Low ecotoxicity of byproducts	4	2	2	3	3	1	1	1
Low human toxicity	4	2	2	3	3	1	1	1
Modification for future concerns								
Adjustment to achieve high level of pathogen reduction	3	3	3	2	3	1	1	1
Compatible with possible future needs for byproduct minimization	3	2	2	3	3	1	1	1
Qualitative economic requirements								
Multiple vendors available for providing equipment	4	2	2	3	2	3	3	2
Training levels for operators	5	2	2	2	2	3	3	3
Sensitivity to cost of electricity	5	2	2	1	1	3	3	2
Indirect environmental and health impacts								
Energy required to operate the facility	4	2	2	1	1	3	3	2
Energy required for processing and production of process chemicals	4	3	3	1	2	1	1	2
Conversion and degradation of process chemicals	4	2	2	3	2	1	1	1
Public Perception	11	2	2	2	3	2	2	1
Weighting Factor Total	100							
Weighting Factor Scores		216	213	221	224	204	205	180

Scoring system: 1 = Poor; 2 = Fair; 3 = Good

For each of the evaluation criteria, each alternative was scored from a scale of 1 (poor) to 3 (good). Table 8-25 presents the scores. These scores were multiplied by the weighting factors assigned to each criterion, to arrive at a total score for that criterion and alternative. The task force and the MWRDGC used these results to determine which technologies were worth pursuing further. Although the scores for the two UV alternatives were similar, only the medium pressure UV systems were considered for detailed cost estimation, due to concerns about handling the large numbers of low intensity UV lamps that would be required by the very large MWRDGC facilities. Similarly, the scores for air-generated ozone were slightly higher than for oxygen-generated ozone, but only the oxygen-based system was considered for detailed cost estimation, because the capital and O&M costs for air-based systems are approximately three to four times higher than for oxygen-based systems. Additionally, there were no known full-scale air-based ozone systems operating at facilities similar in size to those owned by the MWRDGC.

The task force and the MWRDGC therefore decided to perform detailed cost estimates on UV medium pressure and oxygen-generated ozone systems. The results of this study (cost estimates for effluent disinfection and potential public health impacts) will be used by state and federal environmental agencies to help determine whether to require effluent disinfection at the MWRDGC plants in the future.

8.5.2 Roseville, California

The City of Roseville, CA (Roseville) operates a tertiary water reclamation plant with an average dry weather capacity of 18 MGD. Due mainly to safety concerns associated with the use, storage, and transportation of chlorine gas and other concerns related to disinfection byproducts formation and toxicity on aquatic life, Roseville retained a consultant to evaluate its disinfection alternatives (Gillette et al., 2003). The disinfection alternatives evaluated included the following:

- Continued use of chlorine gas with dechlorination;
- Bulk sodium hypochlorite solution with dechlorination;
- Onsite generation of sodium hypochlorite (0.8% purity);
- Peracetic acid (PAA);
- UV disinfection, algae control with launder covers, and onsite generation of sodium hypochlorite to provide residual for recycled water distribution; and
- UV disinfection and onsite generation of sodium hypochlorite to provide residual for recycled water distribution and for algae control.
- Ozone, bromine, hydrogen peroxide, and radiation were also considered but quickly eliminated.

Roseville and its consultants identified important criteria and the associated weighting factors (Table 8-26). The weighting factors were derived though pairwise comparisons of the criteria, by using a multiple-criteria decision-making process called the analytic hierarchy process. These weighting factors were then normalized, so that the sum of the factors was one. Within each criterion, a similar analysis was conducted for each disinfection alternative to provide scores for each alternative for that criterion. Just as with the MWRDGC analysis, these scores were multiplied by the relative weighting factor to give weighted alternative technology scores, as shown in Table 8-26. The option with the highest overall score (0.238) was UV for disinfection, launder and filter covers to control algae, and onsite generated sodium hypochlorite to provide a residual in the reclaimed water. As a result of the analysis, the consultant recommended pilot-scale testing of UV disinfection to help determine design and operational parameters that could be used for full-scale system design.

Table 8-26. Evaluation Criteria, Weighting Factors, and Weighted Alternative Technology Scores for Roseville, CA. (Gillette et al., 2003 with permission from WEF)

Criteria	Normalized Weighting Factor	Cl_2 gas[1]	Cl_2 gas[2]	Bulk NaOCl	Onsite NaOCl	PAA	UV + NaOCl with covers	UV + NaOCl without covers
Employee safety	0.299	0.012	0.005	0.030	0.044	0.044	0.091	0.073
Offsite safety	0.299	0.012	0.005	0.028	0.062	0.062	0.065	0.065
Impacts on receiving water quality	0.142	0.013	0.013	0.007	0.007	0.032	0.041	0.030
System reliability	0.104	0.023	0.023	0.024	0.015	0.003	0.009	0.008
Present and proposed regulations	0.063	0.001	0.001	0.01	0.01	0.014	0.013	0.013
Operation and maintenance	0.053	0.001	0.001	0.011	0.008	0.011	0.012	0.009
Present worth cost (capital + O&M)	0.040	0.007	0.007	0.005	0.006	0.004	0.006	0.006
Weighting Factor Total	1.000							
Weighting Factor Scores		0.069	0.056	0.114	0.151	0.169	0.238	0.204

[1]Assumes a minor leak response
[2]Assumes a moderate leak response

8.5.3 Chlorine Gas Decision Tool

In 2006, the U.S. Department of Homeland Security released a CD entitled "Chlorine Gas Decision Tool for Water and Wastewater Utilities." This program was developed by the National Association of Clean Water Agencies (NACWA) and is available free of charge to water and wastewater treatment plants in the US. The CD can be ordered from the NACWA web address, http://www.nacwa.org/clcd.cfm. The goal of the program was to encourage facilities to consider alternatives to gas chlorine disinfection. The CD contains the decision tool, a report template, and a sample report. It includes four disinfection options: chlorine gas with a scrubber, purchased sodium hypochlorite solution, on-site hypochlorite generation, and UV.

The selection process is similar to those described in Section 8.5.1 and 8.5.2. The decision criteria include cost and 13 non-monetary factors: safety (for operators, for neighbors, permitting requirements, and containment), public acceptance, design/installation requirements (space, traffic, security, and constructability), O&M issues (process reliability, operational complexity, and maintenance requirements), and effluent water quality requirements. The facility using the decision tool must first assign weighting factors to each of these 13 criteria; the weights range from one to ten, and are based on the importance of each criterion in the selection process. The facility then evaluates how well each alternative performs on each of the criteria, again on a scale from one to ten. Using the entered numbers, the program calculates a relative benefit value. The program also estimates the capital cost, operating cost, and net present value of each alternative, based on inputs provided by the utility on plant flow, chlorination and dechlorination parameters, the need to install emergency scrubbers, chemical costs, UV-specific costs, and general costs. From this information, the program can calculate the ratio of the costs to benefits.

The major value of this program is that it provides a standardized template for selecting a disinfection technology. However, it is limited with respect to both the alternatives offered and the decision criteria. As such, it is likely to be helpful to facilities thinking about a change in disinfection technology, but should not be relied upon as a sole resource unless the CD covers all of the alternatives the facility is considering and all of the relevant decision criteria.

CHAPTER 9.0

CONCLUSIONS AND RECOMMENDATIONS

This report examined a wide range of disinfection technologies, their prevalence of use, and information on their implementation at major POTWs (defined as wastewater treatment plants that are state or municipally owned, with an average dry weather design flow of greater than 0.95 MGD). This chapter discusses future research that may help improve, implement, or evaluate the various disinfection alternatives. The mature technologies are discussed individually in Sections 9.1 (Chlorine), 9.2 (UV), and 9.3 (Ozone). The data gaps for the developing technologies are generally quite similar, and are grouped together into Section 9.4. Finally, Section 9.5 discusses the combined disinfection technologies.

Over the next three to five years, most major POTWs that are considering a change in disinfection technology are likely to select one of the mature disinfection technologies, primarily because they are more established and more information is known about them. This means that in the short term, utilities will probably be most interested in resolving the issues listed in Sections 9.1 to 9.3. However, each of these mature technologies has inherent disadvantages with respect to either disinfection of specific organisms, or formation of disinfection byproducts. As a result, a growing number of facilities may consider combining their current disinfection process with an additional technology. These facilities will need information about both the individual alternatives and the issues given in Section 9.5. Over the longer term, research on the developing technologies may prove useful, as wastewater effluents increasingly impact drinking water sources, and the limitations of the current alternatives become apparent. Before these technologies can be widely applied, more information is needed on the issues in Section 9.4.

One common issue for all technologies is the lack of knowledge about the efficacy of these processes against a wide range of pathogens in the wastewater and reclaimed water environment. There are some data for established disinfectants such as chlorine, but even with chlorine, much of the work has traditionally focused on indicator organisms, such as total or fecal coliform. Recent research has indicated that these indicators may not always be appropriate proxies for the pathogens of concern; this may be increasingly true as new technologies with different disinfection mechanisms are developed. More work is needed to establish the efficacy of both established and new disinfection technologies against the pathogens of concern, and to ensure that the surrogates used for compliance accurately reflect the safety and quality of the resulting effluent.

9.1 Chlorine

Chlorine (in the form of chloramines) has been used extensively for wastewater disinfection over the last century. Consequently, there is a wealth of information on the design requirements and O&M experience. However, some issues such as disinfection byproducts have not been fully resolved. In addition, the technology continues to evolve, with advances in on-site chlorine generation and the possible application of free chlorine, now that many plants have incorporated biological nitrogen removal. As a result, questions are continually raised and more research is needed on topics such as the following:

- Chemical reaction intermediate and end products (especially those due to new or newly discovered compounds); as mentioned in Section 3.5.1.1, most disinfection byproducts have not yet been identified
- The reaction kinetics of chlorine and chloramines with emerging chemicals of concern (e.g., naproxen) over a range of water and wastewater conditions
- Performance and economics of on-site generation of hypochlorite
- Re-evaluation of the regulatory CT requirements, to account for the chlorine species used for disinfection and the quality of the effluent
- For free chlorine: methods for residual measurement in wastewater, assessment of the required operating parameters for disinfection (largely unknown for wastewater) over a range of water qualities and operating conditions, and the resulting toxicity
- Dechlorination kinetics, to ensure proper contact time
- Real time dechlorination measurements for assurance of non-toxic discharges

9.2 UV

Although UV use is fairly common, it is still a relatively new technology. As such, there are design, testing and operational guidelines such as those from NWRI, but optimal operation remains as much an art as a science. A number of issues deserve further investigation, to improve the consistency of system performance and control of UV systems:

- Identification and modeling of all of the important parameters that affect disinfection efficacy, such as water quality, flow conditions, and biological characteristics (initial concentration of organisms, disinfection kinetics of organisms, etc.)
- A standardized method to measure, calibrate, and certify how well equipment works before and after installation
- Improved understanding of hydraulic limitations and scale up of closed channel systems
- An accurate method to scale up and validate, from pilot test results to full scale (e.g., are banks additive?)
- Reliable determination of in-situ UV intensity (which varies with lamp age, fouling, distance from lamps, flow paths, temperature, distance along length of lamp)
- Development of better control systems (integration with plant SCADA systems, more user-friendly interface, etc.)
- The effects of water quality and operational conditions on regrowth
- Understanding and optimization of the destruction of emerging compounds of concern; although this requires high doses, it may be an important decision criterion for some plants
- Disinfection byproducts: although UV does not use chemicals, it does apply large amounts of energy to the water (especially medium pressure lamps) and creates radical and excited species that may form byproducts that have not yet been identified
- The performance of developing UV technologies such as pulsed UV, broad-band pulsed xenon lamps, and narrow-band excimer lamps

- ♦ Energy efficiency of UV lamp/ballast technologies; energy use is a major concern with this alternative

9.3 Ozone

Ozone presents a number of advantages relative to chlorine and UV, such as better ability to remove endocrine disrupting and other trace contaminants such as pharmaceuticals and personal care products, and the addition of dissolved oxygen to the effluent. However, of the three mature technologies, it is also the least used for wastewater disinfection. Consequently, there are a number of data gaps on disinfection efficacy, the impacts of ozonation on effluent water quality, and operations. Currently, only limited (or conflicting) data are available on

- ♦ The effectiveness of ozonation for inactivation of specific pathogenic organisms (e.g., protozoan cysts, nematodes) in wastewater effluents
- ♦ The effects of water quality parameters on disinfection efficacy and design
- ♦ The important parameters for pathogen inactivation and process control (ozone concentration vs. contact time vs. transferred ozone dose)
- ♦ The regrowth potential of pathogens following ozonation
- ♦ The reaction mechanisms and fate of disinfection intermediates and byproducts
- ♦ The toxicological relevance of the compounds resulting from ozonation of wastewater (results from experiments conducted to date on the overall effect of ozone on effluent toxicity are variable)
- ♦ Cost data that reflect the recent advances in ozone technology (the current perception is that ozonation is significantly more expensive than other disinfection alternatives, but this is largely based on data from previous generations of ozone generators and contactor equipment)

Filling in these information gaps will help utilities make better-informed decisions regarding ozone disinfection, particularly as a growing number of POTWs consider this technology because of its many benefits.

9.4 Developing Technologies

Of the developing technologies, PAA has the most published information, with data from a number of pilot plant tests, while a small number of pilot tests have been conducted with ferrate and pasteurization. BCDMH has been implemented at a handful of wastewater treatment plants in Japan, but has only been pilot-tested in the U.S. for combined sewer overflows. For broader applicability of these technologies, more information is needed in areas such as

- ♦ The disinfection efficacy against a wide range of organisms. The resistance of microbes to disinfection depends on the disinfection mechanism, which is different for each technology. As a result, each disinfectant has difficulty removing specific microbes. With experience, microbes resistant to the mature technologies have been identified (e.g., cysts and spores for chlorine, adenovirus for UV), but these are largely unknown for the newer technologies.
- ♦ The effect of water quality on disinfection efficacy and/or dose requirements, and the optimization of dose under operating conditions
- ♦ Disinfection byproducts (DBPs), and the conditions (e.g., dose, contact time, water quality) that affect their formation. Many DBPs that are formed during

chlorination have been identified, because chlorine has been used extensively for decades; however, it is likely that not all DBPs have been identified with emerging technologies.

- Other effects of disinfection on the effluent water quality (e.g., TDS)
- The potential for regrowth
- Practical experience with implementation and operation (e.g., maintenance requirements, reliability of disinfection performance)
- Economic evaluations at the full scale

Addressing these issues will facilitate the adoption of these technologies, and clarify the advantages and limitations of each alternative.

9.5 Combined Disinfection

As discussed in Section 6.5, facilities are increasingly interested in the use of multiple disinfection technologies, because they may potentially gain the advantages and mitigate the disadvantages of the individual disinfection methods. As with the other alternatives, the selection of appropriate individual technologies depends on-site-specific constraints and priorities. Several combinations of disinfection alternatives have been investigated recently, but reports remain limited. The data gaps for the combined disinfection alternatives include all of the data gaps for the individual technologies that comprise the combination, but also include the following:

- Synergistic effects on disinfection efficacy, disinfection byproducts, and water quality, i.e., the effects of combined disinfection may not equal the sum of the individual effects of each technology
- The operational complexity added by having multiple disinfection technologies
- The cost impacts: overall costs may increase or decrease

9.6 Concluding Remarks

This report shows that the disinfection landscape of the U.S. is changing, shifting away from chlorine gas toward alternatives such as hypochlorite, UV, and ozone. All of these technologies have distinct advantages relative to chlorine gas, but no alternative is perfect with respect to all of the concerns that a facility might face. As a result, a number of new technologies are under development, and the use of combined disinfection is being explored. All of these alternatives are safer than gas chlorine, both on-site and off-site generation, although most maintain some level of safety requirements for operators.

However, many questions and uncertainties remain on other issues that are key decision criteria for facilities considering these disinfection alternatives:

- Disinfection efficacy against a wide range of pathogens of concern
- Effects of influent water quality on disinfection (e.g., TSS, particle size distribution, organics, pH, temperature, etc.)
- Effects of disinfection on effluent water quality (e.g., formation of disinfection byproducts, addition of TDS, changes in toxicity, removal of compounds of concern, etc.)
- Data for optimizing design, operations, and maintenance
- Costs

Additional research and data on these topics will enable POTWs to better compare and select disinfection alternatives that are appropriate for their facilities, will allow for smoother transition and operation of the technologies, and will help to ensure high quality effluents from wastewater treatment plants.

APPENDIX A

UV SURVEY FORM

GENERAL – OWNER INFO.

Owner:

Address:

City:

State:

Zip:

Owner Contact:

Owner Contact Phone:

Owner Contact Email:

WERF Project Contact:

WERF Project Contact Phone:

WERF Project Contact Fax:

WERF Project Contact Email:

GENERAL – FACILITY INFO.

Owner:
Treatment Plant:
Address:
City:
State:
Zip:
County:
Category:
Region:

TECHNICAL

Average Dry Weather Design Flow:	_____ MGD
Collection System	☐ None ☐ Sanitary Sewer Only ☐ Combined Storm Overflow

WERF

TECHNICAL - (continued)

Types of Treatment:

Primary

Conventional Primary

- ☐ Sedimentation
- ☐ Microstrainer
- ☐ Other

Advanced Primary

- ☐ Alum addition
- ☐ Ferric chloride addition
- ☐ Single Stage lime treatment
- ☐ Two stage lime treatment

Secondary

Activated Sludge (AS)

- ☐ Air
  - ☐ Complete Mix
  - ☐ Conventional
  - ☐ Extended Aeration
  - ☐ High Rate
  - ☐ Step Aeration
  - ☐ Other Mode
- ☐ Pure oxygen
- ☐ Other AS
  - ☐ AS – contact stabilization
  - ☐ Ferric Chloride Addition
- ☐ AS with N &/or P removal
  - ☐ AS - Anaerobic/Anoxic/Oxic
  - ☐ AS with Biological Denitrification
  - ☐ Biological Denitrification – Separate Stage
  - ☐ Biological Nitrification – Separate Stage
  - ☐ Biological Phosphorus Removal
  - ☐ Biological Phosphorus Removal – Modified Bardenpho
  - ☐ Biological Phosphorus Removal – Phostrip
  - ☐ Combined Biological Nitrification and BOD Removal

Trickling filter

- ☐ Rock filter
- ☐ Rock media
- ☐ Plastic media
- ☐ Redwood slats
- ☐ Other media

Non-activated sludge

- ☐ Rotating biological contractor (RBC)
- ☐ Sequence batch reactor (SBR)
- ☐ Other suspended growth process

Other

- ☐ Aerated Lagoon
- ☐ Facultative Lagoon
- ☐ Polishing Lagoon
- ☐ Oxidation ditch
- ☐ Stabilization pond
- ☐ Total containment pond

Tertiary

Clarification

- ☐ Alum Addition
- ☐ Ferric chloride addition
- ☐ Tube settlers
- ☐ Coagulation

Filtration

- ☐ Mixed media filter
- ☐ Moving bed filter
- ☐ Pressure filter
- ☐ Rapid sand filter
- ☐ Sand filter-intermittent
- ☐ Sand filter-recirculating
- ☐ Slow sand filter

Polishing

- ☐ Electrodialysis
- ☐ Ion exchange
- ☐ Reverse osmosis

Physical/chemical

- ☐ Single stage lime treatment
- ☐ Two stage lime treatment

Disinfection

- ☐ Chlorination
- ☐ UV disinfection
- ☐ Dechlorination
- ☐ Ozonation
- ☐ Other
- ☐ None

If you use chlorine as a backup to your UV system, please fill out the following form about Chlorine Treatment.
Please do not include chlorine used elsewhere at your facility.

Chlorine added as:	☐ None ☐ Gas	☐ Liquid ☐ Generated on-site
Average Annual Chlorine Dose:	_____ ☐ mg/L	☐ lbs/day
Average Annual Design Contact Time:	_____ minutes	

Dechlorination:	☐ None	☐ Gas	☐ Liquid

PLEASE FILL OUT THE FOLLOWING FORM ABOUT UV TREATMENT

	System #1	System #2	System #3
Vendor:	________	________	________
Year placed in service (e.g., 1985):	________	________	________
Model:	________	________	________
Type of Lamps:	☐ LPLO ☐ LPHO ☐ MP	☐ LPLO ☐ LPHO ☐ MP	☐ LPLO ☐ LPHO ☐ MP
Total Number of Lamps:	________	________	________
Number of UV Channels in parallel:	________	________	________
Number of UV Units in Series:	☐ 1 ☐ 2 ☐ 3 ☐ 4 ☐ >4	☐ 1 ☐ 2 ☐ 3 ☐ 4 ☐ >4	☐ 1 ☐ 2 ☐ 3 ☐ 4 ☐ >4
Number of Lamps per Channel:	________	________	________
Design Flow of UV System (MGD):	________	________	________
Cleaning:	☐ Manual ☐ Wiper	☐ Manual ☐ Wiper	☐ Manual ☐ Wiper
Configuration:	☐ Open Channel ☐ Pressure	☐ Open Channel ☐ Pressure	☐ Open Channel ☐ Pressure
Lamp Orientation:	☐ Horizontal and Parallel to Flow ☐ Horizontal and Perpendicular to Flow ☐ Vertical to Flow	☐ Horizontal and Parallel to Flow ☐ Horizontal and Perpendicular to Flow ☐ Vertical to Flow	☐ Horizontal and Parallel to Flow ☐ Horizontal and Perpendicular to Flow ☐ Vertical to Flow
Year taken out of service(e.g., 1985):			

INITIAL UV DESIGN ISSUES

Reasons for selecting UV Check all that apply:	☐ Elimination of gas chlorine ☐ Elimination of gas chlorine upgrades to meet building code changes ☐ Elimination of permit requirements such as Risk Management Plans ☐ Elimination of chlorination/dechlorination ☐ Ease of operation ☐ More stringent discharge requirements ☐ More stringent receiving water requirements ☐ Because the water is to be reused/reclaimed ☐ Lower operating costs ☐ Safer work environment ☐ Less O&M Labor ☐ Other		
Select the Average of UVT of the water to be disinfected:	☐ 100-90 ☐ 90-80 ☐ 80-70	☐ 70-60 ☐ 60-50 ☐ <50	☐ Don't know
Select the Minimum of UVT of the water to be disinfected:	☐ 100-90 ☐ 90-80 ☐ 80-70	☐ 70-60 ☐ 60-50 ☐ <50	☐ Don't know
Did conversion of UV eliminate use of chlorine at the plant?	☐ Yes	☐ No	
Was there a bench study done to assist in the design?	☐ Yes	☐ No	
Is a bench study necessary?	☐ Yes	☐ No	
Was a pilot study done to assist the design?	☐ Yes	☐ No	
Is the pilot study necessary?	☐ Yes	☐ No	
Did the design need to meet a guideline or mandated requirements?	☐ Yes	☐ No	
To get a successful project, rank the relative importance of the following: **Using each number only once (i.e. no ties), rank from 1 as most important to 5 as least important:**	Consultants skill or knowledge Own staff skill and knowledge Vendor Available technical resources such as books, articles, etc. Visiting and touring other facilities		____ ____ ____ ____ ____

UV SYSTEM OPERATING ISSUES

Rate the difficulty of the transition to UV:	☐ Difficult ☐ Not too bad ☐ Easier than expected
Rate the complexity of your system compared to the one it is replacing:	☐ More ☐ The same ☐ Less

UV SYSTEM OPERATING ISSUES (Cont.)

How many hours of training is	_____	Hands On

necessary to operate the system?:	_____	Classroom
What is your annual lamp replacement cost?		
What is your typical lamp life (in hours)?		
What is your annual ballast replacement costs?		
What is your annual sleeve replacement costs?		
How many full-time equivalents are needed to maintain your UV system?	_____	☐ FTE/year ☐ FTE/month ☐ hours/month ☐ hours/week

UV VENDOR ISSUES

How often do you need vendor support?	☐ Weekly ☐ Monthly ☐ Every other month ☐ Other	
Would you recommend the UV vendor of your system?	☐ Yes	☐ No
If rules allowed, would you sole source the UV vendor, or put it out for competitive bids?	☐ Sole Source ☐ Competitive bids	
What are the most important issues in selecting a vendor? **(Using each number only once, i.e. no ties, rank from 1 as most important to as least important):**	Technical knowledge Responsiveness Size of Company Sales staff System Design System features	_____ _____ _____ _____ _____ _____

WATER QUALITY PERMITTED VALUES

Enter N/A if Not Applicable:	**Permitted Values**	**Typical Values**
Average BOD:	_____ mg/L	_____ mg/L
Average TSS:	_____ mg/L	_____ mg/L
Average Turbidity:	_____ NTU	_____ NTU

Average Microbial limit:	☐ N/A ☐ Total coliform ☐ Fecal coliform ☐ Fecal streptococcus ☐ Other	☐	☐ MPN/100 mL ☐ CFU/100 mL
Type of receiving water:	☐ River ☐ Ocean	☐ Stream ☐ Estuary	☐ Lake ☐ Other
Designated use of receiving water:			

GENERAL UV TECHNOLOGY QUESTIONS
What is the most unexpected thing that has occurred by changing to UV?
What has been the most unexpected thing that has occurred while operating a UV disinfection system?
What is the most important piece of advice that you can give to someone thinking about converting to UV disinfection?
If you could improve your UV installation, what would it be?
Any other comments?

APPENDIX B

UV SURVEY DATA

B.1 Statistical Analysis of Correlations

The UV survey asked facilities to evaluate a number of transitional and operational parameters: complexity of UV relative to the previous disinfection system, ease of transition to UV, classroom and hands-on training times, maintenance time, lamp life, and lamp replacement costs. The data were analyzed to determine whether these parameters varied significantly with facility size (average dry weather design flow rate), lamp type, or the level of treatment. For the analyses, each of the treatment system characteristics was categorized into "bins," shown in Table B-1. The standard significance level of $P \leq 0.05$ was used; in other words, correlations with $P \leq 0.05$ are considered significant. It should be noted that this analysis may miss correlations that exist or may find significance in data when no correlations exist in reality; the statistics simply indicate the likelihood of correlation.

Table B-1. "Bins" for the Analysis of Variance (ANOVA).

Parameter	Bin 1	Bin 2	Bin 3	Bin 4
Average Dry Weather Design Flow	1-2.5 MGD	> 2.5- 5 MGD	> 5-20 MGD	>20 MGD
Lamp Type	LPLO	LPHO	MP	
Treatment Level	Secondary	Tertiary		

The questions on relative complexity and ease of transition were categorical (i.e., the responses all fell into discrete categories), while the other variables such as times and costs were continuous. For the categorical data, a chi-squared test was applied to determine whether complexity and transition depended on the design flow, lamp type, or treatment level. The null hypothesis was that the complexity and ease of transition were independent of the treatment system characteristics. A P value of ≤ 0.05 indicates that there is $\leq 5\%$ probability of obtaining the observed data if the assumption of independence is true. For example, the significance level for the correlation of ease of transition and lamp type is 0.04. This indicates that there is only a 4% chance of getting the observed distribution of responses if the transition and lamp type are independent; in other words, transition and lamp type are most likely correlated. The results of all of the chi-squared tests are given in Section 7.3.4.

For the continuous variables, three types of analysis of variance (ANOVA) were applied: the standard ANOVA, the Welch test, and the Kruskal-Wallis (KW) test. The requirements for each type of ANOVA are given in Table B-2. The standard ANOVA has the most stringent requirements: the transitional and operational parameters must be normally distributed, and must have similar variances (i.e., standard deviations within a factor of three) across the bins that are being compared. For the standard ANOVA and Welch test, the null hypothesis was that the mean values for each of the bins shown in Table B-1were equal. A P value of ≤ 0.05 indicates that there is $\leq 5\%$ probability that the means are equal. For the KW test, all of the measured data across all bins were first ranked (e.g., from smallest to largest), and then the mean ranking for each bin was compared with the mean ranking of other bins. A P value of ≤ 0.05 indicates that there is $\leq 5\%$ probability that the mean rankings are equal; in other words, the measurements in one bin tend to be lower or higher (and therefore have a significantly lower or higher mean ranking) than measurements in other bins.

Table B-2. Requirements for the ANOVA Tests.

Parameter	Normal Distribution	Equal Variances
Standard ANOVA	Yes	Yes
Welch Test	Yes	No
Kruskal-Wallace Test	No	No

The training and maintenance times, lamp life, and lamp component costs appear to be roughly log-normally distributed, so the logarithm of the data was used for the ANOVA. The variance was evaluated using Levene's test; a result of 0.05 or less indicates that the variances are unequal, and the significance level (P) from the Welch test must be used. Table B-3 summarizes the significance values from all three ANOVAs, for the correlations between the training times and treatment system characteristics. Table B-4 summarizes the significance values from all three ANOVAs, for the correlations between the operational parameters and treatment system characteristics. The bolded values in Tables B-3 and B-4 indicates significant correlations.

Table B-3. Significance Levels (P Values) for Training Times and Treatment System Characteristics.

	Classroom Training Time	Hands on Training Time
POTW Average Dry Weather Design Flow	Levene's: 0.1282 ANOVA: 0.0834 Welch: 0.0571 KW: 0.1032	**Levene's: 0.0203** **ANOVA: 0.0004[a]** **Welch: 5.548e-5[a]** **KW: 0.0006[a]**
Lamp Type	Levene's: 0.1364 ANOVA: 0.0733 Welch: 0.0694 KW: 0.0665	**Levene's: 0.0058** **ANOVA: 0.0362[a]** **Welch: 0.0194[a]** **KW: 0.0344[a]**
Treatment Level	**Levene's: 0.0112** ANOVA: 0.5420 Welch: 0.4784 KW: 0.6127	**Levene's: 0.0271** ANOVA: 0.1888 Welch: 0.1383 KW: 0.2875

[a] P values indicated significant correlation in this one-way ANOVA, but showed no significant correlation with a two-way ANOVA (see Table B-5).

Table B-4. Significance Levels (P Values) for Operational Parameters and Treatment System Characteristics.

	Maintenance Time	Lamp Life	Lamp Costs	Sleeve Costs	Ballast Costs
POTW Average Dry Weather Design Flow	Levene's: 0.4209 **ANOVA: 0.0090** **Welch: 0.0140** **KW: 0.0165**	Levene's: 0.8777 **ANOVA: 0.0178[a]** **Welch: 0.0357[a]** **KW: 0.0130[a]**	Levene's: 0.4764 **ANOVA: 1.213e-9** **Welch: 1.647e-8** **KW: 9.725e-8**	Levene's: 0.0739 **ANOVA: 5.422e-5** **Welch: 1.998e-5** **KW: 0.0003**	Levene's: 0.6138 **ANOVA: 1.453e-7** **Welch: 7.339e-5** **KW: 7.991e-6**
Lamp Type	Levene's: 0.8203 ANOVA: 0.5582 Welch: 0.5459 KW: 0.4990	Levene's: 0.7135 **ANOVA: 0.0001** **Welch: 0.0001** **KW: 2.99e-5**	Levene's: 0.8640 **ANOVA: 0.0004** **Welch: 0.0009** **KW: 0.0017**	Levene's: 0.3438 **ANOVA: 0.0004** **Welch: 0.0004** **KW: 0.0009**	Levene's: 0.4583 **ANOVA: 0.0001** **Welch: 0.0005** **KW: 0.0008**
Treatment Level	Levene's: 0.3653 ANOVA: 0.6400 Welch: 0.6478 KW: 0.6656	Levene's: 0.8528 ANOVA: 0.9766 Welch: 0.9764 KW: 0.8839	Levene's: 0.9942 ANOVA: 0.9615 Welch: 0.9611 KW: 0.9210	Levene's: 0.2650 ANOVA: 0.5206 Welch: 0.4743 KW: 0.6268	Levene's: 0.0969 ANOVA: 0.8464 Welch: 0.8059 KW: 0.9072

[a] P values indicated significant correlation in this one-way ANOVA, but showed no significant correlation with a two-way ANOVA (see Table B-6).

The treatment level (secondary or tertiary) showed no significant correlations with any of the tested transitional or operational parameters, while design flow and lamp type do correlate significantly with some of the tested parameters. However, it is possible that the effects of lamp type and design flow are intertwined, because the POTWs within each category are not evenly distributed. For example, the largest facility size bin (>20 MGD) comprised 32% of all MP plants, but only 8% of all LPLO plants and 11% of all LPHO plants.

To separate these effects, a 2-way (or 2-factor) ANOVA was performed to test the correlations of design flow and lamp type simultaneously with transitional and operational parameters. As with the 1-way analyses, the ANOVAs were run on the logarithms of the data, and if the Levene's test indicated that the variance was not evenly distributed, the Welch test was used to determine the significance levels (P values). Tables B-5 and B-6 provide the significance levels for each of the correlations tested; these are the values reported in Chapter 7.0 for the average dry weather design flow rate and the lamp type. The bolded values had significant correlations.

Table B-5. 2-Way ANOVA. Significance Levels (P Values) for Training Times and Treatment System Characteristics.

	Classroom Training Time	Hands-on Training Time
POTW Average Dry Weather Design Flow	0.0863	0.4544
Lamp Type	0.3980	0.6555

Table B-6. 2-Way ANOVA. Significance Levels (P Values) for Operational Parameter and Treatment System Characteristics.

	Maintenance Time	Lamp Life	Lamp Costs	Sleeve Costs	Ballast Costs
POTW Average Dry Weather Design Flow	**0.0246**	0.4286	**2.63e-7**	**0.0002**	**0.0039**
Lamp Type	0.7914	**0.0038**	**0.012**	**0.0040**	**0.0017**

The results of the 2-way ANOVAs are quite similar to the 1-way results. The only differences are that the correlations are no longer detectably significant between hands-on training times and design flow/lamp type, and between lamp life and design flow.

B.2 Transition Data

Table B-7. Relative Complexity of UV Systems by Facility Size.

Statistic	1-2.5 MGD	> 2.5-5 MGD	> 5-20 MGD	> 20 MGD
Less Complex	4	12	6	3
The Same	0	4	5	2
More Complex	13	10	27	13
Total Responses	17	26	38	18

Table B-8. Relative Complexity of UV Systems by Lamp Type.

Statistic	LPLO	LPHO	MP
Less Complex	9	13	4
The Same	3	2	6
More Complex	14	23	28
Total Responses	26	38	38

Table B-9. Relative Complexity of UV Systems by Upstream Treatment Level.

Statistic	Secondary	Tertiary
Less Complex	13	14
The Same	6	4
More Complex	33	30
Total Responses	52	48

Table B-10. Relative Difficulty of Transition to UV Systems by Facility Size.

Statistic	1-2.5 MGD	> 2.5-5 MGD	> 5-20 MGD	> 20 MGD
Easier Than Expected	10	16	13	4
Not Too Bad	5	6	19	9
Difficult	2	4	6	5
Total Responses	17	26	38	18

Table B-11. Relative Difficulty of Transition to UV Systems by Lamp Type.

Statistic	LPLO	LPHO	MP
Easier Than Expected	17	17	10
Not Too Bad	6	16	19
Difficult	3	5	9
Total Responses	26	38	38

Table B-12. Relative Difficulty of Transition to UV Systems by Upstream Treatment Level.

Statistic	Secondary	Tertiary
Easier Than Expected	21	23
Not Too Bad	22	18
Difficult	9	7
Total Responses	52	48

Table B-13. Summary of Training Times by Facility Size.

	Classroom				Hands-On			
Statistic	1 - 2.5 MGD	> 2.5 – 5 MGD	> 5 – 20 MGD	>20 MGD	1 - 2.5 MGD	> 2.5 – 5 MGD	> 5 – 20 MGD	>20 MGD
Median Hours	4	4	6	8	8	4	10	20
Minimum Hours	1	1	1	3	1	2	1	8
Maximum Hours	24	20	40	24	80	40	50	40
Total Responses	7	16	31	14	13	21	33	18

Table B-14. Summary of Training Times by Lamp Type.

Statistic	Classroom LPLO	Classroom LPHO	Classroom MP	Hands-On LPLO	Hands-On LPHO	Hands-On MP
Median Hours	8	4	8	9	8	12
Minimum Hours	1	1	1	1	1	2
Maximum Hours	30	40	40	80	24	40
Total Responses	16	23	31	20	28	37

Table B-15. Summary of Training Times by Upstream Treatment Level.

Statistic	Classroom Secondary	Classroom Tertiary	Hands-On Secondary	Hands-On Tertiary
Median Hours	5	8	8	10
Minimum Hours	1	1	1	2
Maximum Hours	40	40	40	80
Total Responses	34	36	44	41

B.3 O&M Data

Table B-16. Summary of Weekly Maintenance Times by Facility Size.

Statistic	1 - 2.5 MGD	>2.5 – 5 MGD	>5 – 20 MGD	>20 MGD
Median Hours	1	2	7	10
Minimum Hours	0.2	0.5	0.5	0.5
Maximum Hours	12	35	120	90
Total Responses	9	15	28	15

Table B-17. Summary of Weekly Maintenance Times by Lamp Type.

Statistic	LPLO	LPHO	MP
Median Hours	7	2	5
Minimum Hours	0.2	0.5	0.2
Maximum Hours	120	80	90
Total Responses	18	22	29

Table B-18. Summary of Weekly Maintenance Times by Upstream Treatment Level.

Statistic	Secondary	Tertiary
Median Hours	4	4
Minimum Hours	0.2	0.5
Maximum Hours	120	80
Total Responses	36	33

Table B-19. Summary of UV Lamp Life by Facility Size.

Statistic	1 - 2.5 MGD	> 2.5 – 5 MGD	> 5 – 20 MGD	>20 MGD
Median Hours	8,800	10,000	8,700	5,000
Minimum Hours	1,200	2,000	2,100	2,000
Maximum Hours	18,000	25,000	18,000	18,000
Total Responses	14	20	30	17

Table B-20. Summary of UV Lamp Life by Lamp Type.

Statistic	LPLO	LPHO	MP
Median	10,000	10,000	5,000
Maximum	25,000	18,000	20,000
Minimum	2,100	1,200	2,000
Total Responses	25	25	33

Table B-21. Summary of UV Lamp Life by Upstream Treatment Level.

Statistic	Secondary	Tertiary
Median Hours	8,800	8,000
Minimum Hours	2,500	1,200
Maximum Hours	18,000	25,000
Total Responses	42	41

Table B-22. Summary of Annual Lamp Component Costs by Facility Size.

Statistic	Lamps				Sleeves				Ballasts			
	1 - 2.5 MGD	> 2.5 – 5 MGD	> 5 – 20 MGD	> 20 MGD	1 - 2.5 MGD	> 2.5 – 5 MGD	> 5 – 20 MGD	> 20 MGD	1 - 2.5 MGD	> 2.5 – 5 MGD	> 5 – 20 MGD	> 20 MGD
Median Annual Cost	$6,500	$6,000	$15,000	$48,000	$500	$1,000	$3,000	$15,000	$1,500	$1,000	$3,850	$13,000
Minimum Annual Cost	$1,500	$600	$1,600	$19,000	$360	$130	$250	$1,000	$360	$110	$500	$1,500
Maximum Annual Cost	$20,000	$24,000	$80,000	$250,000	$2,000	$22,000	$25,000	$70,000	$5,000	$25,000	$12,000	$200,000
Total Responses	14	21	29	13	8	18	25	11	9	14	24	12

Table B-23. Summary of Annual Lamp Component Costs by Lamp Type.

Statistic	Lamps			Sleeves			Ballasts		
	LPLO	LPHO	MP	LPLO	LPHO	MP	LPLO	LPHO	MP
Median Costs	$6,500	$11,000	$20,000	$1,000	$880	$5,000	$1,000	$1,500	$5,500
Minimum Costs	$600	$1,600	$2,500	$130	$250	$300	$110	$350	$500
Maximum Costs	$44,000	$65,000	$250,000	$25,000	$10,000	$70,000	$12,000	$10,000	$200,000
Total Responses	24	23	32	20	16	26	16	17	26

Table B-24. Summary of Annual Lamp Component Costs by Treatment Level.

Statistic	Lamps		Sleeves		Ballasts	
	Secondary	Tertiary	Secondary	Tertiary	Secondary	Tertiary
Median Costs	$10,000	$14,000	$2,000	$1,600	$2,600	$3,000
Minimum Costs	$600	$800	$220	$130	$120	$110
Maximum Costs	$250,000	$100,000	$22,000	$70,000	$200,000	$80,000
Total Responses	39	40	29	32	30	29

B.4 Effluent Quality Data

Table B-25. Summary of BOD Values by Facility Size.

Statistic	BOD Permit Limit (mg/L)				Typical BOD Level (mg/L)			
	1 - 2.5 MGD	> 2.5 – 5 MGD	> 5 – 20 MGD	> 20 MGD	1 - 2.5 MGD	> 2.5 – 5 MGD	> 5 – 20 MGD	> 20 MGD
Median BOD Level	18	25	25	20	3	5	4	2
Minimum BOD Level	10	5	4	5	0.5	0.9	0.8	0.5
Maximum BOD Level	45	150	50	60	205	140	15	12
Total Responses	16	25	36	15	14	23	35	15

Table B-26. Summary of TSS Levels and Fecal Coliform Limits by Facility Size.

Statistic	TSS Permit Limit (mg/L)				Typical TSS Levels (mg/L)				Fecal Coliform Limit (MPN/100 mL)			
	1 - 2.5 MGD	> 2.5 – 5 MGD	> 5 – 20 MGD	> 20 MGD	1 - 2.5 MGD	> 2.5 – 5 MGD	> 5 – 20 MGD	> 20 MGD	1 - 2.5 MGD	> 2.5 – 5 MGD	> 5 – 20 MGD	> 20 MGD
Median	21	25	30	30	3	5	5	3	40	50	50	200
Minimum	5	9	4	6	1	0.9	0.01	0.4	1	0.5	1	20
Maximum	45	175	50	60	18	120	15	10	2,000	850	400	2,000
Total Responses	16	25	36	15	14	22	33	14	9	18	25	14

Table B-27. Summary of BOD Values by Lamp Type.

Statistic	BOD Permit Limit (mg/L)			Typical BOD Level (mg/L)		
	LPLO	LPHO	MP	LPLO	LPHO	MP
Median BOD Level	30	25	20	5	4	3
Minimum BOD Level	5	4	4	0.5	1	0.5
Maximum BOD Level	150	45	60	140	15	18
Total Responses	24	33	37	23	27	36

Table B-28. Summary of TSS Levels and Fecal Coliform Limits by Lamp Type.

	TSS Permit Limit (mg/L)			Typical TSS Levels (mg/L)			Fecal Coliform Limit (MPN/100 mL)		
Statistic	LPLO	LPHO	MP	LPLO	LPHO	MP	LPLO	LPHO	MP
Median	30	30	30	4	6	3	16	200	200
Minimum	6	5	4	0.4	0.01	1	0.5	0.5	1
Maximum	175	34	60	120	150	18	200	850	2,000
Total Responses	24	33	37	24	27	33	24	27	33

Table B-29. Summary of BOD Values by Treatment Level.

	BOD Permit Limit (mg/L)		Typical BOD Level (mg/L)	
Statistic	Secondary	Tertiary	Secondary	Tertiary
Median BOD Level	25	14	5	2
Minimum BOD Level	5	4	1	0.5
Maximum BOD Level	150	45	205	117
Responses	49	45	44	42

Table B-30. Summary of TSS Levels and Fecal Coliform Limits by Treatment Level.

	TSS Permit Limit (mg/L)		Typical TSS Levels (mg/L)		Fecal Coliform Limit (MPN/100 mL)	
Statistic	Secondary	Tertiary	Secondary	Tertiary	Secondary	Tertiary
Median	30	30	7	2	75	200
Minimum	4	5	1	0.01	0.5	0.5
Maximum	175	45	120	8	2000	5000
Total Responses	49	45	41	43	37	28

APPENDIX C

UV SURVEY COMMENTS

Flow (MGD)	Lamp Type	Issues During Transition	Issues During Operation	Advice to Others	Ways to Improve UV	Comments
1-2.5	LPHO	O&M time, costs	The cost	Make sure to get a UV system designed for your facility	Less bulbs with the same results, as far as disinfection/water quality	
1-2.5	LPHO			Make sure the software is programmed for your application.	Better programming.	
1-2.5	LPHO	Every once in a while we have a fecal coliform excursion but below our maximum limits	Same	BEFORE you put pen to paper (design the system) you should conduct a performance based evaluated bid for the equipment. In that evaluated bid process, require capital and O&M costs, performance Bio-essay results, and a list of OPERATORS who operate the system.	I'm satisfied with the performance and results of my UV system.	
1-2.5	LPHO	Ease of operation. Also, we like the flexibility of being able to turn on many banks at a time with no problems.	Awkwardness of the light banks and the use of lifting equipment because of the heaviness	Placement of UV in channels, making sure there is enough room to work on the ballast bulbs	Move above over 5 inches	
1-2.5	LPHO	How easy change was.	The cost of replacement parts.	To maintain a very high quality effluent as our SS climb so do our fecal count. Permit value 870-1740. Our high about 600	Better way to clean bulbs and trough.	
1-2.5	LPLO	Nothing surprising - new plant		Do it!		
1-2.5	LPLO	It is easier to maintain.	The cost efficiency, and it's very effective.	It may be less dangerous than chlorine, but protective eyewear is a must.	Add automatic cleaning.	
1-2.5	LPLO	Did not have to clean bulbs as often.		The clearer the effluent the better the kill.	Self cleaning model.	

	Type	Issues During Transition	Issues During Operation	Advice to Others	Ways to Improve UV	Comments
1-2.5	LPLO	No Cl2 residual for Fecal	No Cl2 residual	Make sure they come with wipers	Easier way to remove the quartz sleeves	
1-2.5	LPLO			Would highly recommend UV disinfection	Upgrade to latest technology	
1-2.5	MP	The cost of parts was much greater than expected (esp. with secondary effluent). Most are proprietary, and exorbitantly expensive. All but ~3 parts must come from the vendor - and costs add up quickly. They've replaced the 3-way valve in their UV cabinet 6 times, many cuvettes for the UVT unit, and their touch screen interface 4-5 times, at $4000-5000/screen. At one point, their screen died and "Vendor" didn't have any in stock - but without it, operation is impossible (e.g., can't turn off pumps or lamps, do UVT calibrations, etc.) - they now keep an extra one as a spare part. "Vendor" now says that the screen should not be in the sun. There was a long learning process during the transition to UV. It's hard to say how many hours of training are needed to operate the system, but it took them ~2 years to get it working smoothly.	Random alarms. There are dozens of possible alarms, but the interface with the main plant computer only says "major" or "minor" UV alarm - you have to go to the unit and look at the touch-screen interface to find out what the actual problem is. They once got a "restrike alarm," which shut down the banks. They couldn't find anything in the manual, so they contacted Vendor who had to consult 3 of their techs before they found someone who knew what the alarm was - and even then, the banks could not be restarted for 24 hrs. Current lamp life is 5000 hours, but was ~4000 (or even 3000) with secondary effluent. They've replaced 3-4 ballasts in the 4 years they've had the system. Maintenance now takes 3-4 hours/week, but took ~15 hours/week with secondary effluent.	You need exceptional effluent quality (UVT >60), practically at drinking water quality. The UV system is *very* sensitive to power fluctuations, which can blow lamps and ballasts. People may need to add power conditioning into the design of the unit, if their power source isn't completely stable. He also recommended having an electrical engineer on staff, or at least one in the area who can help with control problems.	Have more banks, so they can work on one and shut it down without worrying about the effectiveness of disinfection.	They opted to install UV rather than build more chlorination and dechlorination units. At the time they installed the system, they only had secondary treatment, and they had all sorts of maintenance problems. They later a tertiary process, and now their water is much cleaner - and the maintenance issues have gone away. Vendor issues: They have not needed vendor support in the last year. They did a lot of research before they bought, and decided that "Vendor" was the best system out there, but they've had a lot of problems.... They have found "Vendor" to be very arrogant, and not very helpful - at one point, it took "Vendor" 2 weeks to get to their plant to fix a problem (they didn't have a local technician, but they hired one ~ 1 year ago). Overall, he says that UV is cleaner than chlorine, and provides good (even better) disinfection. When it works, it's great - they haven't had any problems with violations of their permits. But when there are problems, they are hard to fix....
1-2.5	LPLO	Ease of use	Difficulty in cleaning	Proper design is critical	Automatic cleaning	

Flow (MGD)	Lamp Type	Issues During Transition	Issues During Operation	Advice to Others	Ways to Improve UV	Comments
1-2.5	LPLO	It works as well as the salesman said it would	Maintenance is easier than we thought it would be	Be sure design will handle flow surges through plant	It should have been built with room for the addition of more lamp units, so it could be upgraded for higher flows later on	
1-2.5	MP					
1-2.5	MP		Ease of using the system.	System is a lot safer than chlorine.		
1-2.5	MP		Nothing unexpected as far as operating the system. The unexpected was the installation design for the system. Piping modifications were needed to ensure chambers remained flooded at all times.	Work closely with the vendor and keep the vendor closely involved with the design of systems prior to and after the UV system, which may have any effect on how the UV system operates.	If the UV system must be validated, include in the RFQ for the Vendor to perform the validation.	
1-2.5	MP	Ease of transition.	Less maintenance than anticipated.	Surge protection.	Install surge protection.	
>2.5-5	LPHO	Can't increase dose as you can with Cl2	Not working to specifications	Make sure system is designed with enough bulbs to meet dose requirements and install multiple channels to expand if needed	Install another channel for expansion and have enough lamps to be able to run one channel and do maintenance on the other.	
>2.5-5	LPHO			Clear effluent is needed, because turbidity affects performance. Replacement parts are expensive.		

Flow (MGD)	Lamp Type	Issues During Transition	Issues During Operation	Advice to Others	Ways to Improve UV	Comments
>2.5-5	LPHO	Does not perform well when UVT is less than 55%. We have had 6 weekly excursions in the last 2 years for fecal coliform.	Instrumentation shows excellen doses, but still get high counts occasionally. Downstream channels have biological growth that reoccurs, stalked ciliates mainly and algae. Sloughs off and shows up in final effluent sample. I may do a pilot study with peroxyacetic acid to minimize cleaning and reduce bacteria entering UV.	Analyze effluent for BOD, TSS, any known UV inhibitors i.e., iron. Transmissivity at most always should have 55% plus year round or will have to supplement with chemical.	I would not recommend UV without a filter prior to unit. We do not have one, solids settle in channels, when channels shutdown and start up it changes velocities and stirs up growths and settle solid.	
>2.5-5	LPHO		Initially we had a manual cleaned system. Very labor intensive. Auto-clean is a must.	Make sure there are no UV %T issues pertaining to compliance. Space for worst case scenario.		The old UV system was replaced because the manual cleaning was labor intensive. In addition, the power output was insufficient to handle higher flow.
>2.5-5	LPHO					
>2.5-5	LPHO			Big		
>2.5-5	LPHO	An infestation of midges.	None.	Consider ways to control insects.		
>2.5-5	LPHO	Technical knowledge needed	Ease of breakage and short life of bulbs	Check maintenance cost thoroughly	Not familiar enough with system, yet	Just Completed expansion and upgrade project.
>2.5-5	LPHO	Ease of operation.	Animals in and around bulbs.	Do it!	Have a drain in channel	Recommend UV system installed inside a building. We installed a carport over ours, which helped with P.M. during bad weather and algae growth on bulbs/chamber.
>2.5-5	LPHO	None	None	Make sure the water going in has very low TSS	N/A	The old UV system was taken out of service for the need of increased flow and better performance.
>2.5-5	LPHO		The amount of humidity in the room. Offline bulbs must always be submerged.	Have a spare rack on hand. You always have a backup system.	Need always to test bulbs and ballasts prior to installation. Also the UVT is small and difficult to work.	

Flow (MGD)	Lamp Type	Issues During Transition	Issues During Operation	Advice to Others	Ways to Improve UV	Comments
>2.5-5	LPHO		Ballasts went out constantly at first. They were all under warranty fortunately. "Vendor" did very poorly with customer service at first but are doing a fine job now.	Go with "Other Vendor" for UV disinfection.	Trade it in for a system from another vendor.	
>2.5-5	LPHO			Thoroughly research tech support capabilities of vendor. Talk to people at other plants. We got a lot of excellent feedback on sizing.		
>2.5-5	LPLO	Consistent disinfection, relative to chlorination. Since we exit into a small stream, our chlorine residual had to be almost non-existent; our equipment wasn't too sophisticated and the chlorination basin was small, so it was difficult to maintain the low chlorine residual and consistent disinfection. We probably should have been doing dechlorination, but we weren't far from the construction of the existing plant, so it was decided not to spend the money on equipment and additional basin space.	Low maintenance	Size the system properly	Current system is over-engineered; control board setup too complicated	

Flow (MGD)	Lamp Type	Issues During Transition	Issues During Operation	Advice to Others	Ways to Improve UV	Comments
>2.5-5	LPLO	The plant has always used UV.	Fires from the previous UV system that happened due to design error (electronic cards would catch on fire).	Go and inspect as many systems as you can, and talk to more on the phone, to get a feel on what is the best system for your facility.	Use a channel instead of a vessel, because it would be easier to clean the quartz tubes. Currently we would have to drain and remove the quartz from the vessel for major cleaning or replacement. See comments for more detail.	Their system uses an enclosed, nonpressurized vessel. Operator thinks that an open channel would be easier, for cleaning of the quartz sleeves. Currently they need to drain and remove the quartz from the vessel for major cleaning or replacement. They don't have a wiper system, and rely on in-vessel cleaning with Citric Acid. The Citric Acid works well so they haven't needed to remove the quartz tubes yet. The bulbs are supposed to last for 8000 hours; however, their bulbs last for ~3000 hours. They attribute this to the cold water temperatures. On a good week, maintenance only requires ~2 hrs/week; on a bad week it may be ~25 hrs/week.
>2.5-5	LPLO				Install wipers. However we clean ballasts approximately once per 2 months. So it really isn't a large issue with us.	
>2.5-5	LPLO	Realizing its simplicity.	Low fecal readings.	Clean bulbs weekly.	Cover the entire UV structure.	
>2.5-5	LPLO	None.	Build up of male contraceptives	Make sure you have low turbidity.	Self-cleaning.	
>2.5-5	LPLO			Get auto-cleaning	Auto-cleaning	We run both systems parallel, full time on.
>2.5-5	LPLO	High iron conc. in influent after UV installation		Pretreat iron in influent water		

Flow (MGD)	Lamp Type	Issues During Transition	Issues During Operation	Advice to Others	Ways to Improve UV	Comments
>2.5-5	LPLO				Upgrade to a system with automated cleaning system.	We have a great track record of compliance with both fecal and total coliform limits.
>2.5-5	LPLO					
>2.5-5	MP	The cost of maintaining the system	Nothing	Look at bulb cost and spare parts costs	Install a PLC controlled system. The system we have is a chip and not operator friendly.	
>2.5-5	MP	The low weekly maintenance that the UV system takes to manage.	A continuous hatch of midges in building. We have UV system in building with open channels and the midges hatch all year long.	This is a no-brainer. I would convert in a heartbeat. With UV no deadly gas or chemicals to deal with no chance of accidence to release to river of chlorine or release to community.	When we install unit we didn't bring back all information from UV PLC to plant scada system. This only gives us minor& major alarms. So we couldn't make adjustments without going to building. This is being changed now.	
>2.5-5	MP			Choose a UV system that it is efficient, easy to troubleshoot and maintain.	I am satisfied with the system.	
>2.5-5	MP	Always had UV	System is new, have not had any problems.	No more chemicals		
>2.5-5	MP	Change out condensers in cabinets	Easy to use	Learn everything you can about your system. What makes it work. How does it work.	All good	
>2.5-5	MP	Nothing in particular	Sometimes seeing high coliform plate counts when effluent quality appears very good.	Select the most effluent system using the least amount of lamps - minimize maintenance requirements		
>2.5-5	MP	We had to place another bank of lamps in series to get proper disinfection	Fish being trapped in lamps	Simple is Best	More simple	

WERF

Flow (MGD)	Lamp Type	Issues During Transition	Issues During Operation	Advice to Others	Ways to Improve UV	Comments
>2.5-5	MP	Insects activity around UV structure compared to time when chlorine was used.	Troubleshooting circuit board	Test actual UVT in plant effluent prior to design, if possible.	Install a cover over the system.	
>2.5-5	MP		We did not expect the high cost involved with power	Find a domestic UV manufacturer	A method to test ballast before installation	
>2.5-5	MP		UVT that controlled unit stopped working. We entered a set value as a minimum.	Filter ahead of UV/ cover unit.	Make sure if you use UV that a save filter be used or designed in construction with each installation. We didn't install a filter and that were most of the problems and cost came from. As well the unit *must be covered* due to all of the stainless steel and sun heating up control panels. Major problems if this is not done.	We were the first installation by "Vendor" west of the Mississippi. We have a first generation unit and the new technology is changing every year. "Vendor" has the best support, but the operator and I have visited many plants and talked to a lot of operators and managers, and we feel the "Other Vendor" unit is safer for the operators hands on, for parts replacement. I highly feel that UV is the only way to go. Safer for operators and customers. No chemical storage.
>2.5-5	None		Quality of disinfection has not diminished.	Sleeve cleanliness. Do not use "Vendor".	Control panel installation. Channel gate design. Access to low water cut off probe.	
>5-20	LPHO	Insects in the disinfection building.	Software/control systems.		Include build in air conditioning units on the ballast cabinets.	
>5-20	LPHO	The ease of transition, when it is working right you can just forget about it.	Channel level is controlled by automatic gate, which is the weak link. UV manufacturer supplied cheap components instead of gate UFG we wanted Also ballast cabinets were constantly overheating because cooling units were under sized (cheap).	Do your homework, understand how the different systems work, strengths weaknesses. Tour other facilities to look at systems and ask a lot of questions. Have vendors set-up a test unit.	Add redundancy with two channels use two ranks in one channel. Have a better set-up for controlling channel level. Enclose land area and have overhead lift.	

Flow (MGD)	Lamp Type	Issues During Transition	Issues During Operation	Advice to Others	Ways to Improve UV	Comments
>5-20	LPHO		Look very carefully at each design and head loss associated with each especially on a retrofit	Raise the elevation of the electrical junction box		
>5-20	LPHO			Ensure the system has a good reputation. Visit other sites installed your preferred unit and have been operating it for a few years.		
>5-20	LPHO					
>5-20	LPHO	It was easier than expected	We did experience failure of bulbs and ballast at start-up. The problem was remedied under the warranty.	Choose a quality vendor	Include diagnostic equipment - The ability to test lamps, ballasts and control cards.	Like all new systems, there is a learning curve. This system is somewhat complex, but it is manageable and can be operated and maintained by plant operators.
>5-20	LPHO		Cost of bulbs			
>5-20	LPHO	The width of channel	Wiper failed to torque out when jammed and broke quartz sleeve and ruined lots of bulbs	Insure you have support when first starting units	None now that they are running good	
>5-20	LPHO	The ease of operation. No more chlorine cylinders to change.	High flow flooded units causing extensive damage. Needed to make an emergency run to factory for lamps plus ballasts.	Choose reliable units with O & M costs.	A friendlier MMI.	
>5-20	LPHO	Algae growth in effluent	Sensitivity to power surges, heat and storms	Know your equipment. Have a good factory rep and competent service techs.	Install circuit breakers instead of fuses (because of power surges and storms). Have plenty of spare parts on hand.	
>5-20	LPHO	Insect build up and stuck after chlorine usage went down.	Flooding of UV structure; maintenance for this issue.		Prevent flooding of structure.	

Flow (MGD)	Lamp Type	Issues During Transition	Issues During Operation	Advice to Others	Ways to Improve UV	Comments
>5-20	LPHO	Some unknown bacteria also grows on fecal filter paper, colonies are beige and show up with increased water temp	More man-hours are required to operate UV system	Do your homework. Make sure design engineer has already designed similar systems, then check with these owners of their opinion of project and to see if other improvements can be made	Satisfied with current design and installation	
>5-20	LPHO	Less monitoring	Ease of operation	Check all disinfection options	Provide back-up power; Better human to machine interface; Redesign of screen/flow baffle to facilitate easier cleaning	UV disinfection system was online for only 1 month at time of survey.
>5-20	LPHO		Inability to change modes of operation.	Have operators involved in the design, selection, and commissioning.	Program the PLC to be more friendly.	
>5-20	LPHO					
>5-20	LPHO	Plant has only used UV.	During a large rain event a quartz jacket had broken and due to the increase velocity, the water entered the ballast/LCA/DCA cabinet and shorted out the cell.	Keep control cabinets dry. Any moisture due to broken jackets or seals will cause damage to the control boards and ballasts. These items are costly.	Put the UV system in a building. Our system is out in the open.	
>5-20	LPLO	Can only buy parts from one vendor	UV intensity sensors never worked	Do not buy a cheap system	Choose a different location	
>5-20	LPLO	Not compatible with plant's SCADA system		The effluent has to be really clean. Over design system to allow for upsets.	Include a parallel system to allow for cleaning the channel	Switching to UV was one of the best upgrades done at the plant
>5-20	LPLO	Easy to operate	UVT difficult to use	Visit other plants and talk to their staff. Pilot study recommended. Buy only proven equipment	Keep it simple. Have it work on flow as first option	
>5-20	LPLO					

Flow (MGD)	Lamp Type	Issues During Transition	Issues During Operation	Advice to Others	Ways to Improve UV	Comments
>5-20	LPLO	The violations in bioassays due to DBP byproducts formations stopped significantly	Limit the distance between the PLC and unit to minimize faulty error messages	Keeping the PLC closer to unit and keep the UV unit in an enclosed environment to minimize dust falling and algae formation.	Move the PLC closer to system	
>5-20	LPLO	Unreliability of UV vendor service	Unreliability of UV vendor service	Get your check book out – UV is expensive	Increase size of module connectors, eliminate flow pacing	
>5-20	LPLO	High maintenance requirements	High maintenance requirements	Educate your technical staff so they can troubleshoot system	Install automatic cleaning system, switch to more current system that has fewer lamps	
>5-20	LPLO	Longer bulb life than anticipated.	See above & ease of operation.	I highly recommend the installation of sand filters before the UV system. This will cut down on cleaning and deliver a higher quality of water to the UV system.	Add more automation to the system to control flow to channels starting and stopping during high flow, to control flow gates and add more monitoring failure alarms.	
>5-20	MP	Nothing	Nothing	Research actual O&M Costs	2 Channels	
>5-20	MP	UV was designed into plant.	UV vendors going out of business.	Use vender that will be in business 20 years later.	Ballast	
>5-20	MP	Only one source for parts; Cost is a major issue	Cost	Do your research and look at budgetary impacts	House unit in an enclosure/building for easier maintenance	
>5-20	MP	Much more algae than expected	Originally had system sized for 10 MGD. Needed to upgrade to 17 MGD to handle high flows resulting from heavy rains	Visit other plants and talk to operators that have experience running UV	System has two banks of lamps in series in the same channel. Two banks in parallel in separate channels are preferable from a maintenance standpoint	UV disinfection is much better than using sodium hypochlorite. Vendor support is needed once a year once system is up and running. More frequent vendor support was required at start-up

Flow (MGD)	Lamp Type	Issues During Transition	Issues During Operation	Advice to Others	Ways to Improve UV	Comments
>5-20	MP		It is labor intensive compared to chlorine: filling of the wiper system, replating wiper seals, daily service of the UVT analyzer	1. The best thing we did is request the same service tech each and every time a company rep is needed and 2. Have the UV unit enclosed or within a building for easier maintenance in bad weather.		Overall, they're pretty happy with their UV system. The system requires more cleaning in the summer, when they operate their NDN process; they get more slime buildup in the system then. Operator also noted that having the system enclosed makes for easier maintenance in winter or bad weather. The system has all metal decking and is out in the open, which makes it a potential hazard in thunderstorms. Being inside would also protect the display panel, since it should be kept out of the sun.
>5-20	MP		Lightning hit the system soon after it started up and killed the whole unit (fried some boards or something like that). Fixed it and re-started, and has been good ever since.	Be aware of the cost of material when it comes time to do maintenance on a UV bank.	Our system would have been better if they could see the UV program through a modem, but our system is not equipped with that option.	The UV system works well and does a good job at disinfection, although it is more technical than chlorine because of the programming and controls involved in the system.
>5-20	MP	Still have to use chlorine (bleach) to keep effluent filters clean.	Even when not feeding chlorine still must dechlorinate	Keep your effluent discharge as close as possible to the UV discharge.	Remove all trees from any channels or filters upstream from the UV to prevent fouling the wipers.	
>5-20	MP					
>5-20	MP		Amount of cleaning required was more than expected	Consider energy cost and all other costs	Reduce electrical usage	
>5-20	MP	Still using chlorine (hypochlorite) to aid in disinfection	Early lamp failure during maintenance procedures	Be sure you are ready for the expense in running these types of units	Application correct for your system. Early bulb failure following vendor maintenance program	

Flow (MGD)	Lamp Type	Issues During Transition	Issues During Operation	Advice to Others	Ways to Improve UV	Comments
>20	LPHO	Effluent quality is critical to lower O&M of UV system	heat generation in ballast cabinets and cable trays	Need good quality effluent that is low in TSS; no algae; no solids; no color	automatic chemical cleaning of bulbs	Vendor gave so/so service - problems at start-up. It is important to have *good* vendor support at system startup.
>20	LPHO	No big surprises. The cost for electrical power was higher than for chemicals, but this wasn't expected.	The time for preventative maintenance; however, this is very system dependent. The old UV system required too much maintenance, and the maintenance was too difficult. It didn't have wipers, and there were too many bulbs in a module (28 vs. 8 in the current system) to be able to clean them individually, so cleaning required an acid bath with air scouring. This resulted in large volumes of acid that had to be returned to the plant, and could mess up the treatment processes. Also, the electronics were very complicated relative to the current system, so it was tricky to work on and there were more safety concerns. Finally, the modules were very heavy, and required a crane to lift out.	Find a system with easy maintenance: easy to work on, self-cleaning, not too many parts, etc. Fewer bulbs make it easier to clean, simpler system means fewer parts that can break and fewer parts to stock (they had ~850 parts in stock for the old UV system, vs. 150-200 for the current system), less complicated means easier to find problems and fix them (just disconnecting the power for the old system was complicated, because there were so many switches and connections).	Increase the bandwidth on the discharge gate. Surge protection. Enclose the system to protect it from hot and cold (help maintain the seals on the power distribution panels and modules. Power back up.	Operator has been very happy with their current system. It has a smaller footprint and uses less power than their old system. It also works extremely well, even beyond the system design. Virtually all fecal coliform samples are below detection limit, even at peak flow. The system is designed to reduce fecal coliform to 23 CFU/100 mL for ~15 MGD and 13.6 mg/L TSS. They tested at 23 MGD with 48 mg/L TSS, and still only got a fecal count of 4 CFU/100 mL. Operator also liked several of the automated features with their current UV system, e.g., the fully automated inlet gates (so the operator doesn't have to go out to the unit to start up), and the online UVT monitoring to regulate dose (hasn't required much maintenance). For power: they can use either of 2 sources. They also have battery backup for 5 minutes, which is long enough to allow them to switch power sources if one fails. Their lift & UV systems are close to the power lines, which are configured so that power companies could hook up the units directly to the electrical lines, if necessary. This means that they can get disinfection, as long as there is power on the grid, even if a transformer blows.

Flow (MGD)	Lamp Type	Issues During Transition	Issues During Operation	Advice to Others	Ways to Improve UV	Comments
>20	LPHO	Dealing with head loss at another site owned by this group; much more work for the design, and a more complex control system in order to keep bulbs submerged		UV is very reliable, performs well, and requires only minor maintenance (based on another plant, where UV is already operating)		Chlorine gas is still in service currently, but will come out eventually, and hypochlorite will be used for backwashing the filters.
>20	LPLO	Algae growth in the effluent channels	Nothing	Easy to operate and far less deadly than chlorine	Currently upgrading to "Vendor" because the system uses less bulbs and has easier operation with auto wipes to clean the UV lamps	He also noted that when they first started the system, it worked great and they have very little maintenance for the first ~6 years. At that point the maintenance costs (ballasts and lamps) really started to increase.... Algae is more of a problem in the summer months, and not so much of a problem in the winter months. To deal with the growth they manually clean the effluent flume about once every 3 months.
>20	LPLO	Ease of operation	Sleeve fouling due to ferric (in the first year); solution: cleaned lamps & reduced ferric use, which has solved the problem	Raise as much electrical equipment as possible several feet above the liquid level. Our system has flooded due to tidal issues in the past and the internal ballast and circuit boards, which are located on the top portion of the module have gotten wet, plus there is humidity rising continuously while the lamps are lit. Having all electronics further away from the water surface would help eliminate some of our corrosion issues.	Move all of the power transformers further away from the exit path of the bank removal area. At the floor surface the power transformers and conduits are located several inches away from the UV banks. When raising or lowering a UV bank with the crane it becomes a very sensitive spot for damaging the bulbs and/or bulb banks.	
>20	LPLO					

Flow (MGD)	Lamp Type	Issues During Transition	Issues During Operation	Advice to Others	Ways to Improve UV	Comments
>20	MP	Increased need for routine maintenance.	High flow velocity through the UV channels required that the manufacturer retrofit the units with retainers on the downstream modules. These are used to prevent them from rising out of the water.	Reference check similar size installations with similar design specifications.	Increase the capacity by one unit to eliminate hydraulic restrictions and subsequent headloss.	
>20	MP	That it was inadequate for the application. We disinfect from May to October. The lack of reliable operation forced us to build a hypochlorite disinfection system that we operate routinely; cheaper and more reliably than the UV. We ran UV for the 2001 season and about half of the 2002 season, until we finished construction of the hypochlorite system. UV has not been used since.	It required more attention than expected. Using UV we had a hard time anticipating the dose that we should be using. Measuring UVT is a huge issue and difficult to accomplish. Therefore we just ran the system at 100%+. Even at that we could not comply with NPDES limits reliably.	Make very sure yours is the right application for UV.	Not to have installed it	Up front we knew there was a UVT issue with our waste stream. Our flow is about 75% industrial. Mostly corn wet milling; highly organic in nature with the ability to adversely affect UV operations. We had "normal" UVT's in the vicinity of 50% occasionally deflecting to about 35%. Multiple samples were sent to "Vendor" for analysis at their facility. We also tracked UVT daily on our effluent. "Vendor" assured us that our waste was treatable and that we could meet NPDES requirements. Even so, we required a pilot test before construction began. During the pilot test we found UVT deflections as low as 20%. Several red flags went up but again we were told that the waste was treatable and the pilot demonstrated marginal levels of effectiveness. The pilot did not take into account other variables that existed in the full-scale system, such as algal growth in the open channels that gave shelter to the bacteria. In the end the bench scale and the pilot were not completely representative of the full scale installation

Flow (MGD)	Lamp Type	Issues During Transition	Issues During Operation	Advice to Others	Ways to Improve UV	Comments
>20	MP	We use FeCl3. This has caused some coating of the sleeves, which has resulted in the need for additional maintenance.	We use FeCl3. This has caused some coating of the sleeves, which has resulted in the need for additional maintenance.	We have been very pleased with the results - no Cl_2, good bug kills. Electrical costs have gone up in our area and therefore, the cost of using UV has also increased.	We did improve our installation by putting it inside. The weather here is very variable, and we didn't want to be working on the system in extreme weather	They knew when they installed the system that it would cost more than their chlorine system, and it has - but the costs have been within their budget predictions.
>20	MP	Winter maintenance can be difficult (UV system is very exposed to the elements), and parts replacement.	Hydraulic oil manifold failure rate. The oil helps the wipers go up and down the bulb to clean. The manifolds are located under the UV banks, which makes them hard to get to. They were replacing them ~every 2 months, and were thinking about trying to move all of the parts that are located under the banks to a more accessible location, but it was too much trouble to redesign/re-engineer. Eventually, they got an external source to fabricate one for them - not only was that a cheaper replacement, it worked better, and they haven't had to replace it.	Consider all manufacturers.	Enclose the UV unit, to make maintenance easier (e.g., in winter or rain - can't work on the electronics when it's wet!).	The system does require a lot of maintenance time, but it has worked well for them. They've even exceeded the design flow of 48 MGD, going as high as 65 MGD for ~24 hours, and the system handled it well. The effluent coliform levels did rise, but stayed within permit limits, so they've been pretty happy with the UV performance. Excessive failure rate on ballasts is a problem. They go through 6-12 per year. Engineer did note that although he ranked "technical resources" fairly low in importance, he does find them valuable. He said that reading reports from WEF or other operators in the field helps in understanding the issues and knowing what to ask when you're trying to select a UV system.
>20	MP	Foam generated by high flow conditions as the UV restricts flow and causes foaming issues.	UV bulbs blowing out during start up of the system.	UV is safer and better impact on the environment.	Less man hours and longer lasting bulbs and ballasts.	

Flow (MGD)	Lamp Type	Issues During Transition	Issues During Operation	Advice to Others	Ways to Improve UV	Comments
>20	MP	Dramatic reduction in fecal coliform concentrations. Previously, they had very low permit levels for Cl2, but no dechlorination, so they could only add limited amounts of chlorine and had less disinfection.	Dramatic increase in preventative and corrective maintenance, e.g., bulb replacement (only average life of 2000 hrs, some die immediately after installation).	Install electrical power meter specifically to monitor power usage for the UV system, because it uses so much power (~$100,000/year). This helps with monitoring how efficient the system is, e.g., it draws more power in the summer due to algae growth.	Include the above and invest in reliable lamps.	
>20	MP		Length of the learning process to properly operated and maintain the system; Interrelationship of TSS to dosing	System should be validated for promised kills and capacity (max flow) prior to acceptance. Also, it is useful to request an annotated copy of the control software.	UV system should have a by-pass channel parallel to permit complete cleaning and maintenance of channel; provide canopy over effluent channel to prevent algae growth	The whole plant was upgraded at the time the UV system was installed. The UV system runs seasonally for 6 months/year. Training in sampling techniques is important for assuring accurate and repeatable samples for analysis. Chemical costs have decreased, but are probably equally offset by the increased power costs. Power consumption matches the vendor specs. The 20 hours/week needed to maintain the system breaks down into: 12 hrs maintenance, 4 hrs lab, 4 hrs ops. Vendor support frequency: 2-day annual visit when they startup (though he thinks that an annual visit would be good even for whole-year operation)

Flow (MGD)	Lamp Type	Issues During Transition	Issues During Operation	Advice to Others	Ways to Improve UV	Comments
>20	MP	Higher coliform counts - with Cl2, generally < 10 CFU/100 mL; with UV, avg ~75 CFU/100 mL They did not do any bench or pilot studies, but Vendor did take effluent samples back to their lab to test, and designed the system with their results. They weren't too worried about the lack of prior testing, because Vendor guaranteed the kill. Vendor did do a lot of extra work when the system was started up. For example, they found that the bulbs were rattling, so Vendor came back and provided extra supports for to stop that. Plant and Vendor worked together with the Fe issue as well.	Frequent tube cleaning due to Fe used upstream that deposits on the UV sleeve. As a result, they do ~16 hrs/week of maintenance, to clean the bulbs, replace any burnt out ones, etc. They also do not use the chemical cleaner supplied by the vendor; with the vendor's approval, they've switched to a high strength lime away, which works better for removing the Fe deposits on the bulb. They have very few electrical maintenance problems.	Use two channels, side by side, instead of one. The facility does have 2 banks in parallel, and the banks are redundant, so that they can operate on only one bank. They operate both banks full-time, but can afford to wait until regularly scheduled maintenance to replace bulbs. However, 2 channels in parallel would be a much better configuration for redundancy - e.g., if there is a problem with the weir gate that controls the water level, it affects both banks, while 2 parallel channels would eliminate that problem.	Improve the wipers. When the flow is >90 MGD (design flow is 127 MGD), the wipers don't work, because the pressure is too high for them to move upstream (wipers also work better going with the flow than against). This means that the bulbs don't get automatically cleaned, Fe deposits, and the dose drops. If they have this problem for an extended time (several hours), the coliform count for that day may be ~240; however, this still allows them to meet their geometric mean limits (they'd start to worry if the counts went up to ~700-800, but it never has).	Plant switched from gas to hypochlorite ~10-15 years ago, because of safety concerns, and regulatory requirements. They were interested in moving away from chemical disinfection altogether for the same reasons, and also because it has gotten harder to get the chemicals. In addition, they were upgrading their plant, and they had a new requirement to dechlorinate their water. They did a cost analysis, and UV was essentially the same cost as hypochlorite, given the construction and space requirements for hypochlorite. This prompted them to switch to UV. They sole sourced the system, and were happy with that. They chose MP lamps, because they're very space limited (w/in city limits, and bordering a lake). They have installed it inside a building, because they visited other plants and saw all the algae growth with outdoor systems. Being indoors also makes maintenance during the winter easier - they operate seasonally, from April 1 to Oct 15, and do larger system maintenance during the winter. They've been happy with their UV system, and as they upgrade their smaller facilities, they are switching those to UV as well. They recently installed UV on a 7 MGD plant that they own/operate.

Flow (MGD)	Lamp Type	Issues During Transition	Issues During Operation	Advice to Others	Ways to Improve UV	Comments
>20	MP	Higher operating costs & 100% more man hours for routine maintenance than represented or anticipated for optimal operation, to ensure that they meet regulatory requirements without unexpected failures: weekly inspection of electronics and mechanics; wiping and cleaning of bulbs; and lots of tightening to prevent seal failure/ leakage. Leakage causes electrical failure, premature lamp burnout, and lamp breakage - in 10.5 months, they replaced 25% of their 200 bulbs. They also clean their UV transmitter daily, and at one point the cuvette sea failed and caused leakage into the CPU of the unit. They suspected it was due to the lime-away that the vendor recommends to clean the cuvette. "Vendor" told them the lime-away wasn't a problem - but they've stopped using it anyway, and have not had any leakage problems since. Finally, they check the UVT every shift to calibrate it.	Ballasts subject to flooding during power failure because effluent is pumped. Sleeves and ballasts subject to leakage during normal operation leading to failure	Bulbs are expensive and need to be factored into O&M costs; make sure ballasts are above highest possible water level; have units in parallel instead of series. In a multiple bank installation, should provide the ability to shut down power supply to one bank while keeping the second on. Pay attention to algae build up in the dosing chamber and any loose algae that may escape from upstream units, e.g., final clarifier weirs and launders. Ensure vendor/designer fully incorporates permit disinfection requirements on a daily, weekly, monthly average and design to entire range. Plant had some high coliform concentrations at the beginning and were told "we didn't design to that level," though the superintendent is sure the vendor had a copy of their permits....	Superintendent did note that they were limited to a single channel because of utilities in that area, and that the single channel exacerbates the problems they have. They can't shut the banks down independently - if they shut one bank down, they must shut both down.	Plant superintendent said he "wouldn't have 'Vendor' back if hell froze over." "Vendor's" responses seem to be "CYA" - a typical response is "Well, we have lots of units in the operation and none of them have your problem," rather than an effort to fix the problem. Shipping of parts is slow. He noted that 2 nearby plants that use "Vendor" also had problems (different ones than his plant's), and that at one point, the state environmental agency called in "Vendor" and told them that they were concerned about the problems. "Vendor" did respond and had some more training classes, etc. but superintendent feels that they didn't address the real issue, which is the lack of robustness in the system. He commented that it's as if "Vendor" intended it for an industrial process, where you could shut it down if it wasn't working well - not a WWTP, where you have to run 24/7. This lack of reliability translates into a lot of money and time spent on their part to keep it running. Overall, Superintendent thinks that UV is a fine process and works well - it beats hypochlorite and chlorine because it's safer and cleaner - but at least their unit requires a *lot* of baby-sitting. It's very delicate and can be thrown off-balance (and out of compliance) very easily.

Flow (MGD)	Lamp Type	Issues During Transition	Issues During Operation	Advice to Others	Ways to Improve UV	Comments
>20	MP	More man/hours to maintain the system. We use 2 people for a week at a time just replacing bulbs and ballasts, but we do have a large UV system.	Have had problems with the computer program that runs our system. Also, bulb, ballast and other parts are very expensive, as are electrical costs.	Consider companies that have been in operation for years. We have worked well with "Vendor" in the installation/operation of our UV system they designed for us in 2003. Also consider the expense. It has been much more costly than Cl_2 and SO_2.		
>20	MP	Number of man hours needed. Vendor had estimated 3 hrs/week, but they spend ~20 to clean the bulbs & ballasts, and to clean the secondary weirs (to prevent algae from flaking off and contaminating the UV system)	Cleaning the secondary system to keep algae and other particles out of the UV system. They've also had problems with the automatic wiper cleaning because their effluent isn't clean enough.	Research the project thoroughly. You can't have too many people checking the design. They missed the temperature issue (see below), which has been very expensive for them.	Filters, which would eliminate some of the cleaning that they've had to do. It would also make the automatic cleaning work better. Cooling towers would also help. Their system uses the effluent to cool the system. Vendor says that if the effluent temperature is over ~28-29C, the system should have a cooling tower, so that the water can adequately cool. Their effluent sometimes exceeds that temperature, but they don't have a cooling tower, so they have more ballast and lamp failures, which leads to higher maintenance costs.	Overall, they like their system, and haven't had any issues with meeting their permit requirements. However, this has required vigilance on their part in maintaining and cleaning the secondary system, to minimize particles in the effluent.

Flow (MGD)	Lamp Type	Issues During Transition	Issues During Operation	Advice to Others	Ways to Improve UV	Comments
>20	MP	Higher cost and higher labor intensity	Amount of maintenance required	Prepare to take the needed time to maintain it. If there is another disinfecting system that is less costly and is not sole source, I would suggest looking at that system.	Make it more user friendly.	Their average fecal coliform concentrations in their effluent are 16-63 CFU/100 mL.
>20	MP	The system just never worked. It was designed to be used as the primary disinfection system, but they shut it down in 2003 (after their maintenance contract ran out and it still wasn't working). The contractor that operates the treatment plant says, "In a nutshell, it sucks!"	The system is not very durable and requires a lot of maintenance. Operator said that they needed vendor support constantly when the system was running. Their maintenance contract was for once per month, but the guy would come as often as he could - there was always lot of work to be done. They had a lot of electronics problems, and replaced virtually every electronic part in the system, and it still didn't work. They extended the maintenance contract for 1 more year in the hopes of getting it fixed, but gave up when it still wasn't fixed at the end of the second year.	Owner said, "If you don't need to switch, don't. It's costly, both in terms of capital costs and maintenance. If you must switch to UV, find a design that is *stable*." Operator said, "Research, research, research!" Find out what systems work and which don't (and under what conditions) by talking to people who use the units, looking at the internet, talking to vendors, etc.	Improve the durability - maybe by improving the flow	For a successful project, the owner felt that the equipment was the most important issue. The consultants they hired and their own staff were very competent, so their problems primarily stemmed from the equipment. He highly recommends visiting facilities that have been using the equipment you're considering for at least a couple of years. That way, you can get first-hand experience on the durability of the system. Operator's comment was "This system eats money like candy." He also notes that the system is *very* labor-intensive, and the owner noted that electricity is expensive. Operator liked the vendor, and he felt that they tried really hard to get it to work, but that the system just sucked - so he'd recommend the vendor but not the system. Owner agreed that Vendor worked well with them, but would only recommend them if their product has improved.

APPENDIX D

LITERATURE RESOURCES

Chlorine Technologies

Adam, L.C.; Gordon, G. Hypochlorite Ion Decomposition: Effects of Temperature, Ionic Strength, and Chloride Ion. *Inorg. Chem.* **1999**, *38*(6), 1299-1304.

Aieta, E.M.; Berg, J.D.; Roberts, P.V.; Cooper, R.C. Comparison of Chlorine Dioxide and Chlorine in Wastewater Disinfection. *J. Water Pollut. Control Fed.* **1980**, *52*(4), 810-824.

Ainsworth, L.; Hampton, J. A Safe Option for Disinfection: On-Site Sodium Hypochlorite Generation. *Water Eng. Manage.* **1997**, *144*(12), 25-27.

Baiden, L.H.; Hunter, G.L.; Shurtz, K.B.; Wells, W.J. An Evaluation of UV and NaOCl for Wastewater Disinfection at the Pelham WWTP; *Proc. WEFTEC 2002*, Water Environment Federation, 2002.

Baur, R. Hypochlorite Reality. *Water Environ. Technol.* **2003**, *15*(11), 47-50.

Bedner, M.; MacCrehan, W.A. Transformation of Acetaminophen by Chlorination Produces the Toxicants 1,4-Benzoquinone and N-Acetyl-p-Benzoquinone Imine. *Environ. Sci. Technol.* **2006**, *40*(2), 516-522.

Bedner, M.; MacCrehan, W.A.; Helz, G.R. Making Chlorine Greener: Investigation of Alternatives to Sulfite for Dechlorination. *Water Res.* **2004**, *38*(10), 2505-2514.

Bedner, M.; Maccrehan, W.A.; Helz, G.R. Production of Macromolecular Chloramines by Chlorine-Transfer Reactions. *Environ. Sci. Technol.* **2004**, *38*(6), 1753-1758.

Bergendahl, J.A.; Stevens, L. Oxidation Reduction Potential as a Measure of Disinfection Effectiveness for Chlorination of Wastewater. *Environ. Prog.* **2005**, *24*(2), 214-222.

Bess, J.W.; Casson, L.W.; Navin, T.J.; Lubie, D.B. On-Site Sodium Hypochlorite Generation; *Proc. WEFTEC 2006,* Water Environment Federation, 2006.

Bess, J.W.; Casson, L.W.; Navin, T.J.; Lubie, D.B. The Effect of On-site Generated Ultra Pure Hypochlorite on Water and Wastewater Disinfection; *Proc. WEFTEC 2003*, Water Environment Federation, 2003.

Blatchley, E.R.; Gong, W.L.; Rose, J.B.; Huffman, D.E.; Otaki, M.; Lisle, J.T. Effects of Wastewater Disinfection on Human Health. Report No. 99-HHE-1, 2005.

Bossard, G.; Briggs, G.; Stow, T. Optimizing Chlorination/Dechlorination at a Wastewater Treatment Plant. *Public Works* **1995**, *126*(1), 33-35.

Boyd, G.R.; Zhang, S.; Grimm, D.A. Naproxen Removal from Water by Chlorination and Biofilm Processes. *Water Res.* **2005**, *39*(4), 668-676.

Brunner, T.A.; Stella, D.F.; Vause, K.H. Optimizing Disinfection through Mixing and ORP Control: A Case Study of the Asplund WWTF in Anchorage, Alaska; *Proc. WEFTEC 2004*, Water Environment Federation, 2004.

Butts, E. Chlorination: Is It Really That Important? *Water Well J.* 2004, *58*(6), 30-33.

California Code of Regulations, Title 22 Chapter 3 Water Recycling Criteria; *60301.320,* 2000.

Calmer, J.; Neethling, J.B.; Murbach, D. Nitrification Effects of Chlorination and Disinfection, *Proc. Disinfection 1998,* Water Environment Federation, 1998

Caravelli, A.; Contreras, E.M.; Giannuzzi, L.; Zaritzky, N. Modeling of Chlorine Effect on Floc Forming and Filamentous Micro-organisms of Activated Sludges. *Water Res.* **2003**, *37*(9), 2097-2105.

Casson, L.W.; Bess, J.W. Conversion to on-Site Sodium Hypochlorite Generation; CRC Press: Boca Raton, FL, 2002.

Casson, L.W.; Edgar, W.W.; Bess, J.W. Wastewater Disinfection Using Sodium Hypochlorite Generated on-Site from Process Residuals; *Proc. WEFTEC 2001*, Water Environment Federation, 2001.

Chauret, C.; Liere, J.; Gillem, A.; Smith, C.; Hofmann, R.; Durance, C.; Andrews, R.C. Impact of Particulate Matter on the Inactivation by Chlorine and Monochloramine of *Escherichia Coli* O157:H7 and MS2 Bacteriophages; *Proc. 2004 AWWA Water Quality Technology Conference*, American Water Works Association, 2004.

Chauret, C.; Liere, J.; Gillem, A.; Smith, C.; Hofmann, R.; Durance, C.; Andrews, R.C. Impact of Particulate Matter on the Inactivation by Chlorine and Monochloramine of *Escherichia Coli* O157:H7 and MS2 Bacteriophages; *Proc. 2004 AWWA Water Quality Technology Conference*, American Water Works Association, 2004.

Chen, G.H.; Saby, S.; Djafer, M.; Mo, H.K. New Approaches to Minimize Excess Sludge in Activated Sludge Systems; *Proc. Sludge Management Entering the Third Millenium*, 2001.

Chen, W.; Jensen, J.N. Effect of Chlorine Demand on the Ammonia Breakpoint Curve: Model Development, Validation with Nitrite, and Application to Municipal Wastewater. *Water Environ. Res.* **2001**, *73*(6), 721-731.

Chlorine Institute. Sodium Hypochlorite Manual. Report No. 96, 2000.

Chlorine Institute. Personal Protective Equipment for Chlorine and Sodium Hydroxide. Report No. 65, 1995.

Choi, J.; Valentine, R.L. Formation of N-Nitrosodimethylamine (NDMA) from Reaction of Monochloramine: A New Disinfection By-product. *Water Res.* **2002**, *36*(4), 817-824.

Connell, G.F. Key Operating Strategies for Chlorine Disinfection Operating Systems; *Proc. WEFTEC 2006*, Water Environment Federation, 2006.

Connell, G.F. The Chlorination/Dechlorination Handbook; Water Environment Federation: Alexandria, VA, 2002.

Conrad, D.J.; Huck, P.M. Chlorination By-products from Ferulic Acid (a Lignin Model Compound) and Biologically Treated Thermomechanical Chemithermomechanical Pulp Mill Effluents. *Water Res.* **1996**, *30*(11), 2776-2784.

Cooper, R.C.; Salveson, A.; Sakaji, R.H.; Tchobanoglous, G.; Requa, D.A.; Whitley, R. Comparison of the Resistance of MS-2 and Poliovirus to UV and Chlorine Disinfection; *Proc. WateReuse Association Symposium*, WateReuse Association, 2000.

Darby, J.; Heath, M.; Jacangelo, J.; Loge, F.Swaim, P. Comparison of UV Irradiation to Chlorination: Guidance for Achieving Optimal UV Performance. Report No. WERF 91-WWD-1, 1995.

Delgado-Diaz, S.; Alvarez, M.; Rodriguez-Gomez, L.E.; Aguiar, E. Using Oxidation-Reduction Potential as Septicity Control Parameter during Reclaimed Wastewater Transportation. *Water Environ. Res.* **2000**, *72*(4), 455-459.

Devkota, L.M.; Williams, D.S.; Matta, J.H.; Albertson, O.E.; Grasso, D.; Fox, P. Variation of Oxidation Reduction Potential Along the Breakpoint Curves in Low-Ammonia Effluents. *Water Environ. Res.* **2000**, *72*(5), 610-617.

Devkota, L.; Williams, D.; Matta, J.; Albertson, O.; Grasso, D.; Fox, P. Staged Dosing. *Water Environ. Technol.* **1996**, *8*(9), 53-59.

Diao, H.F.; Li, X. Y.; Gu, J.D.; Shi, H.C.; Xie, Z.M. Electron Microscopic Investigation of the Bactericidal Action of Electrochemical Disinfection in Comparison with Chlorination, Ozonation and Fenton Reaction. *Process Biochemistry* **2004**, *39*(11), 1421-1426.

Dietrich, J.P.; Basagaoglu, H.; Loge, F.J.; Ginn, T.R. Preliminary Assessment of Transport Processes Influencing the Penetration of Chlorine into Wastewater Particles and the Subsequent Inactivation of Particle-Associated Organisms. *Water Res.* **2003**, *37*(1), 139-149.

Dodd, M.C.; Huang, C.H. Transformation of the Antibacterial Agent Sulfamethoxazole in Reactions with Chlorine: Kinetics, Mechanisms, and Pathways. *Environ. Sci. Technol.* **2004**, *38*(21), 5607-5615.

Doorn, M. Environmental Technology Verification Report: On-Site Sodium Hypochlorite Generation and Inactivation of *Pseudomonas* in Raw Drinking Water. Report No. NSF 01/21/EPADW395, 2001.

Dzombak, D.; Deeb, R.; Theis, A.; Young, T.; Kavanaugh, M.; Markowitz, D. Cyanide Formation and Fate in Complex Effluents and Its Relation to Water Quality Criteria - Results from a Three-Year WERF Study; *Proc. WEFTEC 2003*, Water Environment Federation, 2003.

Dzombak, D.; Deeb, R.; Zheng, A.; Anderson, T.; Kavanaugh, M. Cyanide Formation Due to Wastewater Disinfection: Laboratory Issues and Compliance Strategies; *Proc. WEFTEC 2004*, Water Environment Federation, 2004.

Elefritz, R.A. Jr. Comparison of Alternate Methods of Disinfection for Reclaim Effluents. *Florida Water Resource Journal* **2000**, 19-25.

Environment Canada. Review of Municipal Effluent Chlorination/Dechlorination Principles, Technologies and Practices. Report No. En84-36/2006E, 2006.

Esvelt, L.A.; Kaufman, W.J.; Selleck, R.E. Toxicity Removal from Municipal Wastewater. Report No. SERL 71-7 (Berkeley Sanitary Engineering Research Laboratory), 1971.

Florida Administrative Code, *62-600.400,* 1996.

Freese, S.D.; Nozaic, D.J. Chlorine: Is It Really So Bad and What Are the Alternatives? *Water SA* **2004**, *30*(5), 566-572.

Freese, S.D.; Nozaic, D.J.; Bailey, I.; Trollip, D.L. Alternative Disinfectants for Wastewater Effluents: Viable or Prohibitively Expensive? *Water SA* **2002**, *28*(SPEC ISS), 23-32.

Gillette, R.; Marks, K.; Pawluk, D. City of Roseville Alternative Disinfection Study; *Proc. WEFTEC 2003*, Water Environment Federation, 2003.

Gordon, G.; Bubnis, B. Sodium Hypochlorite Speciations; *Proc. 2000 AWWA Water Quality Technology Conference*, American Water Works Association, 2000.

Gordon, G.L.; Adam, L.; Bubnis, B. Minimizing Chlorate Ion Formation in Drinking Water When Hypochlorite Ion Is the Chlorinating Agent; *Proc. 1995 AWWA* , American Water Works Association, 1995.

Great Lakes - Upper Mississippi River Board of State and Provincial Public Health and Environmental Managers. Ten State Standards: Recommended Standards for Wastewater Facilities. 2004.

Groninger, T.J.; Apicella, G.A.; Delva, E. Evaluation of Enhanced Mixing and ORP Control for Wastewater Disinfection; *Proc. WEFTEC 2002*, Water Environment Federation, 2002.

Hamby, S. A Comparison of Operating Strategies - Chlorine Vs UV; *Proc. WEFTEC 2006*, Water Environment Federation, 2006.

Hanzon, B.; Hartfelder, J.; O'Connell, S.; Murray, D. Comparing UV and Chlorine for Wastewater Disinfection; *Proc. Disinfection 2005*, Water Environment Federation, 2005.

Harp, D.L. Specific Determination of Inorganic Monochloramine in Chlorinated Wastewaters. *Water Environ. Res.* 2000, *72*(6), 706-713.

Hassen, A.; Heyouni, A.; Shayeb, H.; Cherif, M.; Boudabous, A. Inactivation of Indicator Bacteria in Wastewater by Chlorine - a Kinetics Study. *Bioresour. Technol.* **2000**, *72*(1), 85-93.

Hernlem, B.J.; Tsai, L. Titration of Chlorine: Amperometric Versus Potentiometric. *J. Am. Water Works Assoc.* **2000**, *92*(12), 101-107.

Huitric, S.J.; Kuo, J.; Creel, M.; Tang, C.C.; Snyder, D.; Horvath, R.W.; Stahl, J.F. Reclaimed Water Disinfection Alternatives to Avoid NDMA and THM Formation; *Proc. WEFTEC 2006*, Water Environment Federation, 2006.

Isaac, R.A. Disinfection Chemistry. *Water Environ. Technol.* **1996**, *8*(9), 47-50.

Jameel, R.H.; Helz, G.R. Organic Chloramines in Disinfected Wastewaters: Rates of Reduction by Sulfite and Toxicity. *Environmental Toxicology and Chemistry* **1999**, *18*(9), 1899-1904.

Jensen, J.S.; Lam, Y.; Helz, G.R. Role of Amide Nitrogen in Water Chlorination: Proton NMR Evidence. *Environ. Sci. Technol.* **1999**, *33*(20), 3568-3573.

Jensen, J.S.; Helz, G.R. Rates of Reduction of N-Chlorinated Peptides by Sulfite: Relevance to Incomplete Dechlorination of Wastewaters. *Environ. Sci. Technol.* **1998**, *32*(4), 516-522.

Jimenez Cisneros, B.; Chavez Mejia, A. Chlorine Disinfection of Advanced Primary Effluent for Reuse in Irrigation in Mexico; *Proc. 2000 WateReuse Research Conference*, WateReuse Foundation, 2000.

Jimenez, B.; Chavez, A.; Hernandez, C. Alternative Treatment for Wastewater Destined for Agricultural use. *Water Sci. Technol.* **1999**, *40*(4-5), 355-362.

Kavanaugh, M.; Dzombak, D.A.; Theis, T.L.; Young, T.C.; Luthy, R.G. Cyanide Formation and Fate in Complex Effluents and Its Relation to Water Quality Criteria. Report No. 98-HHE-5, 2003.

Khoury, J.; Pang, M.; Young, C.; Pandit, A.; Carr, S.; Fischer, D. Cyanide Generation during Preservation of Chlorinated Wastewater Effluent Samples for Total Cyanide Analysis; *Proc. WEFTEC 2005*, Water Environment Federation, 2005.

Kim, H.; Kwon, S.; Han, S.; Yu, M.; Gong, S.; Colosimo, M.F. New Process Control Strategy for Wastewater Chlorination and Dechlorination Using ORP/pH. *Water Sci. Technol.* 2006, *53*(4-5), 431-438.

Kim, H.; Kwon, S.; Han, S.; Yu, M.; Kim, J.; Gong, S.; Colosimo, M.F. New ORP/pH Based Control Strategy for Chlorination and Dechlorination of Wastewater: Pilot Scale Application. *Water Sci. Technol.* 2006, *53*(6), 145-151.

Kim, Y.H.; Hensley, R. Effective Control of Chlorination and Dechlorination at Wastewater Treatment Plants Using Redox Potential. *Water Environ. Res.* 1997, *69*(5), 1008-1014.

Kim, Y.; Strand, R. Effective Control of Wastewater Chlorination and Dechlorination; *Proc. 1996 International Conference on Advances in Instrumentation and Control. Part 2 (of 2), Oct 6-11 1996*, Instrument Society of America, Research Triangle Park, NC, USA: Chicago, IL, USA, 1996.

Kim, Y.H.; Strand, R.L. Oxidation Reduction Potential Cuts Wastewater Chemicals. *INTECH* **1997**, *44*(3), 40-44.

Kopchynski, D.M.; Palmer, T.M.; Ross, M.C. Comparisons of On-line ORP and Chlorine Residual Monitoring/Control Systems for Wastewater Treatment Plant Final Effluent Chlorination; *Proc. WEFTEC 2001*, Water Environment Federation, 2001.

Koukouraki, E.; Diamadopoulos, E. Modeling the Formation of THM during Chlorination of Treated Municipal Wastewater; *Proc. IWA Regional Symposium on Water Recycling in the Mediterranean Region*, 2002.

Koukouraki, E.; Diamadopoulos, E. THM Formation during Chlorination of Treated Municipal Wastewater; *Proc. World Water Congress of the International Water Association*, 2001.

Krasner, S.W.; Weinberg, H.S.; Richardson, S.D.; Pastor, S.J.; Chinn, R.; Sclimenti, M.J.; Onstad, G.D.; Thurston, A.D., Jr. Occurrence of a New Generation of Disinfection Byproduct. *Environ. Sci. Technol.* **2006**, *40*(23), 7175-85.

Kuo, J.; Stahl, J.F.; Burton, D.; El Jack, Z.; Horvath, R.W.; Tang, C.C. Chloramination of N/DN Effluent - Meeting the Ammonia, Coliform and THM Limits; *Proc. WEFTEC 2003*, Water Environment Federation, 2003.

Kuo, J.; Smith, S.O. Disinfection. *Water Environ. Res.* **1996**, *68*(4), 503-510.

Lehmann, F.M. The use of Bromine Chemistry for Sand Filter Disinfection; *Proc. AESF/EPA Conference for Environmental Excellence, Jan 17-19 2000*, American Electroplaters and Surface Finishers Soc. Inc: Orlando, FL, United States, 2000.

Lind, C. Wastewater Dechlorination Options. *Public Works* **1995**, *126*(10), 69-70.

Macauley, J.J.; Qiang, Z.; Adams, C.D.; Surampalli, R.; Mormile, M.R. Disinfection of Swine Wastewater Using Chlorine, Ultraviolet Light and Ozone. *Water Res.* **2006**, *40*(10), 2017-2026.

MacCrehan, W.A.; Bedner, M.; Helz, G.R. Making Chlorine Greener: Performance of Alternative Dechlorination Agents in Wastewater. *Chemosphere* **2005**, *60*(3), 381-388.

MacCrehan, W.A.; Jensen, J.S.; Helz, G.R. Detection of Sewage Organic Chlorination Products That Are Resistant to Dechlorination with Sulfite. *Environ. Sci. Technol.* **1998**, *32*(22), 3640-3645.

Marks, D. Automated chlor/dechlor Control Protects Fishery, Reduces Chemical and Labor Costs. *Water/Engineering and Management* **1998**, *145*(11), 4.

Marx, S. Sodium Hypochlorite Safer than Chlorine, but Must Be Approached with Respect. *J. New England Water Environment Association* **1996**, *30*(1), 49-52.

Massachusetts Department of Environmental Protection. The Occurrence and Sources of Perchlorate in Massachusetts. Updated April 2006.

Mitch, W.A.; Oelker, G.L.; Hawley, E.L.; Deeb, R.A.; Sedlak, D.L. Minimization of NDMA Formation during Chlorine Disinfection of Municipal Wastewater by Application of Pre-Formed Chloramines. *Environ. Eng. Sci.* **2005**, *22*(6), 882-890.

Mitch, W.A.; Sedlak, D.L. Characterization and Fate of N-Nitrosodimethylamine Precursors in Municipal Wastewater Treatment Plants. *Environ. Sci. Technol.* **2004**, *38*(5), 1445-1454.

Mitch, W.A.; Sedlak, D.L. Fate of N-Nitrosodimethylamine (NDMA) Precursors during Municipal Wastewater Treatment; *Proc. AWWA Annual Conference*, American Water Works Association, 2003.

Moffa, P.E.; Somerlot, C.A Comparison between High Rate Chlorine and Bromine Disinfection; *Proc. Disinfection 2007*, Water Environment Federation, 2007.

Moore, L.W.; Baker, W.L. Solving a Complex Chlorination Problem. *Water Environ. Technol.* **1996**, *8*(12), 41-46.

Na, C.; Olson, T.M. Mechanism and Kinetics of Cyanogen Chloride Formation from the Chlorination of Glycine. *Environ. Sci. Technol.* **2006**, *40*(5), 1469-1477.

Naber, J. Switching to Sodium Hypochlorite Disinfection. *Water Environ. Technol.* **2003**, *15*(9), 100-102.

Nagel, W.H. Importance of Response Time in Chlorination Control Systems. *Public Works* **1995**, *126*(2), 42-43.

Najm, I.; Trussell, R.R. NDMA Formation in Water and Wastewater. *J. Am. Water Works Assoc.* **2001**, 92-99.

Narkis, N.; Kott, Y. Comparison between Chlorine Dioxide and Chlorine for use as a Disinfectant of Wastewater Effluents. *Water Sci. Technol. WSTED 4,* 1992, *26*(7-8).

Ormeci, B.; Linden, K.G. Comparison of UV and Chlorine Inactivation of Particle and Non-Particle Associated Coliform. *Water Sci. Technol.: Water Supply* **2002**, *2*(5-6), 403-410.

Pandit, A.; Young, C.; Pang, M.; Khoury, J.; Carr, S.; Fischer, D.; Stahl, J. Factors Affecting Cyanide Generation in Chlorinated Wastewater Effluent Matrix; *Proc. WEFTEC 2006*, Water Environment Federation, 2006.

Pinkston, K.E.; Sedlak, D.L. Transformation of Aromatic Ether- and Amine-Containing Pharmaceuticals during Chlorine Disinfection. *Environ. Sci. Technol.* **2004**, *38*(14), 4019-4025.

Poliakova, E.E.; Zholdakova, Z.I.; Khamidoulina, K.K.; Lebedev, A.T. Formation of Organochlorine By-products during Chlorination of Wastewaters Containing Cyclohexene and Butanol; *Proc. Sixth International Conference on the Modelling, Monitoring and Management of Water Pollution, Water Pollution VI, Sep 17-19 2001*, WIT Press, Southampton, United Kingdom: Rhodes, Greece, 2001.

Prentiss, D. Safety Training for Chlorine Gas. *Water Environ. Technol.* **2003**, *15*(8), 64-65.

Ramirez, G.W.; Alonso, J.L.; Villanueva, A.; Guardino, R.; Basiero, J.A.; Bernecer, I.; Morenilla, J.J. Rapid, Direct Method for Assessing Chlorine Effect on Filamentous Bacteria in Activated Sludge. *Water Res.* **2000**, *34*(15), 3894-3898.

Rebhun, M.; Heller-Grossman, L.; Manka, J. Formation of Disinfection Byproducts during Chlorination of Secondary Effluent and Renovated Water. *Water Environ. Res.* **1997**, *69*(6), 1154-1161.

Reddy, S. Computer Simulation: Reducing Cost of Chlorine in Wastewater Plant. *Chemical Engineering World* **2003**, *38*(4), 66-67.

Reisch, M. Assessing Rita, *Chemical and Engineering News* **2005**, 83(40), 9.

Ryder, R.A.; Faller, J.A. Disinfection Challenges with Partially Nitrifying Fixed Film Reactor Effluent, *Proc. Planning, Design, & Operations of Effluent Disinfection Systems*, Water Environment Federation, 1993.

Ryder, R.A. Testing & Correcting Short- Circuiting of Chlorine Contact Chambers, *Proc Disinfecting Wastewater for Discharge & Reuse*, Water Environment Federation, 1996.

Ryder, R.A. Lowering pH of Hypochlorinated Trickling Filter Effluent Assure Coliform Discharge Compliance, *Proc. Disinfection 2000*, Water Environment Federation, 2000.

Ryder, RA. The Necessity of Absolute Reliability In Chlorine Disinfection & Dechlorination Systems *Proc Disinfection 2002,* Water Environment Federation. 2002.

Ryu, H.; Alum, A.; Abbaszadegan, M. Microbial Characterization and Population Changes in Nonpotable Reclaimed Water Distribution Systems. *Environ. Sci. Technol.* **2005**, *39*(22), 8600-8605.

Sanitation Districts of Los Angeles County. Pomona Virus Study - Final Report. 1977.

Schreiber, I.M.; Mitch, W.A. Influence of the Order of Reagent Addition on NDMA Formation during Chloramination. *Environ. Sci. Technol.* **2005**, *39*(10), 3811-3818.

Scruggs, C.; Hunter, G.; Wallis-Lage, C. Could EDC Control Change Disinfection Practices at WWTPs? *Proc. Disinfection 2005,* Water Environment Federation, 2005.

Shayeb, H.; Riabi, T.; Roustan, M.; Hassan, A. Etude Experimentale Et Modelisation De La Desinfection Par Le Chlore Des Eaux Usees Epurees; Experimental Study of Wastewater Chlorination and Application to the Design of Chlorine Contact Facilities. *Rev. Sci. Eau/J. Water Sci.* **1999**, *11*(4), 517-536.

Shayeb, H.; Riabi, T.; Roustan, M.; Hassan, A. Étude Expérimentale Et Modélisation De La Désinfection Par Le Chlore Des Eaux Usées Épurées. *Revue Des Sciences De l'Eau (Paris)* **1998**, *11*(4), 517-536.

Siddiqui, M.; Atasi, K.Z. N-Nitrosodimethylamine: A Disinfectant Byproduct and Its Occurrence in Wastewater. *Water Environ. Res.* **2004**, *76*(4), 316-326.

Singer, P.C. Disinfection Byproducts in Drinking Water: Additional Science and Policy Considerations in the Pursuit of Public Health Protection. *The 2006 Clarke Lecture*, National Water Research Institute, 2006.

Sloane, J.R.; Branch, B. Considerations Involved in Selecting the on-Site Sodium Hypochlorite Process for Water Disinfection at the Daytona Beach Brennan Water Treatment Plant; *Proc. 2001 Annual Conference;* American Water Works Association, 2001.

Snyder, C.H. The Extraordinary Chemistry of Ordinary Things; John Wily & Sons: New York, 1995.

Snyder, S.A.; Wert, E.C.; Lei, H.; Westerhoff, P.; Yoon, Y. Removal of EDCs and Pharmaceuticals in Drinking and Reuse Treatment Processes. Report No. 2758 (AWWARF), 2007.

Stella, D.F.; Brunner, T.A.; Vause, K.H. Optimizing Disinfection through Induction Mixing and ORP Control. *Water Environ. Technol.* **2005**, *17*(8), 45-49.

Swift, J.L.; Wilson, J.P.; Johnson, M. Assessment of Operation and Maintenance Costs for Ultraviolet Disinfection Systems; *Proc. WEFTEC 2000*, Water Environment Federation, 2000.

Tang, C.C.; Huitric, S.J.; Kuo, J.F.; Horvath, R.W.; Stahl, J.F. Disinfection Alternatives to Avoid NDMA Formation. *Wastewater Professional* **2006**, *42*(3), 14-23.

Tchobanoglous, G.; Burton, F.L.; Stensel, D.H. Wastewater Engineering Treatment and Reuse; McGraw Hill: Boston, 2003.

Tikkanen, M., Schroeter, J.H., Leong, L.Y.C.; Ganesh, R. Guidance Manual for Disposal of Chlorinated Water, AWWA Research Foundation, 2001.

Tobin, A.; Burror, J.; Yahya, M.; Wistrom, A.; Nedic, S. Alternatives for Disinfection Effluent Generated at Orange County Sanitation District Wastewater Treatment Plants; *Proc. WEFTEC 2005*, Water Environment Federation, 2005.

Tree, J.A.; Adams, M.R.; Lees, D.N. Chlorination of Indicator Bacteria and Viruses in Primary Sewage Effluent. *Appl. Environ. Microbiol.* **2003**, *69*(4), 2038-2043.

Tree, J.A.; Adams, M.R.; Lees, D.N. Virus Inactivation during Disinfection of Wastewater by Chlorination and UV Irradiation and the Efficacy of F^+ Bacteriophage as a 'Viral Indicator'. *Water Sci. Technol.* **1997**, *35*(11-12), 227-232.

Tsai, C.T.; Kuo, C.T.; Lin, S.T. Analysis of Organic Halides in Hospital Waste Sludge Disinfected Using Sodium Hypochlorite (NaOCl). *Water Res.* **1999**, *33*(3), 778-784.

Ueno, H.; Nakamuro, K.; Moto, T.; Sayato, Y. Disinfection By-products in the Chlorination of Organic Nitrogen Compounds: Possible Pathways for the Formation of Disinfection By-products. *Water Supply* **1995**, *13*(3-4), 171-176.

U.S. Department of Homeland Security, Chlorine Gas Decision Tool for Water and Wastewater Utilities; 2006.

U.S. EPA. Alternative Disinfectants and Oxidants Guidance Manual. Report No. EPA 815/R-99/014, 1999.

U.S. EPA. Design Manual: Municipal Wastewater Disinfection. Report No. EPA 625/1-86/021, 1986.

Wang, H.; Shao, X.; Falconer, R.A. Flow and Transport Simulation Models for Prediction of Chlorine Contact Tank Flow-through Curves. *Water Environ. Res.* **2003**, *75*(5), 455-471.

Wang, L.; Zhang, T.; Hu, H. Comparison of the Quality and Toxicity of Wastewater After Chlorine and Chlorine Dioxide Disinfections. *Huanjing Kexue/Environmental Science* **2005**, *26*(6), 75-78.

WEF. Wastewater Disinfection – Training Manual. 2006.

WEF. Water Environment Federation Disinfection Process Survey Summary. *Water Environ. Technol.* **2004**, *16*(7), 47-49.

WEF. Wastewater Disinfection - MOP FD-10. 1996.

Wei, D.; Wang, L.; Wei, J.; Hu, H. Toxicity Screening and Evaluating in Chlorination Disinfection of Wastewater Reclamation Processes. *Water Sci. Technol.* **2006**, *53*(9), 239-246.

White, G.C. Handbook of Chlorination and Alternative Disinfection; Wiley & Sons: New York, 1999.

Wilczak, A.; Hoover, L.L.; Hubert Lai, H. Effects of Treatment Changes on Chloramine Demand and Decay. *J. Am. Water Works Assoc.* **2003**, *95*(7), 94-106.

Woolschlager, J.; Soucie, W.J. Chorine Reactivity with Concrete Pipe Surfaces; *Proc. World Water and Environmental Resources Congress 2003*, American Society of Civil Engineers, 2003.

Yang, X.; Shang, C.; Huang, J. DBPFormation in Breakpoint Chlorination of Wastewater. *Water Res.* **2005**, *39*(19), 4755-4767.

Yoon, J.; Jensen, J.N. Chlorine Transfer from Inorganic Monochloramine in Chlorinated Wastewater. *Water Environ. Res.* **1995**, *67*(5), 842-847.

Yu, R.F.; Cheng, W.P. Determination of Chlorine Demand in Water and Wastewater Chlorination by Oxidation-Reduction Potential. *Water Sci. Technol.: Water Supply* **2003**, *3*(1-2), 313-320.

Yu, R. Feed-Forward Dose Control of Wastewater Chlorination Using on-Line pH and ORP Titration. *Chemosphere* **2004**, *56*(10), 973-980.

Zenz, D.R.; Law, K.; Lanyon, R.; Sobanski, J.; Kunetz, T.; Bouchard, A.B.; Hass, C.; Schwab, K.; Marinas, B. Study of Effluent Disinfection for Urban Rivers in Chicago; *Proc. Disinfection 2007*, 2007.

Zheng, A.; Dzombak, D.A.; Luthy, R.G. Effects of Thiocyanate on the Formation of Free Cyanide during Chlorination and Ultraviolet Disinfection of Publicly Owned Treatment Works Secondary Effluent. *Water Environ. Res.* **2004**, *76*(3), 205-212.

UV Technologies

Ahn, K.H.; Park, K.Y.; Maeng, S.K.; Song, K.G.; Kim, K.P.; Lee, S.H.; Kweon, J.H. Color Removal and Disinfection with UV/H_2O_2 System for Wastewater Reclamation and Reuse. *Water Sci. Technol.: Water Supply* **2005**, *5*(1), 51-57.

Allbaugh, T.A.; Kang, S.J.; Olmstead, K.P.; Erickson, T.L.; Thomas, P.; Thomas, L. Design and Operation of an Ultraviolet Disinfection System for A Municipal Wastewater Treatment Plant in Ypsilanti, Michigan; *Proc. WEFTEC 2004*, Water Environment Federation, 2004.

Amirhor, P. Overview of Wastewater Effluent Disinfection by Ultraviolet Irradiation. *J. New England Water Environment Association* **1999**, *33*(1), 27-38.

Amos, S.A.; Davey, K.R.; Thomas, C.J. A Comparison of Predictive Models for the Combined Effect of UV Dose and Solids Concentration on Disinfection Kinetics of *Escherichia Coli* for Potable Water Production. *Process Safety and Environmental Protection: Transactions of the Institution of Chemical Engineers, Part B* May, 2001, *79*(3), 174-182.

Andreadakis, A.; Mamais, D.; Christoulas, D.; Kabylafka, S. Ultraviolet Disinefection of Secondary and Tertiary Effluent in the Mediterranean Region; *Proc. 1998 6th International Conference on Advanced Wastewater Treatment, Recycling and Reuse*, Elsevier Science, 1999.

Anon. Gold Bar Centre of Excellence. *Canadian Consulting Engineer* **2005**, *46*(4), 27.

Anon. High-Rate Treatment System Minimizes Bremerton's CSOs. *Water Environ. Technol.* **2003**, *15*(5), 78.

Anon. Military Point Regional Advanced Wastewater Treatment Facility. *Water Environ. Technol.* **2002**, *14*(8), 74.

Asano, T.; Burton, F.L.; Leverenz, H.L.; Tshuchihashi, R.; Tchobanoglous, G. Water Reuse: Issues, Technologies, and Applications; McGraw-Hill: New York, 2007.

Awad, J. Ultraviolet Disinfection for Water Reuse. Planning, Design, and Operation of Effluent Disinfection Systems. *Proc. WEF Specialty Conference - Effluent Disinfection Systems*, Water Environment Federation, 1993.

Babcock Jr., R.W.; McNair, D.A.; Edling, L.A.; Nagato, H. Evaluation of a System for Residential Treatment and Reuse of Wastewater. *J. Environ. Eng.* **2004**, *130*(7), 766-773.

Baiden, L.H.; Hunter, G.L.; Shurtz, K.B.; Wells, W.J. An Evaluation of UV and NaOCl for Wastewater Disinfection at the Pelham WWTP; *Proc. WEFTEC 2002*, 2002.

Balchunas, B.M.; Hentz, L.H. Situation Normal during a UV Disinfection Retrofit. *Water Engineering and Management* **2002**, *149*(11), 24-28.

Baron, J. Repair of Wastewater Microorganisms After Ultraviolet Disinfection Under Seminatural Conditions. *Water Environ. Res.* **1997**, *69*(5), 992-998.

Baron, J.; Bourbigot, M. Repair of *Escherichia Coli* and Enterococci in Sea Water After Ultraviolet Disinfection Quantification Using Diffusion Chambers. *Water Res.* **1996**, *30*(11), 2817-2821.

Batch, L.F.; Schulz, C.R.; Linden, K.G. Evaluating Water Quality Effects on UV Disinfection of MS2 Coliphage. *J. Am. Water Works Assoc.* **2004**, *96*(7), 75-87.

Bigham, R.; Fisher, S. Unsung Heroes. *Pollut. Eng.* **2004**, *36*(6), 30-33.

Blatchley, E.R.; Dumoutier, N.; Halaby, T.N.; Levi, Y.; Laine, J.M. Bacterial Responses to Ultraviolet Irradiation. *Water Sci. Technol.* **2001**, *43*(10), 179-186.

Blatchley, E.R.; Wood, W.L.; Schuerch, P. UV Pilot Testing: Intensity Distributions and Hydrodynamics. *J. Environ. Eng.* **1995**, *121*(3), 258-262.

Blatchley III, E.R.; Bastian, K.C.; Duggirala, R.K.; Alleman, J.E.; Moore, M.; Schuerch, P. Ultraviolet Irradiation and chlorination/dechlorination for Municipal Wastewater Disinfection: Assessment of Performance Limitations. *Water Environ. Res.* **1996**, *68*(2), 194.

Blume, T.; Neis, U. Improved Wastewater Disinfection by Ultrasonic Pre-Treatment. *Ultrason. Sonochem.* **2004**, *11*(5), 333-336.

Bolton, J.R.; Linden, K.G.; Kuo, J.; Chen, C.; Nellor, M. Discussion of "Standardized Collimated Beam Testing Protocol for water/wastewater Ultraviolet Disinfection" by Kuo,Jeff; Chen,Ching-Lin; Nellor,Margaret. *J. Environ. Eng.* **2005**, *131*(5), 827-829.

Bolton, J.R.; Linden, K.G. Standardization of Methods for Fluence (UV Dose) Determination in Bench-Scale UV Experiments. *J. Environ. Eng.* **2003**, *129*(3), 209-215.

Bolton, J.R.; Fitzsimmons, S.R.; Patel, M.; Lem, W.; Gigliotti, J. Experimental Determination of the Fluence (UV Dose) Required in the UV Photodegradation of NDMA in a Treated Waste Water; *Proc. 2002 AWWA Water Quality Technology Conference*, 2002.

Borroum, Y.; Emerick, R.W. Bioassay Comparison of Similar Pilot- and Full-Scale UV Disinfection Systems; *Proc. WEFTEC 2005*, Water Environment Federation, 2005.

Bourgeous, K.; Salveson, A.; Narayanan, B. Pesky Particles. *Water Environ. Technol.* **2005**, *17*(2), 35-40.

Bourrouet, A.; Garcia, J.; Mujeriego, R.; Penuelas, G. Faecal Bacteria and Bacteriophage Inactivation in a Full-Scale UV Disinfection System used for Wastewater Reclamation; IWA Publishing, 2001.

Bradley, B.R.; Braband, S. Advanced Onsite Treatment and Dispersal Offers New Solution for Mobile Home Parks; *10th National Symposium on Individual and Small Community Sewage*

Systems, Mar 21-24 2005, American Society of Agricultural Engineers, St. Joseph, MI 49085-9659, United States: Sacramento, CA, United States, 2004.

Brownell, S.A.; Nelson, K.L. Inactivation of Single-Celled Ascaris Suum Eggs by Low-Pressure UV Radiation. *Appl. Environ. Microbiol.* **2006**, *72*(3), 2178-2184.

Brunet, R.; Sakamoto, G.; Wright, H.; Cairns, W. UV Disinfection of Wastewater: Key Factors for Effective System Design; *2000 Annual Conference - Canadian Society for Civil Engineering, Jun 7-10 2000*, Canadian Society for Civil Engineering, Montreal, H3H 2R9, Canada: London, Ont., Canada, 2000.

Bukhari, Z.; LeChevallier, M. Assessing UV Reactor Performance for Treatment of Finished Water. *Water Sci. Technol.* **2003**, *47*(3), 179-184.

Bush, A. We Found the Solution, and We Worked Out the Bugs, Start-Up of a Major POTW Ultraviolet Disinfection System; *Proc. Disinfection 2005*, Water Environment Federation, 2005.

Cairns, B.; Petri, B.; Aldin, S.; Mao, T.; Lu, D.; Gowman, L. UV Disinfection of Wastewater - the First Barrier in a Multiple Barrier Strategy for Drinking Water Protection. *J. Harbin Institute of Technology (New Series)* **2005**, *12*(SUPPL 1), 178-183.

Cairns, W.L. UV Technology for Water Supply Treatment. *Water Supply* **1995**, *13*(3-4), 211-214.

Carrasco, L.; Turner, C.D. Evaluation of Disinfection Techniques in the Treatment of Advanced Primary Treated Wastewater for Ciudad Juarez, Mexico. *Water Environ. Res.* **2006**, *78*(1), 49-58.

Catley, G.; Jones, W. Combined MBR Process Recycles Industrial Wastewater for Potable use. *Water and Wastewater International* 2004, *19*(9), 16-17.

Cavanagh, H.; Hunter, G.; Wood, J. Implementing a Fast Track Wastewater Reuse Program to Meet Water Demands at a Power Plant. *2002 Proc. Water Sources Conference*, 2002.

Chen, J. Application of Ultraviolet Disinfection Technology in Water and Wastewater Treatment. *Zhongguo Jishui Paishui/China Water & Wastewater* 2002, *18*(7), 29.

Chiu, K.; Lyn, D.A.; Savoye, P.; Blatchley III, E.R. Integrated UV Disinfection Model Based on Particle Tracking. *J. Environ. Eng.* **1999**, *125*(1), 7-16.

Chiu, K.; Lyn, D.A.; Savoye, P.; Blatchley III, E.R. Effect of UV System Modifications on Disinfection Performance. *J. Environ. Eng.* **1999**, *125*(5), 459-469.

Chmiel, H.; Kaschek, M.; Blocher, C.; Noronha, M.; Mavrov, V. Concepts for the Treatment of Spent Process Water in the Food and Beverage Industries. *Desalination* **2003**, *152*(1-3), 307-314.

Close, J.; Ip, J.; Lam, K.H. Water Recycling with PV-Powered UV-LED Disinfection. *Renewable Energy* **2006**, *31*(11), 1657-1664.

Collivignarelli, C.; Bertanza, G.; Pedrazzani, R.A Comparison among Different Wastewater Disinfection Systems: Experimental Results. *Environ. Technol.* **2000**, *21*(1), 1-16.

Cooper, R.C.; Salveson, A.; Sakaji, R.H.; Tchobanoglous, G.; Requa, D.A.; Whitley, R. Comparison of the Resistance of MS-2 and Poliovirus to UV and Chlorine Disinfection; *Proc. WateReuse Association Symposium*, WateReuse Association, 2000.

Cotton, C.; Passantino, L.; Owen, D.; Bishop, M.; Valade, M.; Becker, W.; Joshi, R.; Young, J.; LeChevallier, M.; Hubel, R. Integrating UV Disinfection into Existing Water Treatment Plants. Report No. 91086 (AWWA), 2005.

Cotton, C.A.; Owen, D.M.; Cline, G.C.; Brodeur, T.P. UV Disinfection Costs for Inactivating *Cryptosporidium. J. Am. Water Works Assoc.* 2001*, 93*(6), 82-94.

Craik, S.A.; Leinan, B.E.; Smith, D.W.; Belosevic, M. Impact of Particles on Dose Delivery in an Ultraviolet Light Reactor for Drinking Water Disinfection; *Proc. Canadian Society for Civil Engineering - 31st Annual Conference: 2003 Building our Civilization*, Canadian Society for Civil Engineering, 2003.

Craik, S.A.; Finch, G.R.; Bolton, J.R.; Belosevic, M. Inactivation of *Giardia Muris* Cysts Using Medium-Pressure Ultraviolet Radiation in Filtered Drinking Water. *Water Res.* **2000**, *34*(18), 4325-4332.

Cutler, D.; Chalmers, R.B.; Ng, H. Implementation of Emerging Technology UV Irradiation of NDMA for Recycled Water Projects; *Proc. 2002 Annual Conference*, American Water Works Association, 2002.

Daigger, G.T. Tools for Future Success. *Water Environ. Technol.* **2003**, *15*(12), 38-45.

Darby, J.; Emerick, R.; Loge, F.; Tchobanoglous, G. The Effect of Upstream Treatment Processes on UV Disinfection Performance; Water Environment Research Foundation: 1999.

Darby, J.; Heath, M.; Jacangelo, J.; Loge, F.Swaim, P. Comparison of UV Irradiation to Chlorination: Guidance for Achieving Optimal UV Performance. Report No. WERF 91-WWD-1, 1995.

Diamant, B.Z. Water Treatment: The Ultra-Violet Option. *Int Water Irrig Rev* **1996**, *16*(3), 6.

Dinkloh, L. Sizing and Monitoring UV Systems for Wastewater Reuse. *Water and Wastewater International* **2002**, *17*(6), 29-30.

Durance, C.; Hofmann, R.; Andrews, R.C.; Brown, M. Application of Ultraviolet Light for Inactivation of Adenovirus. *Proc. Disinfection2005,* Water Environment Federation, 2005.

Dykstra, T.S.; Chauret, C.; Gagnon, G.A. Hydraulic Calibration and Fluence Determination of Model Ultraviolet Disinfection System. *J. Environ. Eng.* **2002**, *128*(11), 1046-1055.

Elefritz, R.A. J. Comparison of Alternate Methods of Disinfection for Reclaim Effluents. *Florida Water Resource Journal* **2000**, 19-25.

Emerick, R.W.; Loge, F.J.; Thompson, D.; Darby, J.L. Factors Influencing Ultraviolet Disinfection Performance. Part II: Association of Coliform Bacteria with Wastewater Particles. *Water Environ. Res.* **1999**, *71*(6), 1178-1187.

Freeman, S.; Leitner, G.F.; Crook, J.; Vernon, W.A Clear Advantage. *Water Environ. Technol.* **2002**, *14*(1), 16-21.

Freese, S.D.; Nozaic, D.J.; Bailey, I.; Trollip, D.L. Alternative Disinfectants for Wastewater Effluents: Viable or Prohibitively Expensive? *Water SA* 2002, *28*(SPEC ISS), 23-32.

Fries, A. Ultraviolet Disinfection at the Regina Waste Water Treatment Plant; *Proc. 48th Annual Conference of the Western Canada Water and Wastewater Association*, 1996.

Gehr, R.; Wright, H. UV Disinfection of Wastewater Coagulated with Ferric Chloride: Recalcitrance and Fouling Problems. *Water Sci. Technol.* **1998**, *38*(3), 15-23.

Gehr, R.; Nicell, J. Pilot Studies and Assessment of Downstream Effects of UV and Ozone Disinfection of a Physicochemical Wastewater. *Water Qual. Res. J. Can.* **1996**, *31*(2), 263-281.

Getter, R.; Loveland, J.; Patel, M. Disinfection with UV Irradiation for Water Reuse in California. *Proc. 2002 Water Sources Conference*, 2002.

Giese, N.; Darby, J. Sensitivity of Microorganisms to Different Wavelengths of UV Light: Implications on Modeling of Medium Pressure UV Systems. *Water Res.* **2000**, *34*(16), 4007-4013.

Giiler, H.; Kostioutchenko, S. St. Petersburg Upgrades Water Treatment Plants with UV Systems. *Water and Wastewater International* 2005, *20*(7), 38.

Gillette, R., Marks City of Roseville Alternative Disinfection Study; *Proc. WEFTEC 2003*, Water Environment Federation, 2003.

Goodman, G.V.; Mills, J.A. Let There Be Light! *Water Environ. Technol.* **2002**, *14*(1), 42-46.

Gori, R.; Lubello, C. Pilot Plant for Reclaimed Wastewater Reuse in Nurseries. *Water Sci. Technol.* **2000**, *42*(1), 221-226.

Green, R.G.; Equels, T.; Stadnyckyj, M.S. Welcome Neighbor. *Water Environ. Technol.* **2003**, *15*(11), 38-39.

Greyshock, A.E.; Vikesland, P.J. Triclosan Reactivity in Chloraminated Waters. *Environ. Sci. Technol.* **2006**, *40*(8), 2615-2622.

Hamby, S.A Comparison of Operating Strategies - Chlorine vs. UV; *Proc. WEFTEC 2006*, Water Environment Federation, 2006.

Hancock, G.G.; Davis, E.M. Regrowth Potential of Coliforms After UV Disinfection of Municipal Wastewater. *J. Environmental Science and Health, Part A: Toxic/Hazardous Substances and Environmental Engineering* 1999, *34*(9), 1737-1743.

Hanzon, B.D.; Vigilia, R. Two Experts Offer Practical Guidance in Designing and Operating Ultraviolet Disinfection Systems. *Water Environ. Technol.* **1999**, *11*, 35-42.

Hanzon, B.; Hartfelder, J.; O'Connell, S.; Murray, D. Comparing UV and Chlorine for Wastewater Disinfection; *Proc. Disinfection 2005*, Water Environment Federation, 2005.

Hartz, K. Pilot Trial shows UV Disinfection Inactivates Microbes. *Water/Engineering and Management* **1999**, *146*(8), 21-23.

Hassen, A.; Mahrouk, M.; Ouzari, H.; Cherif, M.; Boudabous, A.; Damelincourt, J.J. UV Disinfection of Treated Wastewater in a Large-Scale Pilot Plant and Inactivation of Selected Bacteria in a Laboratory UV Device. *Bioresour. Technol.* **2000**, *74*(2), 141-150.

Havelaar, A.H.; Nieuwstad, T.J.; Meulemans, C.C.E.; Olphen, M.F-Specific RNA Bacteriophages as Model Viruses in UV Disinfection of Wastewater. *Water Sci. Technol.* **1991**, *24*(2), 347-352.

Hijnen, W.A.M.; Beerendonk, E.F.; Medema, G.J. Inactivation Credit of UV Radiation for Viruses, Bacteria and Protozoan (Oo)Cysts in Water: A Review. *Water Res.* **2006**, *40*(1), 3-22.

Hill, V.R.; Kantardjieff, A.; Sobsey, M.D.; Westerman, P.W. Reduction of Enteric Microbes in Flushed Swine Wastewater Treated by a Biological Aerated Filter and UV Irradiation. *Water Environ. Res.* **2002**, *74*(1), 91-99.

Ho, C.H.; Pitt, P.; Mamais, D.; Chiu, C.; Jolis, D. Evaluation of UV Disinfection Systems for Large-Scale Secondary Effluent. *Water Environ. Res.* **1998**, *70*(6), 1142-1150.

Hunter, G. The Light at the End of the Tunnel – How to Successfully Operate a UV Disinfection System; *Proc. WEFTEC 2006*, Water Environment Federation, 2006.

Hunter, G.; Hulsy, B.; Mackey, H.; Neeman, J. Progress of Wastewater UV Systems - A Survey of Large Systems; *Proc. WEFTEC 2005,* Water Environment Federation, 2005.

Hunter, G.; Long, J. Ultraviolet Blues. *Water Environ. Technol.* **2003**, *15*(11), 51-53.

Hunter, G.; Cummings, T. Low Pressure/High Output UV: Is It the Wave of the Future? *Proc. WEFTEC 2002*, Water Environment Federation, 2002.

Hunter, G. The History of UV Disinfection in the Last 20 Years; *Proc. UV – A Technical Symposium*, National Water Research Institute, 2000.

Hunter, G.L.; O'Brien, W.J.; Hulsey, R.A.; Carns, K.E.; Ehrhard, R. Emerging Disinfection Technologies. *Water Environ. Technol.* **1998**, *10*(6), 40-44.

Hymel, R. Modified UV Disinfection System Meets Specifications. *Public Works* **1998**, *129*(2), 50,52.

Iranpour, R.; Garnas, G.; Moghaddam, O.; Taebi, A. Hydraulic Effects on Ultraviolet Disinfection: Modification of Reactor Design. *Water Environ. Res.* **1999**, *71*(1), 114-118.

Jacangelo, J.G.; Loughran, P.; Petrik, B.; Simpson, D.; McIlroy, C. Removal of Enteric Viruses and Selected Microbial Indicators by UV Irradiation of Secondary Effluent. *Water Sci. Technol.* 2003, *47*(9), 193-198.

Jalali, Y.; Huitric, S.J.; Kuo, J.; Tang, C.C.; Garcia, A.; Thompson, S.; Horvath, R.W.; Stahl, J.F. A Large-Scale UV Pilot-Plant Study: Tertiary Effluent Disinfection and Effect on NDMA and Cyanide; *Proc. WEFTEC 2005*, Water Environment Federation, 2005.

Janex, M.L.; Savoye, P.; Do-Quang, Z.; Blatchley III, E.; Laine, J.M. Impact of Water Quality and Reactor Hydrodynamics on Wastewater Disinfection by UV, use of CFD Modeling for Performance Optimization. *Water Sci. Technol.* **1998**, *38*(6 pt 5), 71-78.

Jensen, J.N.; Madge, B.A. UV Disinfection Modeling: Implications of Bacterial Colonization of Wastewater Solids; *Proc. WEFTEC 2004*, Water Environment Federation, 2004.

Jeyanayagam, S.; Cotton, C. Practical Considerations in the Use of UV Light for Drinking Water Disinfection; *Proc. CASE/ASCE Joint Conference: An International Perspective on Environmental Engineering*, 2002.

Jin, S.; Ferran, B.; Kelly, R.F.; Webber, E.T.; Schieble, O.K. Validation Testing of a UV Disinfection System to Achieve Regulatory Approval for Wastewater Reuse; *Proc. WEFTEC 2004*, Water Environment Federation, 2004.

Jin, S.; Mofidi, A.A.; Linden, K.G. Polychromatic UV Fluence Measurement Using Chemical Actinometry, Biodosimetry, and Mathematical Techniques. *J. Environ. Eng.* **2006**, *132*(8), 831-841.

Job, G.D.; Trengove, R.; Realey, G.J. Trials Using a Mobile Ultraviolet Disinfection System in South West Water. *J. Chartered Institution of Water and Environment Management* **1995**, *9*(3), 257-263.

Johns, F.; Lichtwardt, M.; Grundemann, P.; Gallegos, D. Selection and Installation of a UV Disinfection System as a Retrofit to an Existing Wastewater Treatment Plant; *Proc. WEFTEC 2002*, Water Environment Federation, 2002.

Jolis, D.; Lam, C.; Pitt, P. Particle Effects on Ultraviolet Disinfection of Coliform Bacteria in Recycled Water. *Water Environ. Res.* **2001**, *73*(2), 233-236.

Jolis, D.; Hirano, R.; Pitt, P. Tertiary Treatment Using Microfiltration and UV Disinfection for Water Reclamation. *Water Environ. Res.* **1999**, *71*(2), 224-231.

Jolis, D.; Lam, C.; Pitt, P.; Hirano, R. Particle Size Effects on UV Light Disinfection of Filtered Reclaimed Wastewater; *Proc. 1996 WateReuse Conference*, WateReuse Foundation, 1996.

Jones, D.E. Florida's First Permitting Experience with the New NWRI/AWWARF UV Disinfection Guidelines; *World Water and Environmental Resources Congress 2003, Jun 23-26 2003*, American Society of Civil Engineers: Philadelphia, PA, United States, 2003.

Kalisvaart, B.F. UV Disinfection Clears Up *Giardia* Outbreak in Bergen Waterworks. *Water and Wastewater International* **2005**, *20*(5), 36.

Kalisvaart, B.F. Re-use of Wastewater: Preventing the Recovery of Pathogens by Using Medium-Pressure UV Lamp Technology. *Water Sci. Technol.* **2004**, *50*(6), 337-344.

Kalisvaart, B.F. Photobiological Effects of Polychromatic Medium Pressure UV Lamps. *Water Sci. Technol.* **2001**, *43*(4), 191-197.

Kang, S.J.; Allbaugh, T.A.; Reynhout, J.W.; Erickson, T.L.; Olmstead, K.P.; Thomas, L.; Thomas, P. Selection of an Ultaviolet Disinfection System for a Municipal Wastewater Treatment Plant. *Water Sci. Technol.* **2004**, *50*(7), 163-169.

Karanfil, T.; Schlautman, M.A.; Erdogan, I. Survey of DOC and UV Measurement Practices with Implications for SUVA Determination. *J. Am. Water Works Assoc.* **2002**, *94*(12), 68-80.

Kashimada, K.; Kamiko, N.; Yamamoto, K.; Ohgaki, S. Assessment of Photoreactivation Following Ultraviolet Light Disinfection. *Water Sci. Technol.* **1996**, *33*(10-11), 261-269.

Keller, J.; Kobylinski, E.A.; Hunter, G.H. Meeting Compliance during Wet Weather Events; *Proc. Disinfection 2005*, Water Environment Federation, 2005.

Keller, R.; Passamani-Franca, R.F.; Passamani, F.; Vaz, L.; Cassini, S.T.; Sherrer, N.; Rubim, K.; Sant'Ana, T.D.; Goncalves, R.F. Pathogen Removal Efficiency from UASB+BF Effluent Using Conventional and UV Post-Treatment Systems. *Water Sci. Technol.* **2004**, *50*(1), 1-6.

Keller, R.; Passamani, F.; Vaz, L.; Cassini, S.T.; Goncalves, R.F. Inactivation of *Salmonella* Sp. from Secondary and Tertiary Effluents by UV Irradiation. *Water Sci. Technol.* **2003**, *47*(3), 147-150.

Khan, S.; Wintgens, T.; Sherman, P.; Zarlcky, J.; Schafer, A.A Performance Comparison of Individual and Combined Treatment Modules for Water Recycling. *Environ. Prog.* **2005**, *24*(4), 383-391.

Kolch, A. Disinfecting Drinking Water with UV Light. *Pollut. Eng.* **1999**, *31*(10), 34-36.

Konig, R. UV Wastewater Disinfection: The Key to the Future. *Water 21* **2001**(APR), 30.

Kuo, J.; Chen, C.; Nellor, M. Standardized Collimated Beam Testing Protocol for Water/Wastewater Ultraviolet Disinfection. *J. Environ. Eng.* **2003**, *129*(8), 774-779.

Kuo, J.; Smith, S.O. Disinfection. *Water Environ. Res.* **1996**, *68*(4), 503-510.

Kwan, A.; Archer, J.; Mohammed, A.; Freeman, A. Technical Evaluation of UV Disinfection Alternatives for Gold Bar WWTP; *Proc. 47th Annual Conference of the Western Canada Water and Wastewater Association*, 1995.

Laine, S.; Poujol, T.; Dufay, S.; Baron, J.; Robert, P. Treatment of Stormwater to Bathing Water Quality by Dissolved Air Flotation, Filtration and Ultraviolet Disinfection. *Water Sci. Technol.* **1998**, *38*(10 pt 8), 99-105.

Lau, P.J. Applying Disinfection Alternatives to Wastewater Treatment. *Pollut. Eng.* 1997, *29*(9), 64-66.

Lawryshyn, Y.A.; Cairns, B. UV Disinfection of Water: The Need for UV Reactor Validation. *Water Sci. Technol.: Water Supply* **2003**, *3*(4), 293-300.

Lazarova, V.; Sheilds, P.; Levine, B.; Savoye, P.; Hranisavljevic, D.; Renaud, P. Production of High Quality Water for Reuse Purposes: The West Basin Experience. *Water Sci. Technol.: Water Supply* **2003**, *3*(3), 167-175.

Lazarova, V.; Savoye, P.; Janex, M.L.; Blatchley III, E.R.; Pommepuy, M. Advanced Wastewater Disinfection Technologies: State of the Art and Perspectives. *Water Sci. Technol.* **1999**, *40*(4-5), 203-213.

Lazarova, V.; Janex, M.L.; Fiksdal, L.; Oberg, C.; Barcina, I.; Pommepuy, M. Advanced Wastewater Disinfection Technologies: Short and Long Term Efficiency. *Water Sci. Technol.* **1998**, *38*(12), 109-117.

Lazarova, V.; Savoye, P. Technical and Sanitary Aspects of Wastewater Disinfection by UV Irradiation for Landscape Irrigation. *Water Sci. Technol.* **2004**, *50*(2), 203-209.

Liberti, L.; Notarnicola, M.; Petruzzelli, D. Advanced Treatment for Municipal Wastewater Reuse in Agriculture. UV Disinfection: Parasite Removal and By-product Formation. *Desalination* **2003**, *152*(1-3), 315-324.

Liberti, L.; Notarnicola, M.; Boghetich, G.; Lopez, A. Advanced Treatment for Municipal Wastewater Reuse in Agriculture. UV Disinfection: Bacteria Inactivation. *J. Water Supply: Research and Technology - AQUA* **2001**, *50*(5), 275-285.

Liberti, L.; Lopez, A.; Notarnicola, M.; Barnea, N.; Pedahzur, R.; Fattal, B. Comparison of Advanced Disinfecting Methods for Municipal Wastewater Reuse in Agriculture. *Water Sci. Technol.* **2000**, *42*(1), 215-220.

Lin, L.S.; Blatchley, III, E.R. UV Dose Distribution Characterization Using Fractal Concepts for System Performance Evaluation. *Water Sci. Technol.* **2001**, *43*(11), 181-188.

Lin, L.; Johnston, C.T.; Blatchley, III, E.R. Inorganic Fouling at Quartz:Water Interfaces in Ultraviolet Photoreactors - I. Chemical Characterization. *Water Res.* **1999**, *33*(15), 3321-3329.

Lin, L.; Johnston, C.T.; Blatchley, III, E.R. Inorganic Fouling at Quartz:Water Interfaces in Ultraviolet Photoreactors - II. Temporal and Spatial Distributions. *Water Res.* **1999**, *33*(15), 3330-3338.

Lin, L.; Johnston, C.T.; Blatchley, III, E.R. Inorganic Fouling at Quartz:Water Interfaces in Ultraviolet Photoreactors - III. Numerical Modelling. *Water Res.* **1999**, *33*(15), 3339-3347.

Linden, K.G.; Shemer, H.; Reckhow, D.A.; Makdissy, G. Ultraviolet Light Induced Disinfection Byproducts: Realities and Challenges; *Proc. Disinfection 2007*, 2007.

Linden, K.G.; Shin, G.; Faubert, G.; Cairns, W.; Sobsey, M.D. UV Disinfection of *Giardia Lamblia* Cysts in Water. *Environ. Sci. Technol.* **2002**, *36*(11), 2519-2522.

Linden, K.G. UV Dose Verification Using Chemical Actinometry and Biodosimetry Methods; *Proc. UV 2000: A Technical Symposium*, 2000.

Linden, K.G.; Darby, J.L. Ultraviolet Disinfection of Marginal Effluents: Determining Ultraviolet Absorbance and Subsequent Estimation of Ultraviolet Intensity. *Water Environ. Res.* **1998**, *70*(2), 214-223.

Loge, F.J.; Bourgeous, K.; Emerick, R.W.; Darby, J.L. Variations in Wastewater Quality Parameters Influencing UV Disinfection Performance: Relative Impact of Filtration. *J. Environ. Eng.* **2001**, *127*(9), 832-836.

Loge, F.J.; Emerick, R.W.; Thompson, D.E.; Nelson, D.C.; Darby, J.L. Factors Influencing Ultraviolet Disinfection Performance Part I: Light Penetration to Wastewater Particles. *Water Environ. Res.* **1999**, *71*(3), 377-381.

Loge, F.J.; Emerick, R.W.; Heath, M.; Jacangelo, J.; Tchobanoglous, G.; Darby, J.L. Ultraviolet Disinfection of Secondary Wastewater Effluents: Prediction of Performance and Design. *Water Environ. Res.* **1996**, *68*(5), 900-916.

Loge, F.J.; Darby, J.L.; Tchobanoglous, G. UV Disinfection of Wastewater: Probabilistic Approach to Design. *J. Environ. Eng.* **1996**, *122*(12), 1078-1084.

Lyn, D.A.; Chiu, K.; Blatchley, III, E.R. Numerical Modeling of Flow and Disinfection in UV Disinfection Channels. *J. Environ. Eng.* **1999**, *125*(1), 17-26.

Lyssandridou, A.A.; Lyn, D.A.; Blatchley, III, E.R. Numerical Modeling of Process Behavior in Ultraviolet Disinfection Systems; *Proc. Disinfection 2002*, Water Environment Federation, 2002.

Mackey, E.D.; Wright, H.; Hargy, T.; Fonseca, A.C.; Cabaj, A. Reducing UV Design Safety Factors - Optimization of UV Reactor Validation; *Proc. 2004 AWWA Water Quality Technology Conference*, American Water Works Association, 2004.

Mahar, E.; Eaton, C.E.; Salveson, A. Velocity Profiling Versus Checkpoint Bioassay for Performance Testing of UV Disinfection Systems; *Proc. WEFTEC 2006*, Water Environment Federation, 2006.

Malley, J.P.; Shaw, J.P.; Ropp, J.R. Evaluation of By-products Produced by Treatment of Groundwaters with Ultraviolet Irradiation. Report No. 90685 (AWWARF), 1995.

Mamane, H.; Linden, K.G. Impact of Particle Aggregated Microbes on UV Disinfection. II: Proper Absorbance Measurement for UV Fluence. *J. Environ. Eng.* **2006**, *132*(6), 607-615.

Mann, M.A.; Cramer, J.A. Disinfecting with Ultraviolet Radiation. *Water Environ. Technol.* **1992***, 4*(12), 40-45.

Markus, M.; Sevenandt, W. Replenishing Groundwater. *Public Works* **2005***, 136*(2), 32-35.

Marsden, D. Asian UV Market Offers Tremendous Potential. *Water and Wastewater International* **2005***, 20*(7), 40.

Marshall, T.H. On the Pulse. *Water Environ. Technol.* **2002***, 14*(6), 51-54.

Marshall, T. Deadly Pulses. *Water Environ. Technol.* **1999***, 11*(4), 37-41.

Martin, N.; Gehr, R. Reduction of Photoreactivation with the Combined UV/Peracetic Acid Process or by Delayed Exposure to Visible Light , *Water Environ. Res.* **2007**, 79 (9), 991-999.

Mason, I.G.; Li, Y. UV Transmittance Characteristics of a Low-Quality Wastewater. *Transactions of the American Society of Agricultural Engineers* **2001***, 44*(2), 397-405.

Maya, C.; Beltran, N.; Jimenez, B.; Bonilla, P. Evaluation of the UV Disinfection Process in Bacteria and Amphizoic Amoebae Inactivation. *Water Sci. Technol.: Water Supply* **2003***, 3*(4), 285-291.

McCann, B. Disinfection Developments. *Water 21* **2002**(APR), 25-26.

McDonald, K.; Clevenger, T.; Curry, R.; Golden, J. Ultraviolet Light Technologies for Water Treatment; *Proc. ASCE National Conference on Environmental and Pipeline Engineering 2000, Jul 23-26 2000*, American Society of Civil Engineers: Kansas City, MO, United States, 2000.

McMartin, D.W. Two Is Better than One. *Resource: Engineering and Technology for Sustainable World* 2004*, 11*(1), 13-14.

Melin, G., Ed.; In *Ultraviolet Disinfection Guidelines for Drinking Water and Water Reuse;* National Water Research Institute, in collaboration with the AWWA Research Foundation: Fountain Valley, CA, 2003.

Mitch, W.A.; Sharp, J.O.; Trussell, R.R.; Valentine, R.L.; Alvarez-Cohen, L.; Sedlak, D.L.N-Nitrosodimethylamine (NDMA) as a Drinking Water Contaminant: A Review. *Environ. Eng. Sci.* **2003***, 20*(5), 389-404.

Mofidi, A.A.; Baribeau, H.; Green, J.F.; De Leon, R.; Wolfe, R.L. Disinfection Using High-Intensity Pulsed-Ultraviolet Irradiation; *Proc. 1998 AWWA Water Quality Technology Conference*, American Water Works Association, 1998.

Moreland, V.; Ono, P. Primary Effluent Disinfection - Necessary Procedures in UV System Design for Low Water Quality Applications (to Pilot or Not to Pilot?); *Proc. WEFTEC 2001*, Water Environment Federation, 2001.

Moreland, V.; Rijal, G.K.; Fujioka, R. UV Disinfection of Six Microbial Indicators at a 3 MGD Water Reuse Project; *Proc. WateReuse*, WateReuse Foundation, 1996.

Moreno, B.; Goni, F.; Fernandez, O.; Martinez, J.; Astigarraga, M. The Disinfection of Wastewater by Ultraviolet Light. *Water Sci. Technol.* **1997**, *35*(11), 233-235.

Moreno, B.; Goni, F.; Fernandez, O.; Martinez, J.A.; Astigarraga, M. Disinfection of Wastewater by Ultraviolet Light. *Water Sci. Technol.* **1997**, *35*(11-12), 233-235.

Morita, S.; Namikoshi, A.; Hirata, T.; Oguma, K.; Katayama, H.; Ohgaki, S.; Motoyama, N.; Fujiwara, M. Efficacy of UV Irradiation in Inactivating *Cryptosporidium Parvum* Oocysts. *Appl. Environ. Microbiol.* **2002**, *68*(11), 5387-5393.

Morowitz, H.J. Absorption Effects in Volume Irradiation of Microorganisms. *Science* **1950**, *111*(2879), 229-230.

Nelson, K.L. Ultraviolet Light Disinfection of Wastewater Stabilization Pond Effluents. *Water Sci. Technol.* **2000**, *42*(10-11), 165-170.

Noling, C. Taking a Swing at Industrial Wastewater Reuse. *Pollut. Eng.* **2004**, *36*(4), 48-51.

Norman, J.E.; Patel, M.V.; Leslie, G.L. UV Disinfectability of Membrane Treated Wastewater; *Proc. 1999 Membrane Technology Conference*, American Water Works Association, 1999.

Noronha, M.; Britz, T.; Mavrov, V.; Janke, H.; Chmiel, H. Treatment of Spent Process Water from a Fruit Juice Company for Purposes of Reuse: Hybrid Process Concept and on-Site Test Operation of a Pilot Plant. *Desalination* **2002**, *143*(2), 183-196.

NSF International. Protocol for Equipment Verification Testing for Inactivation of Microbiological Contaminants. Report No. 03/9204/EPADWCTR, 2003.

O'Brien, W.J.; Hunter, G.L.; Rosson, J.J.; Hulsey, R.A.; Carns, K.E. Ultraviolet System Design: Past, Present, and Future; *Proc. 1995 AWWA Water Quality Technology Conference, Part I*, American Water Works Association, 1996.

Oguma, K.; Katayama, H.; Ohgaki, S. Photoreactivation of *Legionella Pneumophila* After Inactivation by Low- or Medium-Pressure Ultraviolet Lamp. *Water Res.* **2004**, *38*(11), 2757-2763.

Oguma, K.; Katayama, H.; Ohgaki, S. Photoreactivation of *Escherichia Coli* After Low-Or Medium-Pressure UV Disinfection Determined by an Endonuclease Sensitive Site Assay. *Appl. Environ. Microbiol.* **2002**, *68*(12), 6029-6035.

Oguma, K.; Katayama, H.; Mitani, H.; Morita, S.; Hirata, T.; Ohgaki, S. Determination of Pyrimidine Dimers in *Escherichia Coli* and *Cryptosporidium Parvum* during UV Light Inactivation, Photoreactivation, and Dark Repair. *Appl. Environ. Microbiol.* **2001**, *67*(10), 4630-4637.

Oliver, M. Validation of Cleaning Methods for UV Disinfection Systems; *Proc. WEFTEC Disinfection 2002*, 2002.

Oppenheimer, J.A.; Jacangelo, J.G.; Laine, J.; Hoagland, J.E. Testing the Equivalency of Ultraviolet Light and Chlorine for Disinfection of Wastewater to Reclamation Standards. *Water Environ. Res.* **1997**, *69*(1), 14-24.

Oppenheimer, J.A.; Jacangelo, J.G.; Laine, J.; Hoagland, J.E. Equivalency Testing of Ultraviolet Disinfection for Wastewater Reclamation; *Proc. 1995 AWWA Water Quality Technology Conference, Part I*, American Water Works Association, 1995.

Ormeci, B.; Linden, K.G. Comparison of UV and Chlorine Inactivation of Particle and Non-Particle Associated Coliform. *Water Sci. Technol.: Water Supply* **2002**, *2*(5-6), 403-410.

Otaki, M.; Takizawa, S.; Ohgaki, S. Control and Modeling of Membrane Fouling due to Microorganism Growth by UV Pretreatment. *Water Sci. Technol.* **1998**, *38*(4-5 pt 4), 405-412.

Parker, J.A.; Darby, J.L. Particle-Associated Coliform in Secondary Effluents: Shielding from Ultraviolet Light Disinfection. *Water Environ. Res.* **1995**, *67*(7), 1065-1075.

Passantino, L.; Malley Jr., J.; Knudson, M.; Ward, R.; Kim, J. Effect of Low Turbidity and Algae on UV Disinfection Performance. *J. Am. Water Works Assoc.* **2004**, *96*(6), 128-137.

Patoczka, J. UV Disinfection of Low Transmittance Pharmaceutical Wastewater; *Proc. Disinfection 2005*, Water Environment Federation, 2005.

Peng, J.; Qiu, Y.; Gehr, R. Characterization of Permanent Fouling on the Surfaces of UV Lamps Used for Wastewater Disinfection. *Water Environ. Res.* **2005**, *77*(4), 309-322.

Perrot, J.; Baron, J. The Disinfection of Municipal Wastewater by Ultraviolet Light: A French Case Study. *Water Sci. Technol.* **1995**, *32*(7), 167-174.

Perrot, J.Y.; Baron, J. Disinfection of Municipal Wastewater by Ultraviolet Light: A French Case Study. *Water Sci. Technol.* **1995**, *32*(7), 167-174.

Petri, B.; Cairns, B.; Mao, T.; Lu, D.; Gowman, L. Safeguarding Public and Environmental Health with UV Disinfection: Validation and Safe System Design. *J. Harbin Institute of Technology (New Series)* **2005**, *12*(SUPPL 1), 155-159.

Petri, B.M.; Lawryshyn, Y.A. Validating UV Reactors for Low Dose Wastewater Applications; *Proc. Disinfection 2005*, Water Environment Federation, 2005.

Petri, B.; Olson, D. Bioassay Validation of Computational Disinfection Models used for UV Reactor Design and Scaleup. *Proc. Disinfection: Health and Safety Achieved through Disinfection WEF* 2002.

Qualls, R.G.; Johnson, J.D. Bioassay and Dose Measurement in UV Disinfection. *Appl. Environ. Microbiol.* **1983**, *45*(3), 872-877.

Rajala, R.L.; Pulkkanen, M.; Pessi, M.; Heinonen-Tanski, H. Removal of Microbes from Municipal Wastewater Effluent by Rapid Sand Filtration and Subsequent UV Irradiation. *Water Sci. Technol.* **2003**, *47*(3), 157-162.

Raju, V.A. Ultraviolet Disinfection: Wastewater & Process Water Applications. *Chemical Engineering World* **2002**, *37*(8), 60.

Richter, A.Y.; Weaver, R.W. Ultraviolet Disinfection of Effluent from Subsurface Flow Constructed Wetlands. *Environ. Technol.* 2003, *24*(9), 1175-1182.

Rijal, G.K.; Fujioka, R.S. Use of Reflectors to Enhance the Synergistic Effects of Solar Heating and Solar Wavelengths to Disinfect Drinking Water Sources. *Water Sci. Technol.* **2003**, *48*(11-12), 481-488.

Robertson, J.M.C.; Robertson, P.K.J.; Lawton, L.A.A Comparison of the Effectiveness of TiO_2 Photocatalysis and UVA Photolysis for the Destruction of Three Pathogenic Micro-Organisms. *J. Photochemistry and Photobiology.A, Chemistry* **2005**, *175*(1), 51-56.

Safe Drinking Water Committee, Ed.; In *Safe Drinking Water and Health;* National Academy Press: Washington, D.C., 1980.

Sakaji, R.H. Reclaimed Water: The California Experience; *Proc. UV – A Technical Symposium*, National Water Research Institute, 2000.

Sakamoto, G.; Schwartzel, D.; Tomowich, D. UV Disinfection for Reuse Applications in North America. *Water Sci. Technol.* **2001**, *43*(10), 173-178.

Salgot, M.; Folch, M.; Huertas, E.; Tapias, J.; Avellaneda, D.; Giros, G.; Brissaud, F.; Verges, C.; Molina, J.; Pigem, J. Comparison of Different Advanced Disinfection Systems for Wastewater Reclamation. *Water Sci. Technol.: Water Supply* **2002**, *2*(3), 213-218.

Samel, L.S.; Norton, J.M.; McLelland, J. Implementation of Best-Value UV Systems through Competitive Life Cycle Cost Bidding Procedures; *Proc. WEFTEC 2003*, Water Environment Federation, 2003.

Savoye, P.; Janex, M.L.; Lazarova, V. Wastewater Disinfection by Low-Pressure UV and Ozone: A Design Approach Based on Water Quality. *Water Sci. Tech.* **2001**, *43*(10), 163-171.

Scheible, O.K.; Lawryshyn, Y.A. UV Reactor Validation: Matching the Microbe with the Target Dose; *Proc. WEFTEC 2004*, Water Environment Federation, 2004.

Scheible, O.K.; Mignone, E.J. Final Draft (Version 3.4) of the Generic Verification Protocol for Secondary Effluent and Water Reuse Disinfection Applications. 2002.

Sellami, M.H.; Hassen, A.; Sifaoui, M. S. Modelling of UV Radiation Field Inside a Photoreactor Designed for Wastewater Disinfection Experimental Validation. *J. Quantitative Spectroscopy and Radiative Transfer* **2003**, *78*(3-4), 269-287.

Shaw, J.P.; Malley Jr., J.P.; Willoughby, S.A. Effects of UV Irradiation on Organic Matter. *J. Am. Water Works Assoc.* **2000**, *92*(4), 157-167.

Sheriff, M.; Gehr, R. Laboratory Investigation of Inorganic Fouling o Flow Pressure UV Disinfection Lamps. *Water Qual. Res. J. Can.* **2001**, *36*(1), 71-92.

Simpson, D.; Jacangelo, J.; Loughran, P.; McIlroy, C. Investigation of Potential Surrogate Organisms and Public Health Risk in UV Irradiated Secondary Effluent. *Water Sci. Technol.* **2003**, *47*(9), 37-43.

Smith, T.; Brown, D. Ultraviolet Treatment Technology for the Henderson Water Reclamation Facility. *Proc. 2002 Water Sources Conference*, 2002.

Snowden-Swan, L.; Piatt, J.; Lesperance, A. Disinfection Technologies for Potable Water and Wastewater Treatment: Alternative to Chlorine Gas. 1998.

Snyder, S.A.; Wert, E.C.; Lei, H.; Westerhoff, P.; Yoon, Y. Removal of EDCs and Pharmaceuticals in Drinking and Reuse Treatment Processes. Report No. 2758 (AWWARF), 2007.

Sommer, R.; Cabaj, A.; Hirschmann, G.; Pribil, W.; Haider, T. Perspectives of UV Drinking Water Disinfection; *Proc. Disinfection 2002*, Water Environment Federation, 2002.

Soroushian, F.; Shen, Y.; Patel, M.; Wehner, M. Evaluation and Pilot Testing of Advanced Treatment Processes for NDMA Removal and Reformation Prevention; *Proc. AWWA Annual Conference*, American Water Works Association, 2001.

Soroushian, F. Knowledge Gaps: What Is Required for Reliable UV Application; *Proc. UV – A Technical Symposium*, National Water Research Institute, 2000.

Soroushian, F.; Norman, J.; Patel, M.; Leslie, G.; Tchobanoglous, G. Comparison of UV Disinfection Technologies for Meeting California Wastewater Reclamation Criteria; *Proc. 2000 WateReuse Conference*, WateReuse Association, 2000.

Soroushian, F.; Abramson, C.; Ferris, M.; Mohammed, A. Pilot-Scale Studies of High-Intensity UV Disinfection By-products; *Proc. WEFTEC 1996*, Water Environment Federation, 1996.

State of North Carolina. Ultraviolet Disinfection Systems – Reliability Guidelines. 2001.

Stedman, L.A Unanimous Vote of Confidence? *Water 21* **2004**(Feb), 31-32.

Stefan, M.I.; Bolton, J.R. UV Direct Photolysis of N-Nitrosodimethylamine(NDMA): Kinetic and Product Study. *Helv. Chim. Acta* **2002**, *85*(5), 1416.

Steinberg, L.J.; Albrecht, J.M.; Basolo, V. The Diffusion of Environmental Technology: A Case Study of Ultraviolet Disinfection for Wastewater; *Proc. WEFTEC 2000*, Water Environment Federation, 2000.

Swaim, P.D.; Jeffcoat, S.B.; Roberson, J.A.; Via, S.; Pihera, G.; Bellamy, W.D. Application of the Proposed UVDGM Requirements to an Existing UV Disinfection Installation. *J. Am. Water Works Assoc.* **2004**, *96*(10), 83-96.

Swift, J.; Emerick, R.; Scheible, K.; Soroushian, F.; Putnam, L.B.; Sakaji, R. Treat, Disinfect, Reuse. *Water Environ. Technol.* **2002**, *14*(11), 21-25.

Swift, J.L.; Wilson, J.P.; Johnson, M. The Impact of UV-Absorbing Wastewater from a Printed Circuit Board Manufacturing Facility on the Performance of a Municipal UV Disinfection System; *Proc. WEFTEC 2001*, Water Environment Federation, 2001.

Swift, J.L.; Wilson, J.P.; Johnson, M. Assessment of Operation and Maintenance Costs for Ultraviolet Disinfection Systems; *Proc. WEFTEC 2000*, Water Environment Federation, 2000.

Taghipour, F. Ultraviolet and Ionizing Radiation for Microorganism Inactivation. *Water Res.* **2004**, *38*(18), 3940-3948.

Tang, C.C.; Kuo, J.; Huitric, S.J.; Jalali, Y.; Horvath, R.W.; Stahl, J.F. UV Systems for Reclaimed Water Disinfection – from Equipment Validation to Operation; *Proc. WEFTEC.06*, Water Environment Federation, 2006.

Tchobanoglous, G.; Burton, F.L.; Stensel, D.H. Wastewater Engineering Treatment and Reuse; McGraw Hill: Boston, 2003.

Tchobanoglous, G.; Loge, F.; Darby, J.; Devries, M. UV Design: Comparison of Probabilistic and Deterministic Design Approaches. *Water Sci. Technol.* **1996**, *33*(10-11), 251-260.

Temmer, B.; Soroushian, F.; Noesen, M. UV Emerges as Bright Spot for Wastewater Disinfection. *Pollut. Eng.* **2000**, *32*(9), 46-49.

Templeton, M.R.; Hofmann, R.; Andrews, R.C.; Whitby, G.E. Biodosimetry Testing of a Simplified Computational Model for the UV Disinfection of Wastewater. *J. Environ. Eng. Sci.* **2006**, *5*(1), 29-36.

Tessele, F.; Monteggia, L.O.; Rubio, J. Treatment of Municipal Wastewater UASB Reactor Effluent by Unconventional Flotation and UV Disinfection. *Water Sci. Technol.* **2005**, *52*(1-2), 315-322.

Thomas, L.; Kalinowski, J.; Cook, C.; Verduzco, L. Fieldbus Instruments Support a Larger Vision. *INTECH* 2004, *51*(7), 22-26.

Thomas, L.; Hajda, P.; Zehner, S. To UV or Not to UV, Is That the Question? Or Is It: When Is Clear Water Not Transparent? *Proc. WEFTEC 2002*, Water Environment Federation, 2002.

Thompson, S.S.; Jackson, J.L.; Suva-Castillo, M.; Yanko, W.A.; El Jack, Z.; Kuo, J.; Chen, C.L.; Williams, F.P.; Schnurr, D.P. Detection of Infectious Human Adenoviruses in Tertiary-Treated and Ultraviolet-Disinfected Wastewater. *Water Environ. Res.* **2003**, *75*(2), 163-170.

Tobin, A.; Burror, J.; Yahya, M.; Wistrom, A.; Nedic, S. Alternatives for Disinfection Effluent Generated at Orange County Sanitation District Wastewater Treatment Plants; *Proc. WEFTEC 2005*, Water Environment Federation, 2005.

Tone, K.; Tamburini, R.; Ramey, W.W. So You're Thinking about an UV Disinfection System: Here Are Some Considerations for You; *Proc. 1998 WEFTEC*, Water Environment Federation, 1998.

Topnik, J.M.; Buchanan, I.D.; Smith, D.W. The Effect of Iron in Wastewater on UV Lamp Sleeve Fouling; *Proc. WEFTEC 2000 Conference*, Water Environment Federation, 2000.

Tosa, K.; Hirata, T. Photoreactivation of Enterohemorrhagic *Escherichia Coli* Following UV Disinfection. *Water Res.* **1999**, *33*(2), 361-366.

Tree, J.A.; Adams, M.R.; Lees, D.N. Virus Inactivation during Disinfection of Wastewater by Chlorination and UV Irradiation and the Efficacy of F^+ Bacteriophage as a 'Viral Indicator'. *Water Sci. Technol.* **1997**, *35*(11-12), 227-232.

U.S. Department of Homeland Security, Chlorine Gas Decision Tool for Water and Wastewater Utilities; 2006.

U.S. EPA. Ultraviolet Disinfection Guidance Manual for the Final Long Term 2 Enhanced Surface Water Treatment Rule. Report No. EPA 815-R-06-007, 2006.

U.S. EPA. Wastewater Technology Fact Sheet – Ultraviolet Disinfection. Report No. EPA 832-F-99-064, 1999.

U.S. EPA. User's Manual for UVDIS, Version 3.1 UV Disinfection Process Design Manual. 1992.

U.S. EPA. Design Manual: Municipal Wastewater Disinfection. Report No. EPA 625/1-86/021, 1986.

Valenti, M. Lighting the Way to Improved Disinfection. *Mech. Eng.* **1997**, *119*(7), 82-86.

Van Der Walt, E. The Effect of Ultraviolet Light, Cavitational Flow and Ultrasound on Protozoan Cysts and Oocysts, Bacteriophages and Clostridium. *Water SA* **2002**, *28*(SPEC ISS), 16-22.

Van Esch, R. UV Combats Microorganisms in Municipal and Dairy Industries. *Water and Wastewater International* **2005**, *20*(7), 35-36.

Veenstra, M. Plant Renovation Brings City to Forefront of Water Reclamation; *Proc. 2002 Water Sources Conference: Reuse, Resources, Conservation*, 2002.

Wallis-Lage, C.; Scruggs, C.; Hunter, G.; Huber, R. ONE Size Doesn't Fit all. *Water Environ. Technol.* **2004**, *16*(12), 32-36.

Watson, R. Banff Upgrade. *Canadian Consulting Engineer* **2003**, *44*(6), 82.

Weber, E.; Scheible, K. Special Section: Security – Pulsed-UV Unit May Inactivate Biological Agents. *J. Am. Water Works Assoc.* **2003**, 34-46.

WEF. Wastewater Disinfection – Training Manual. 2006.

WEF. Water Environment Federation Disinfection Process Survey Summary. *Water Environ. Technol.* **2004**, *16*(7), 47-49.

WEF. Wastewater Disinfection - MOP FD-10. 1996.

Whitby, E. The History of UV Through Patents; *Proc. Disinfection 2002*, Water Environment Federation, 2002.

Williamson, C. UV-Oxidation Disinfects Water and Treats Chemicals Simultaneously. *Water and Wastewater International* **2002**, *17*(6), 33.

Wolfe, R.L. Ultraviolet Disinfection of Potable Water. *Environ. Sci. Technol.* **1990**, *24*(6), 768-773.

Wood, P.; Hunter, G.; Kobylinski, E. To PLC or Not to PLC UV Systems – Alternatives for UV System Control; *Proc. Disinfection 2005*, Water Environment Federation, 2005.

Wright, H.B.; Mackey, E.; Cushing, R.A Comparison of UV Disinfection for Drinking Water, Wastewater, and Reclaimed Wastewater *Proc. WEFTEC 2002*, Water Environment Federation, 2002.

Wright, H.B. Comparison and Validation of UV Dose Calculations for Low- and Medium-Pressure Mercury Arc Lamps. *Water Environ. Res.* **2000**, *72*(4), 439-443.

Xin, Z. Disinfection Development: The Rise of UV in China. *Water 21* **2004**(CT), 18-22.

Yamagiwa, K.; Tsujikawa, M.; Yoshida, M.; Ohkawa, A. Disinfection Kinetics of Legionella Pneumophila by Ultraviolet Irradiation. *Water Sci. Technol.* **2002**, *46*(11-12), 311-317.

Yoon, C.G.; Chung, K.; Ham, J.; Jeon, J. Feasibility Study of UV Disinfection for Agricultural Reuse of Secondary Level Effluent in Korea. *J. Environ. Sci. and Health - Part A Toxic/Hazardous Substances and Environ. Eng.* **2004**, *39*(3), 835-846.

York, D.W.; Potts, E.A. UV Disinfection: Florida's Perspective; *Proc. UV – A Technical Symposium*, National Water Research Institute, 2000.

Zenz, D.R.; Law, K.; Lanyon, R.; Sobanski, J.; Kunetz, T.; Bouchard, A.B.; Hass, C.; Schwab, K.; Marinas, B. Study of Effluent Disinfection for Urban Rivers in Chicago; *Proc. Disinfection 2007*, Water Environment Federation, 2007.

Ozone Technology

Anonymous. Ozone Disinfection - A Technical Overview. Report No. WWFSOM22, 1991.

Ahn, K.H.; Park, K.Y.; Maeng, S.K.; Hwang, J.H.; Lee, J.W.; Song, K.G.; Choi, S. Ozonation of Wastewater Sludge for Reduction and Recycling. *Water Sci. Technol.* **2002**, *46*(10), 71-77.

Amy, G.; Debroux, J.F.; Arnold, R.; Wilson, L.-.Preozonation for Enhancing the Biodegradability of Wastewater Effluent in a Potable-Recovery Soil Aquifer Treatment (SAT) System. *Rev. Sci. Eau/J. Water Sci.* **1996**, *9*(3), 365-380.

Balakrishnan, P.A.; Arunagiri, A.; Rao, P.G. Ozone Generation by Silent Electric Discharge and Its Application in Tertiary Treatment of Tannery Effluent. *J. Electrostatics* **2002**, *56*(1), 77-86.

Baron, J.; Langlais, B.; Perez-Parra, J.; Del Castillo, E.; Vallverdru, A. The use of Ozone as Tertiary Treatment of Wastewaters before Reuse: Practical Application in Almeria (Spain); *Proc. 1998 WateReuse Conference*, WateReuse Association, 1998.

Blatchley, III, E.R.; Hunt, B.A.; Duggirala, R.; Thompson, J.E.; Zhao, J.; Halaby, T.; Cowger, R.L.; Straub, C.M.; Alleman, J.E. Effects of Disinfectants on Wastewater Effluent Toxicity. *Water Res.* **1997**, *31*(7), 1581-1588.

Boyce, D.S.; Sproul, O.J.; Buck, C.E. The Effect of Bentonite Clay on Ozone Disinfection of Bacteria and Viruses in Water. *Water Res.* **1981**, *15*(6), 759-767.

Buffle, M.; Schumacher, J.; Salhi, E.; Jekel, M.; von Gunten, U. Measurement of the Initial Phase of Ozone Decomposition in Water and Wastewater by Means of a Continuous Quench-Flow System: Application to Disinfection and Pharmaceutical Oxidation. *Water Res.* **2006**, *40*(9), 1884-1894.

Bünning, G.; Hempel, D.C. Vital-Fluorochromization of Microorganisms Using 3', 6'-Diacetylfluorescein to Determine Damages of Cell Membranes and Loss of Metabolic Activity by Ozonation. *Ozone: Sci. Eng.* **1996**, *18*(2), 173-181.

Camacho, P.; Paul, E.; Salhi, M.; Ginestet, P. Does Reduction of Sludge Processes/Ozonation, Thermal Treatment have an Impact on Wastewater Treatment Efficacy? *Proc. Technology 2005 Second Joint Specialty Conference for Sustainable Management of Water Quality Systems for the 21st Century: Working to Protect Public Health*, 2005.

Carr, S.A.; Baird, R.B. Mineralization as a Mechanism for TOC Removal: Study of ozone/ozone-Peroxide Oxidation Using FT-IR. *Water Res.* **2000**, *34*(16), 4036-4048.

Case, P. Applying Ozone Technology to Treat Urban Runoff to Santa Monica Bay; *California and the World Ocean '02: Revisiting and Revising California's Ocean Agenda - Proc. Conference, CWO '02, Oct 27-30 2004*, American Society of Civil Engineers, Reston, VA 20191-4400, United States: Santa Barbara, CA, United States, 2005.

Chiang, C.; Tsai, C.; Lin, S.; Huo, C.; Lo, K.V. Disinfection of Hospital Wastewater by Continuous Ozonization. *J. Environ. Sci. and Health - Part A Toxic/Hazardous Substances and Environ. Eng.* **2003**, *38*(12), 2895-2908.

Cho, M.; Chung, H.; Yoon, J. Disinfection of Water Containing Natural Organic Matter by Using Ozone-Initiated Radical Reactions. *Appl. Environ. Microbiol.* **2003**, *69*(4), 2284-2291.

Clark, J.P. Ozone - Cure for some Sanitation Problems. *Food Technol.* 2004, *58*(4), 75-76.

Collivignarelli, C.; Bertanza, G.; Pedrazzani, R.A Comparison among Different Wastewater Disinfection Systems: Experimental Results. *Environ. Technol.* **2000**, *21*(1), 1-16.

Contreras, S.; Ollis, D.F.; Esplugas, S. Sequential Ozonation and Biological Oxidation of Wastewaters: A Model Including Biomass Inhibition by Residual Oxidant. *Ozone: Sci. Eng.* **2003**, *25*(2), 95-105.

Diao, H.F.; Li, X.Y.; Gu, J.D.; Shi, H.C.; Xie, Z.M. Electron Microscopic Investigation of the Bactericidal Action of Electrochemical Disinfection in Comparison with Chlorination, Ozonation and Fenton Reaction. *Process Biochemistry* **2004**, *39*(11), 1421-1426.

Drago, J.; Thompson, C. Historical Ozone use at Municipal Wastewater Treatment Plants in the United States; *Proc. Pan-American Group Annual Conference*, International Ozone Association, 2005.

Drury, D.D.; Snyder, S.A.; Wert, E.C. Using Ozone Disinfection for EDC Removal; *Proc. WEFTEC 2006*, Water Environment Federation, 2006.

Freese, S.D.; Nozaic, D.J.; Bailey, I.; Trollip, D.L. Alternative Disinfectants for Wastewater Effluents: Viable or Prohibitively Expensive? *Water SA* **2002**, *28*(SPEC ISS), 23-32.

Gilmore, L.A.; Howkins, I.; Bowen, P. Wastewater Reclamation: Ozone Treatment for Organic and Pathogen Control - Design Criteria; *Proc. 1998 WateReuse Conference*, WateReuse Foundation, 1998.

Goi, M.; Odagawa, K.; Nishimura, T.; Okoch, T.; Yuzawa, H. Gordona Scum Suppression Mechanism of Ozone Added in Wastewater Treatment Plants. *Water Sci. Technol.* **1998**, *38*(3 pt 3), 87-94.

Hatatatsu, T.; Suzuki, Y. Evaluation of Ozone Disinfection from View Points of Inactivation Disinfection Effects and Byproduct Formation; *Proc.* WEFTEC *2000*, Water Environment Federation, 2000.

Hoigne, J. Chemistry of Aqueous Ozone and Transformation of Pollutants by Ozonation and Advanced Oxidation Processes. *The Handbook of Environmental Chemistry: Quality and Treatment of Drinking Water II* 1998.

Hu, W.R. Study on the Microflocculation-Direct filtration/ozone Pre-Disinfection Process for Treatment of Micro-Polluted Reservoir Water. *Zhongguo Jishui Paishui/China Water & Wastewater* **2001**, *17*(9), 6.

Huber, M.M.; Gobel, A.; Joss, A.; Hermann, N.; Loffler, D.; McArdell, C.S.; Ried, A.; Siegrist, H.; Ternes, T.A.; von Gunten, U. Oxidation of Pharmaceuticals during Ozonation of Municipal Wastewater Effluents: A Pilot Study. *Environ. Sci. Technol.* **2005**, *39*(11), 4290-4299.

Huber, M.M.; Ternes, T.A.; Von Gunten, U. Removal of Estrogenic Activity and Formation of Oxidation Products during Ozonation of 17α-Ethinylestradiol. *Environ. Sci. Technol.* **2004**, *38*(19), 5177-5186.

Hunter, G.; Hulsey, B.; Coughenour, J.; Walz, T.; Snyder, S. Out in the Zone - the Return of Ozone and Hydoxyl Radical to Wastewater Disinfection; *Proc.* WEFTEC.*06 79th Annual Technical Exhibition and Conference*, Water Environment Federation, 2006.

Janex, M.L.; Savoye, P.; Roustan, M.; Do-Quang, Z.; Laine, J.M.; Lazarova, V. Wastewater Disinfection by Ozone: Influence of Water Quality and Kinetics Modeling. *Ozone: Sci. Eng.* **2000**, *22*(2), 113-121.

Janex, M.L.; Xu, P.; Savoye, P.; Laine, J.M.; Lazarova, V. Ozonation as a Wastewater Disinfection Process to Meet Reuse Standards; *Proc. 14Th World Congress*, International Ozone Association: Stamford, CT, 1999.

Jin, P.K.; Wang, X.C.; Hu, G.A Dispersed-Ozone Flotation (DOF) Separator for Tertiary Wastewater Treatment. *Water Sci. Technol.* **2006**, *53*(9), 151-157.

Khan, S.; Wintgens, T.; Sherman, P.; Zarlcky, J.; Schafer, A. A Performance Comparison of Individual and Combined Treatment Modules for Water Recycling. *Environ. Prog.* **2005**, *24*(4), 383-391.

Kuo, J.; Smith, S.O. Disinfection. *Water Environ. Res.* **1996**, *68*(4), 503-510.

Lazarova, V.; Savoye, P.; Janex, M.L.; Blatchley III, E.R.; Pommepuy, M. Advanced Wastewater Disinfection Technologies: State of the Art and Perspectives. *Water Sci. Technol.* **1999**, *40*(4-5), 203-213.

Lazarova, V.; Janex, M.L.; Fiksdal, L.; Oberg, C.; Barcina, I.; Pommepuy, M. Advanced Wastewater Disinfection Technologies: Short and Long Term Efficiency. *Water Sci. Technol.* **1998**, *38*(12), 109-117.

Lee, B.; Kim, S.; Kim, S. DOF (Dissolved Ozone Flotation) Technology in Wastewater Treatment; *8th Korea-Russia International Symposium on Science and Technology, KORUS 2004, Jun 26-Jul 3 2004*, Institute of Electrical and Electronics Engineers Computer Society, Piscataway, NJ 08855-1331, United States: Tomsk, Russian Federation, 2004.

Lee, S.; Lee, K.; Wan, W.M.; Choi, Y. Comparison of Membrane Permeability and a Fouling Mechanism by Pre-Ozonation Followed by Membrane Filtration and Residual Ozone in Membrane Cells. *Desalination* **2005**, *178*(1-3 SPEC ISS), 287-294.

Liberti, L.; Notarnicola, M.; Lopez, A. Advanced Treatment for Municipal Wastewater Reuse in Agriculture. III - Ozone Disinfection. *Ozone: Sci. Eng.* **2000**, *22*(2), 151-166.

Liberti, L.; Lopez, A.; Notarnicola, M.; Barnea, N.; Pedahzur, R.; Fattal, B. Comparison of Advanced Disinfecting Methods for Municipal Wastewater Reuse in Agriculture. *Water Sci. Technol.* **2000**, *42*(1), 215-220.

Loeb, B.L. Ozone: Science and Engineering: The First 23 Years. *Ozone: Sci. Eng.* **2002**, *24*(6), 399-412.

Monarca, S.; Feretti, D.; Collivignarelli, C.; Guzzella, L.; Zerbini, I.; Bertanza, G.; Pedrazzani, R. Influence of Different Disinfectants on Mutagenicity and Toxicity of Urban Wastewater. *Water Res.* **2000**, *34*(17), 4261-4269.

Ni, C.H.; Chen, J.N.; Tsai, Y.C.; Chen, W.B.; Chen, C.H. Ozonation of Domestic Secondary Effluent for Recycling and Reuse--a Pilot Plant Study. *Water Sci. Technol.* **2002**, *46*(4-5), 361-366.

Paraskeva, P.; Graham, N.J. Ozonation of Municipal Wastewater Effluents. *Water Environ. Res.* **2002**, *74*(6), 569-581.

Paraskeva, P.; Lambert, S.D.; Graham, N.J.D. Influence of Ozonation Conditions on the Treatability of Secondary Effluent. *Ozone: Sci. Eng.* **1998**, *20*(2), 133-150.

Paraskeva, P. The Treatment of a Secondary Municipal Effluent by Ozone. Ph.D. Thesis, Imperial College, University of London, 1998.

Rein, D.A.; Jamesson, G.M.; Monteith, R.A. Toxicity Effects of Alternate Disinfection Processes; *Proc. WEFTEC 1992*, Water Environment Federation, 1992.

Rice, R.G. Ozone in the United States of America: State-of-the-Art. *Ozone: Sci. Eng.* **1999**, *21*(2), 99-118.

Robson, C.M.; Rice, R.G. Wastewater Ozonation in the USA.--History and Current Status--1989. *Ozone: Sci. Eng.* **1991**, *13*(1), 23-40.

Salgot, M.; Folch, M.; Huertas, E.; Tapias, J.; Avellaneda, D.; Giros, G.; Brissaud, F.; Verges, C.; Molina, J.; Pigem, J. Comparison of Different Advanced Disinfection Systems for Wastewater Reclamation. *Water Sci. Technol.: Water Supply* **2002**, *2*(3), 213-218.

Savoye, P.; Janex, M.L.; Lazarova, V. Wastewater Disinfection by Low-Pressure UV and Ozone: A Design Approach Based on Water Quality. *Water. Sci. Tech.* **2001**, *43*(10), 163-171.

Snowden-Swan, L.; Piatt, J.; Lesperance, A. Disinfection Technologies for Potable Water and Wastewater Treatment: Alternative to Chlorine Gas. 1998.

Snyder, S.A.; Wert, E.C.; Lei, H.; Westerhoff, P.; Yoon, Y. Removal of EDCs and Pharmaceuticals in Drinking and Reuse Treatment Processes. Report No. 2758 (AWWARF), 2007.

Snyder, S.; Wert, E.; Zegers, R.; Drury, D. Oxidation of EDCs and Pharmaceuticals in Reuse Water; *Proc. Disinfection 2007*, Water Environment Federation, 2007.

Snyder, S.A.; Wert, E.C.; Rexing, D.J.; Zegers, R.E.; Drury, D.D. Ozone Oxidation of Endocrine Disruptors and Pharmaceuticals in Surface Water and Wastewater. *Ozone: Sci. Eng.* **2006**, *28*(6), 445-460.

Stedman, L. O_3 Research: Developments in Disinfection. *Water 21* 2005(FEB), 34-36.

Takanashi, H.; Mayumi, M.; Kato, M.; Hirata, M.; Hano, T. Removal of Mutagen Precursor from Wastewater by Activated Sludge and Oxidation Treatment. *Water Sci. Technol.* **2002**, *46*(11-12), 389-394.

Tchobanoglous, G.; Burton, F.L.; Stensel, D.H. Wastewater Engineering Treatment and Reuse; McGraw Hill: Boston, 2003.

Ternes, T.A.; Stuber, J.; Herrmann, N.; McDowell, D.; Ried, A.; Kampmann, M.; Teiser, B. Ozonation: A Tool for Removal of Pharmaceuticals, Contrast Media and Musk Fragrances from Wastewater? *Water Res.* **2003**, *37*(8), 1976-1982.

Thompson, C.; Leong, L.Y.C. 30 Years of Experience Using Ozone in the United States Disinfecting Wastewater; *Proc. Disinfection 2007*, Water Environment Federation, 2007.

Thompson, C.; Drago, J. Ozone use Trends in the Water Treatment Industry in the United States; *Proc. Pan-American Group Annual Conference*, International Ozone Association, 2005.

U.S. EPA. Long-Term 2 Enhanced Surface Water Treatment Rule. *Federal Register* **2006**, *71*(3), 653-702.

U.S. EPA. Alternative Disinfectants and Oxidants Guidance Manual. Report No. EPA 815/R-99/014, 1999.

U.S. EPA. Wastewater Technology Fact Sheet – Ozone Disinfection. Report No. EPA 832/F-99-063, 1999.

U.S. EPA Interim Enhanced Surface Water Treatment. *Federal Register* **1998**, *63*(241), 69477-69521.

U.S. EPA. Design Manual: Municipal Wastewater Disinfection. Report No. EPA 625/1-86/021, 1986.

Walsh, D.S.; Sproul, O.J.; Buck, C.E. Ozone Inactivation of Floc Associated Viruses and Bacteria. *J. Environ. Eng. Division, ASCE* **1980**, *106*(4), 711-726.

WEF. Wastewater Disinfection – Training Manual. 2006.

WEF. Water Environment Federation Disinfection Process Survey Summary. *Water Environ. Technol.* 2004, *16*(7), 47-49.

WEF. Wastewater Disinfection - MOP FD-10. 1996.

Wert, E.; Snyder, S.; Zegers, R.; Drury, D. Formation of Ozonation DBPs in Tertiary Wastewater; *Proc. Disinfection 2007*, Water Environment Federation: Pittsburgh, Pennsylvania, 2007.

Whitby, E. The History of UV through Patents; *Proc. Disinfection 2002*, Water Environment Federation, 2002.

White, G.C. *Handbook of Chlorination and Alternative Disinfection*; Wiley & Sons: New York, 1999.

Wojtenko, I.; Stinson, M.K.; Field, R. Performance of Ozone as a Disinectant for Combined Sewer Overflow. *Crit. Rev. Environ. Sci. Technol.* 2001, *31*(4), 295-309.

Wu, J.; Doan, H. Disinfection of Recycled Red-Meat-Processing Wastewater by Ozone. *J. Chemical Technology and Biotechnology* **2005**, *80*(7), 828-833.

Xu, P.; Janex, M.L.; Savoye, P.; Cockx, A.; Lazarova, V. Wastewater Disinfection by Ozone: Main Parameters for Process Design. *Water Res.* **2002**, *36*(4), 1043-1055.

Zenz, D.R.; Law, K.; Lanyon, R.; Sobanski, J.; Kunetz, T.; Bouchard, A.B.; Hass, C.; Schwab, K.; Marinas, B. Study of Effluent Disinfection for Urban Rivers in Chicago; *Proc. Disinfection 2007*, 2007.

Chlorine Dioxide

Aieta, E.M.; Berg, J.D.; Roberts, P.V.; Cooper, R.C. Comparison of Chlorine Dioxide and Chlorine in Wastewater Disinfection. *J. Water Pollut. Control Fed.* **1980**, *52*(4), 810-824.

Collivignarelli, C.; Bertanza, G.; Pedrazzani, R.A Comparison among Different Wastewater Disinfection Systems: Experimental Results. *Environ. Technol.* **2000**, *21*(1), 1-16.

Connell, G.F. The Chlorination/Dechlorination Handbook; Water Environment Federation: Alexandria, VA, 2002.

Dernat, M.; Pouillot, M. Theoretical and Practical Approach to the Disinfection of Municipal Waste Water Using Chlorine Dioxide. *Water Sci. Technol. WSTED 4,* **1992**, *25*(12).

International Consultants and Malcolm-Pirnie. Technologies and Costs for Control of Disinfection By-products. Report No. EPA-PB99-111486, 1998.

Kawamura, S. Integrated Design and Operation of Water Treatment Plant Facilities (2nd Ed.); John Wiley & Sons: New York, 2000.

Narkis, N.; Kott, Y. Comparison between Chlorine Dioxide and Chlorine for use as a Disinfectant of Wastewater Effluents. *Water Sci. Technol. WSTED 4,* 1992, *26*(7-8).

Narkis, N.; Armon, R.; Offer, R.; Orshansky, F.; Friedland, E. Effect of Suspended Solids on Wastewater Disinfection Efficiency by Chlorine Dioxide. *Water Res.* **1995**, *29*(1), 227-236.

Oguma, K.; Katayama, H.; Ohgaki, S. Photoreactivation of *Escherichia Coli* After Low-Or Medium-Pressure UV Disinfection Determined by an Endonuclease Sensitive Site Assay. *Appl. Environ. Microbiol.* **2002**, *68*(12), 6029-6035.

Roberts, P.V.; Aieta, E.M.; Berg, J.D.; Chow, B.M. Chlorine Dioxide for Wastewater Disinfection: A Feasibility Evaluation. Report No. EPA-600/2-81-092, 1981.

Salgot, M.; Folch, M.; Huertas, E.; Tapias, J.; Avellaneda, D.; Giros, G.; Brissaud, F.; Verges, C.; Molina, J.; Pigem, J. Comparison of Different Advanced Disinfection Systems for Wastewater Reclamation. *Water Sci. Technol.: Water Supply* **2002**, *2*(3), 213-218.

Shen, X.; Wang, F.; Xie, J.; Han, Q.; Song, J.; Wang, D. Disinfection and Decoloration Effects of Chlorine Dioxide in Wastewater Reuse. *Zhongguo Jishui Paishui/China Water & Wastewater* **2000**, *16*(9), 53-55.

Tchobanoglous, G.; Burton, F.L.; Stensel, D.H. Wastewater Engineering Treatment and Reuse; McGraw Hill: Boston, 2003.

U.S. EPA. Alternative Disinfectants and Oxidants Guidance Manual. Report No. EPA 815/R-99/014, 1999.

U.S. EPA. Design Manual: Municipal Wastewater Disinfection. Report No. EPA 625/1-86/021, 1986.

Wang, L.; Zhang, T.; Hu, H. Comparison of the Quality and Toxicity of Wastewater After Chlorine and Chlorine Dioxide Disinfections. *Huanjing Kexue/Environmental Science* **2005**, *26*(6), 75-78.

WEF. Wastewater Disinfection – Training Manual. 2006.

White, G.C. *Handbook of Chlorination and Alternative Disinfection*; Wiley & Sons: New York, 1999.

You, Z.; Zhang, J. Standard and Development Trend on the Control of Chlorine Dioxide Disinfection of Drinking Water. *Zhongguo Jishui Paishui/China Water and Wastewater* **2003**, *19*(2), 29.

Peracetic Acid

Baldry, M.G.C.; Cavadore, A.; French, M.S.; Massa, G.; Rodrigues, L.M.; Schirch, P.F.T.; Threadgold, T.L. Effluent Disinfection in Warm Climates with Peracetic Acid. *Water Sci. Technol.* **1995**, *31*(5-6), 161-164.

Booth, R.A.; Lester, J.N. The Potential Formation of Halogenated By-products during Peracetic Acid Treatment of Final Sewage Effluent. *Water Res.* **1995**, *29*(7), 1793-1801.

Caretti, C.; Lubello, C. Wastewater Disinfection with PAA and UV Combined Treatment: A Pilot Plant Study. *Water Res.* **2003**, *37*(10), 2365-2371.

Collivignarelli, C.; Bertanza, G.; Pedrazzani, R.A Comparison among Different Wastewater Disinfection Systems: Experimental Results. *Environ. Technol.* **2000**, *21*(1), 1-16.

Crebelli, R.; Conti, L.; Monarca, S.; Feretti, D.; Zerbini, I.; Zani, C.; Veschetti, E.; Cutilli, D.; Ottaviani, M. Genotoxicity of the Disinfection By-products Resulting from Peracetic Acid- or Hypochlorite-Disinfected Sewage Wastewater. *Water Res.* **2005**, *39*(6), 1105-1113.

Dell'Erba, A.; Falsanisi, D.; Liberti, L.; Notarnicola, M.; Santoro, D. Disinfecting Behaviour of Peracetic Acid for Municipal Wastewater Reuse. *Desalination* **2004**, *168*(1-3), 435-442.

Freese, S.D.; Nozaic, D.J.; Bailey, I.; Trollip, D.L. Alternative Disinfectants for Wastewater Effluents: Viable or Prohibitively Expensive? *Water SA* **2002**, *28*(SPEC ISS), 23-32.

Gehr, R.; Wagner, M.; Veerasubramanian, P.; Payment, P. Disinfection Efficiency of Peracetic Acid, UV and Ozone After Enhanced Primary Treatment of Municipal Wastewater. *Water Res.* **2003**, *37*(19), 4573-4586.

Gehr, R.; Cochrane, D.; French, M. Peracetic Acid (PAA) as a Disinfectant for Municipal Wastewaters: Encouraging Performance Results from Physicochemical as Well as Biological Effluents; *Proc. Disinfection 2002*, Water Environment Federation 2002.

Gillette, R., Marks City of Roseville Alternative Disinfection Study; *Proc. WEFTEC 2003*, Water Environment Federation, 2003.

Kitis, M. Disinfection of Wastewater with Peracetic Acid: A Review. *Environ. Int.* **2004**, *30*(1), 47-55.

Koivunen, J.; Heinonen-Tanski, H. Peracetic Acid (PAA) Disinfection of Primary, Secondary and Tertiary Treated Municipal Wastewaters. *Water Res.* **2005**, *39*(18), 4445-4453.

Koivunen, J.; Heinonen-Tanski, H. Inactivation of Enteric Microorganisms with Chemical Disinfectants, UV Irradiation and Combined Chemical/UV Treatments. *Water Res.* **2005**, *39*(8), 1519-1526.

Koivunen, J.; Heinonen-Tanski, H. Inactivation of Enteric Microorganisms in Secondary Treated Municipal Wastewaters with Peracetic Acid (PAA) Disinfection; *Proc. Disinfection 2005*, Water Environment Federation, 2005.

Lazarova, V.; Janex, M.L.; Fiksdal, L.; Oberg, C.; Barcina, I.; Pommepuy, M. Advanced Wastewater Disinfection Technologies: Short and Long Term Efficiency. *Water Sci. Technol.* **1998**, *38*(12), 109-117.

Liberti, L.; Lopez, A.; Notarnicola, M.; Barnea, N.; Pedahzur, R.; Fattal, B. Comparison of Advanced Disinfecting Methods for Municipal Wastewater Reuse in Agriculture. *Water Sci. Technol.* **2000**, *42*(1), 215-220.

Liberti, L.; Notarnicola, M. Advanced Treatment and Disinfection for Municipal Wastewater Reuse in Agriculture. *Water Sci. Technol.* **1999**, *40*(4-5), 235-245.

Lubello, C.; Caretti, C.; Gori, R. Comparison between PAA/UV and H_2O_2/UV Disinfection for Wastewater Reuse. *Water Sci. Technol.: Water Supply* **2002**, *2*(1), 205-212.

Monarca, S.; Feretti, D.; Collivignarelli, C.; Guzzella, L.; Zerbini, I.; Bertanza, G.; Pedrazzani, R. Influence of Different Disinfectants on Mutagenicity and Toxicity of Urban Wastewater. *Water Res.* **2000**, *34*(17), 4261-4269.

National Research Council. Prudent Practices in the Laboratory: Handling and Disposal of Chemicals; National Academy Press: Washington, D.C., 2000.

Nurizzo, C.; Bonomo, L.; Malpei, F. Some Economic Considerations on Wastewater Reclamation for Irrigation, with Reference to the Italian Situation. *Water Sci. Technol.* **2001**, *43*(10), 75-81.

Rossi, S.; Antonelli, M.; Nurizzo, C. PAA Disinfection: A Feasible Alternative to Wastewater Chlorination; *Proc. Disinfection 2005*, Water Environment Federation, 2005.

Salgot, M.; Folch, M.; Huertas, E.; Tapias, J.; Avellaneda, D.; Giros, G.; Brissaud, F.; Verges, C.; Molina, J.; Pigem, J. Comparison of Different Advanced Disinfection Systems for Wastewater Reclamation. *Water Sci. Technol.: Water Supply* **2002**, *2*(3), 213-218.

Sanchez-Ruiz, C.; Martinez-Royano, S.; Tejero-Monzon, I. Evaluation of the Efficiency and Impact of Raw Wastewater Disinfection with Peracetic Acid Prior to Ocean Discharge. *Water Sci. Technol.* **1995**, *32*(7), 159-166.

Santoro, D.; Bartrand, T.A.; Greene, D.J.; Farouk, B.; Haas, C.N.; Notarnicola, M.; Liberti, L. Use of CFD for Wastewater Disinfection Process Analysis: *E. Coli* Inactivation with Peroxyacetic Acid (PAA). *International Journal of Chemical Reactor Engineering* **2005**, *3*, 14.

Santoro, D.; Gehr, R.; Bartrand, T.A.; Liberti, L.; Notarnicola, M.; Dell'Erba, A.; Falsanisi, D.; Haas, C.N. Wastewater Disinfection by PAA: Relating Residual Measurements and Inactivation; *Proc. Disinfection 2005*, 2005.

U.S. EPA. Combined Sewer Overflow Technology Fact Sheet: Alternative Disinfection Methods. Report No. EPA 832-F-99-033, 1999.

U.S. EPA. RED Facts: Peroxy Compounds. Report No. EPA-738-F-93-026, 1993.

Veschetti, E.; Cutilli, D.; Bonadonna, L.; Briancesco, R.; Martini, C.; Cecchini, G.; Anastasi, P.; Ottaviani, M. Pilot-Plant Comparative Study of Peracetic Acid and Sodium Hypochlorite Wastewater Disinfection. *Water Res.* 2003, *37*(1), 78-94.

Wagner, M.; Brumelis, D.; Gehr, R. Disinfection of Wastewater by Hydrogen Peroxide or Peracetic Acid: Development of Procedures for Measurement of Residual Disinfectant and Application to a Physicochemically Treated Municipal Effluent. *Water Environ. Res.* **2002**, *74*(1), 33-50.

Ferrate

Alig, C. Corporate Literature from Ferrate Treatment Technologies, LLC. Orlando, FL, 2007.

Fagan, J.; Waite, T.D. Biofouling Control with Ferrate (VI). *Environ. Sci. Technol.* **1983**, *17*(2), 123-125.

Jessen, A.; Randall, A.; Reinhard, D.; Daly, L. Effectiveness and Kinetics of Ferrate as a Disinfectant for Ballast Water; *Proc. WEFTEC 2006*, Water Environment Federation, 2006.

Jiang, J.; Wang, S.; Panagoulopoulos, A. The Exploration of Potassium Ferrate(VI) as a disinfectant/coagulant in Water and Wastewater Treatment. *Chemosphere* **2006**, *63*(2), 212-219.

Jiang, J.; Wang, S. Enhanced Coagulation with Potassium Ferrate(VI) for Removing Humic Substances. *Environ. Eng. Sci.* **2003**, *20*(6), 627-633.

Jiang, J.; Lloyd, B. Progress in the Development and Use of Ferrate(VI) Salt as an Oxidant and Coagulant for Water and Wastewater Treatment. *Water Res.* **2002**, *36*(6), 1397-1408.

Kazama, F. Viral Inactivation by Potassium Ferrate. *Water Sci. Technol.* **1995**, *31*(5-6), 165-168.

Kazama, F. Inactivation of Coliphage Qβ by Potassium Ferrate. *FEMS Microbiol. Lett.* **1994**, *118*(3), 345-349.

Murmann, R.K.; Robinson, P.R. Experiments Utilizing FeO_4^{2-} for Purifying Water. *Water Res.* **1974**, *8*(8), 543-547.

Reimers, R.S.; Reinhard, D.R.; Sharma, V.K.; Austin, G.C. The Application of the Green Oxidant Ferrate for Wastewater Disinfection and Reuse to Be Utilized for Wetland Restoration, Irrigation, and Groundwater Recharge; *Proc. Disinfection 2007*, Water Environment Federation, 2007.

Safe Drinking Water Committee, Ed.; In *Safe Drinking Water and Health;* National Academy Press: Washington, D.C., 1980.

Schink, T.; Waite, T.D. Inactivation of f2 Virus with Ferrate(VI). *Water Res.* **1980**, *14*(12).

Sharma, V.K.; Kazama, F.; Jiangyong, H.; Ray, A.K. Ferrates (Iron(VI) and Iron(V)): Environmentally Friendly Oxidants and Disinfectants. *J. Water Health* **2005**, *3*(1), 45-58.

Sharma, V.K. Use of Iron(VI) and Iron(V) in Water and Wastewater Treatment. *Water Sci. Technol.* **2004**, *49*(4), 69-74.

Sharma, V.K. Potassium Ferrate(VI): An Environmentally Friendly Oxidant. *Adv. Environ. Res.* **2002**, *6*(2), 143-156.

Waite, T.D. Feasibility of Wastewater Treatment with Ferrate. *J. Environ. Eng. Division* **1979**, *105*(6).

Bromine Compounds

Boner, M.; Kim, J.Y.; Muller, R.J. Is Bromine Disinfection a Viable Wet Weather Solution; *Proc. Disinfection 2002*, Water Environment Federation, 2002.

Freese, S.D.; Nozaic, D.J.; Bailey, I.; Trollip, D.L. Alternative Disinfectants for Wastewater Effluents: Viable or Prohibitively Expensive? *Water SA* **2002**, *28*(SPEC ISS), 23-32.

Fuchu, Y.; Watanabe, S.; Yasuhara, Y.; Okada, S. BCDMH Disinfection for CSO; *Proc. Disinfection 2005*, Water Environment Federation, 2005.

Lehmann, F.M. The use of Bromine Chemistry for Sand Filter Disinfection; *Proc. AESF/EPA Conference for Environmental Excellence, Jan 17-19 2000*, American Electroplaters and Surface Finishers Soc. Inc: Orlando, FL, United States, 2000.

Moffa, P.E.; David, D.P.; Somerlot, C.; Sharek, D.; Gresser, B.; Smith, T. Going for the Kill: Alternative Disinfection Technology Demonstrates Advantages for Wet Weather Applications. *Water Environ. Tech.* **2007**, *19*(1), 46-51.

Moffa, P.E.; Somerlot, C.A Comparison Between High Rate Chlorine and Bromine Disinfection; *Proc. Disinfection 2007*, Water Environment Federation, 2007.

Moffa, P.E.; Davis, D.P.; Somerlot, C.; Sharek, D.; Gresser, B.; Smith, T. Alternative Disinfection Technology Demonstrates Advantages for Wet Weather Applications - a Pilot Study of Powdered Bromine Technology; *Proc. WEFTEC 2006*, Water Environment Federation, 2006.

U.S. EPA. Design Manual: Municipal Wastewater Disinfection. Report No. EPA 625/1-86/021, 1986.

WEF. Wastewater Disinfection – Training Manual. 2006.

Yamamoto, T.; Mikami, K.W.,S. Feasibility Study of Bromine Disinfection to Combined Sewer Overflows by Full-Scale Experiment; *Proc. WEFTEC 2003*, Water Environment Federation, 2003.

Pasteurization

Bidawid, S.; Farber, J.M.; Sattar, S.A.; Hayward, S. Heat Inactivation of Hepatitis A Virus in Dairy Foods. *J. Food Prot.* **2000**, *63*(4), 522-528.

Borges, E.S. M.; Godinho, V.M.; Chernicharo, C.A.L. Thermal Hygienization of Excess Anaerobic Sludge: A Possible Self-Sustained Application of Biogas Produced in UASB Reactors. *Water Sci. Technol.* **2005**, *52*(10-11), 227-234.

Keller, R.; Passamani-Franca, R.F.; Cassini, S.T.; Goncalves, F.R. Disinfection of Sludge Using Lime Stabilisation and Pasteurisation in a Small Wastewater Treatment Plant. *Water Sci. Technol.* **2004**, *50*(1), 13-17.

Madigan, M.T.; Martinko, J.M.; Parker, J. Biology of Microorganisms; Prentice Hall: Upper Saddle River, NJ, 1997.

Moce-Llivina, L.; Muniesa, M.; Pimenta-Vale, H.; Lucena, F.; Jofre, J. Survival of Bacterial Indicator Species and Bacteriophages After Thermal Treatment of Sludge and Sewage. *Appl. Environ. Microbiol.* **2003**, *69*(3), 1452-1456.

Ryan Pasteurization & Power. RP&P Wastewater Pasteurization System Validation Report, Submitted to the California Department of Health Services. 2006.

Salveson, A.; Reina, J.; Ryan, G. Pasteurization – A Sustainable Approach to Wastewater Disinfection; *Proc. WateReuse Foundation 10th Annual Water Reuse Research Conference*, WateReuse Foundation, 2006.

Tchobanoglous, G.; Burton, F.L.; Stensel, D.H. Wastewater Engineering Treatment and Reuse; McGraw Hill: Boston, 2003.

U.S. EPA. Part 503 Implementation Guidance. Report No. 833-R-95-001, 1995.

Ultrasonic Cavitation

Blume, T.; Neis, U. Improving Chlorine Disinfection of Wastewater by Ultrasound Application. *Water Sci. Technol.* **2005**, *52*(10-11), 139-144.

Blume, T.; Martinez, I.; Neis, U. Wastewater Disinfection Using Ultrasound and UV Light. *TU Hamburh-Harburg Reports on Sanitary Engineering* **2002**, *35*, 117-128.

Blume, T.; Neis, U. Improved Wastewater Disinfection by Ultrasonic Pre-treatment. *Ultrason. Sonochem.* **2004**, *11*(5), 333-336.

Dehghani, M.H. Effectiveness of Ultrasound on the Destruction of *E. Coli*. *Am. J. Environ. Sci.* **2005**, *1*(3), 187-189.

Gaines, W.A.; Kim, B.R.; Drews, A.R.; Bailey, C.; Loch, T.; Frenette, S. Hydrodynamic Cavitation Pilot Study for Controlling Cooling Water Quality; *Proc. WEFTEC 2006*, Water Environment Federation, 2006.

Hua, I.; Thompson, J.E. Inactivation of *Escherichia Coli* by Sonication at Discrete Ultrasonic Frequencies. *Water Res.* **2000**, *34*(15), 3888-3893.

Joyce, E.; Phull, S.S.; Lorimer, J.P.; Mason, T.J. The Development and Evaluation of Ultrasound for the Treatment of Bacterial Suspensions. A Study of Frequency, Power and Sonication Time on Cultured *Bacillus* Species. *Ultrason. Sonochem.* 2003, *10*(6), 315-318.

Lohse, D. Bubble Puzzles. *Physics Today* **2003**, *56*, 36-41.

Ma, Y.; Lin, J. Effect of Pre-Sonication on Removal of Organic Matters Resulting from Chlorinated Humic Acids. *Water Sci. Technol.* **1998**, *38*(6 pt 5), 253-260.

Madge, B.A.; Jensen, J.N. Disinfection of Wastewater Using a 20-kHz Ultrasound Unit. *Water Environ. Res.* **2002**, *74*(2), 159-169.

Mahvi, A.H.; Dehghani, M.H.; Vaezi, F. Ultrasonic Technology Effectiveness in Total Coliforms Disinfection of Water. *J. Appl. Sci* **2005**, *5*(5), 856-858.

Mason, T.J.; Duckhouse, H.; Joyce, E.; Lorimer, J.P. Uses of Ultrasound in the Biological Decontamination of Water; *Proc. World Congress on Ultrasonics*, 2003.

Mason, T.J.; Joyce, E.; Phull, S.S.; Lorimer, J.P. Potential Uses of Ultrasound in the Biological Decontamination of Water. *Ultrason. Sonochem.* **2003**, *10*(6), 319-323.

Neis, U.; Blume, T. Ultrasonic Disinfection of Wastewater Effluents for High-Quality Reuse. *Water Sci. Technol.: Water Supply* **2003**, *3*(4), 261-267.

Phull, S.S.; Newman, A.P.; Lorimer, J.P.; Pollet, B.; Mason, T.J. The Development and Evaluation of Ultrasound in the Biocidal Treatment of Water. *Ultrason. Sonochem.* **1997**, *4*(2), 157-164.

Plummer, J.D.; Long, S.C. Sonication as an Alternative Disinfectant for *Escherichia Coli* and MS2 Coliphage in Drinking Waters. *J. New England Water Works Assoc.* **2004**, *118*(4), 240-252.

Scherba, G.; Weigel, R.M.; O'Brien, W.D. Quantitative Assessment of the Germicidal Efficacy of Ultrasonic Energy. *Appl. Environ. Microbiol.* **1991**, *57*(7), 2079-2084.

Weavers, L.; Pee, G.Y.; Frim, A.; Yang, L.; Rathman, J.F. Ultrasonic Destruction of Surfactants: Application to Industrial Wastewaters; *Proc. WEFTEC 2002*, Water Environment Federation, 2002.

Electron Beam and Gamma Irradiation

Basfar, A.A.; Abdel Rehim, F. Disinfection of Wastewater from a Riyadh Wastewater Treatment Plant with Ionizing Radiation. *Radiat. Phys. Chem.* **2002**, *65*(4-5), 527-532.

Borrely, S.I.; Cruz, A.C.; del Mastro, N.L.; Sampa, M.H.O.; Somessari, E.S. Radiation Processing of Sewage and Sludge. A Review. *Prog. Nuclear Energy* 1998, *33*(1-2), 3-21.

Cooper, W.J.; Mezyk, S.P.; Cole, S.K.; Fox, R.V.; Mincher, B.J.; O'Shea, K.E.; Gardinali, P.R. Free Radical Chemistry and Treatment of Halonitromethanes: Possible Reaction By-products in Water Intended for Reuse; *Proc. Annual WateReuse Symposium*, WateReuse Foundation, 2006.

Cooper, W.J.; Nickelsen, M.G.; Mezyk, S.P.; Leslie, G.; Tornatore, P.M.; Hardison, W.; Hajali, P.A. MTBE and Priority Contaminant Treatment with High Energy Electron Beam Injection. *Radiat. Phys. Chem.* 2002, *65*(4-5), 451-460.

Cooper, W.J.; Meacham, D.E.; Nickelsen, M.G.; Lin, K.; Ford, D.B.; Kurucz, C.N.; Waite, T.D. The Removal of Tri-(TCE) and Tetrachloroethylene(PCE) from Aqueous Solution Using High Energy Electrons. *J. Air Waste Manage. Assoc.* 1993, *43*(10), 1358-1366.

de St. Remy, E. Electron Injection Technology for Environmental, Industrial and Municipal Applications. 1995.

Electric Power Research Institute. A Need for Disinfection Alternatives. Report No. TB-114560, 1999.

Farooq, S.; Kurucz, C.N.; Waite, T.D.; Cooper, W.J. Disinfection of Wastewaters: High-Energy Electron vs. Gamma Irradiation. *Water Res.* **1993**, *27*(7), 1177-1184.

Hu, J.; Wang, J. Radiation-Induced Degradation of Chlorophenols in Aqueous Solution. *Fushe Yanjiu Yu Fushe Gongyi Xuebao/Journal of Radiation Research and Radiation Processing* 2005, *23*(3), 135-139.

Kurucz, C.N.; Waite, T.D.; Cooper, W.J. Miami Electron Beam Research Facility: A Large Scale Wastewater Treatment Application. *Radiat. Phys. Chem.* **1995**, *45*(2), 299-308.

Kurucz, C.N.; Waite, T.D.; Cooper, W.J.; Nickelson, M.J., Eds.; In *High Energy Electron-Beam Irradiation of Water, Wastewater, and Sludge;* Lewins, J., Becker, M., Eds.; Advances in Nuclear Science and Technology; Springer: New York, 1991.

Landsman, N.A.; Swancutt, K.L.; Bradford, C.N.; Cox, C.R.; Mezyk, S.P.; Kiddle, J.J. The Free Radical Chemistry of Advanced Oxidation Process Removal of Nitrosamines in Waters Intended for Reuse; *Proc. Annual WateReuse Symposium*, WateReuse Foundation, 2006.

Larson, L.E. Water and Wastewater Technology Demonstration Projects. Report No. P600-00-010, 1999.

Martin, D.I.; Margaritescu, I.; Cirstea, E.; Togoe, I.; Ighigeanu, D.; Nemtanu, M.R.; Oproiu, C.; Iacob, N. Application of Accelerated Electron Beam and Microwave Irradiation to Biological Waste Treatment; *Proc. 7th International Conference on Electron Beam Technologies, Jun 2-5 2003*, Elsevier Ltd, Oxford, OX5 1GB, United Kingdom: Varna, Bulgaria, 2005.

Miyata, T.; Kondoh, M.; Minemura, T.; Arai, H.; Hosono, M.; Nakao, A.; Seike, Y.; Tokunaga, O.; Machi, S. High Energy Electron Disinfection of Sewage Wastewater in Flow Systems. *Radiat. Phys. Chem.* **1990**, *35*(1), 440-444.

Pachan, M. Use of Accelerators for Environmental Protection. 1996.

Pikaev, A.K.; Podzorova, E.A.; Bakhtin, O.M. Combined Electron-Beam and Ozone Treatment of Wastewater in the Aerosol Flow. *Radiat. Phys. Chem.* **1997**, *49*(1), 155-157.

Safe Drinking Water Committee, Ed.; In *Safe Drinking Water and Health;* National Academy Press: Washington, D.C., 1980.

Schuetz, M.N.; Vroom, D.A.; St. Remy, E. Electron Injection System for Treatment of Hazardous Waste. *Transactions of the American Nuclear Society* 1995, *72*.

Swinwood, J.F.; Waite, T.D.; Kruger, P.; Rao, S.M. Radiation Technologies for Waste Treatment: A Global Perspective. *IAEA Bull.* 1994, *36*(1).

Thompson, J.E.; Blatchley, III, E.R. Gamma Irradiation for Inactivation of *C. Parvum*, *E. Coli*, and Coliphage MS-2. *J. Environ. Eng.* **2000**, *126*(8), 761-768.

Thompson, J.E.; Blatchley, III, E.R. Toxicity Effects of γ-Irradiated Wastewater Effluents. *Water Res.* **1999**, *33*(9), 2053-2058.

U.S. EPA. Electron Beam Treatment for the Removal of Benzene and Toluene from Aqueous Streams and Sludge. Report No. EPA/540/F-93/502, 1993.

U.S. EPA. Electron Bream Treatment for Removal of Trichloroethylene and Tetrachloroethylene from Streams and Sludge. Report No. EPA/540/M5-92/009, 1992.

Photocatalysis/TiO_2

Al-Bastaki, N.M. Treatment of Synthetic Industrial Wastewater with UV/TiO_2 and RO Using Benzene as Model Hydrocarbon. *Desalination* **2003**, *156*(1-3), 193-197.

Arana, J.; Herrera Melian, J.A.; Dona Rodriguez, J.M.; Gonzalez Diaz, O.; Viera, A.; Perez Pena, J.; Marrero Sosa, P.M.; Espino Jimenez, V. TiO_2-Photocatalysis as a Tertiary Treatment of Naturally Treated Wastewater. *Catalysis Today* 2002, *76*(2-4), 279-289.

Dillert, R.; Siemon, U.; Bahnemann, D. Photocatalytic Disinfection of Municipal Wastewater. *Chem. Eng. Technol.* **1998**, *21*(4), 356-358.

Gogniat, G.; Thyssen, M.; Denis, M.; Pulgarin, C.; Dukan, S. The Bactericidal Effect of TiO_2 Photocatalysis Involves Adsorption onto Catalyst and the Loss of Membrane Integrity. *FEMS Microbiol. Lett.* **2006**, *258*(1), 18-24.

Herrera Melian, J.A.; Dona Rodriguez, J.M.; Viera Suarez, A.; Tello Rendon, E.; Valdes do Campo, C.; Arana, J.; Perez Pena, J. Photocatalytic Disinfection of Urban Waste Waters. *Chemosphere* **2000**, *41*(3), 323-327.

Kuo, J.; Smith, S.O. Disinfection. *Water Environ. Res.* **1996**, *68*(4), 503-510.

Lee, S.; Nishida, K.; Otaki, M.; Ohgaki, S. Photocatalytic Inactivation of Phage Qβ by Immobilized Titanium Dioxide Mediated Photocatalyst. *Water Sci. Technol.* **1997**, *35*(11), 101-106.

Rincon, A.; Pulgarin, C. Use of Coaxial Photocatalytic Reactor (CAPHORE) in the TiO_2 Photo-Assisted Treatment of Mixed *E. Coli* and *Bacillus* Sp. and Bacterial Community Present in Wastewater. *Catalysis Today* 2005, *101*(3-4 SPEC ISS), 331-344.

Rincon, A.; Pulgarin, C. Bactericidal Action of Illuminated TiO_2 on Pure *Escherichia Coli* and Natural Bacterial Consortia: Post-Irradiation Events in the Dark and Assessment of the Effective Disinfection Time. *Applied Catalysis B: Environmental* **2004**, *49*(2), 99-112.

Rincon, A.G.; Pulgarin, C. Photocatalytical Inactivation of *E. Coli*: Effect of (Continuous-Intermittent) Light Intensity and of (Suspended-Fixed) TiO_2 Concentration. *Applied Catalysis B: Environmental* **2003**, *44*(3), 263-284.

Rincón, A.G.; Pulgarin, C.; Adler, N.; Peringer, P. Interaction Between *E. coli* Inactivation and DBP-precursors – Dihydroxybenzene Isomers – in the Photocatalytic Process of Drinking-Water Disinfection with TiO_2. *J. Photochem. Photobiol. A.* **2001**, *139*(2), 233-241.

Robertson, J.M.C.; Robertson, P.K.J.; Lawton, L.A.A Comparison of the Effectiveness of TiO_2 Photocatalysis and UVA Photolysis for the Destruction of Three Pathogenic Micro-Organisms. *J. Photochem. and Photobiol. A, Chem* **2005**, *175*(1), 51-56.

Seven, O.; Dindar, B.; Aydemir, S.; Metin, D.; Ozinel, M.A.; Icli, S. Solar Photocatalytic Disinfection of a Group of Bacteria and Fungi Aqueous Suspensions with TiO_2, ZnO and Sahara Desert Dust. *J. Photochem. Photobiol. A Chem.* **2004**, *165*, 103-107.

Watts, R.J.; Kong, S.; Orr, M.P.; Miller, G.C.; Henry, B.E. Photocatalytic Inactivation of Coliform Bacteria and Viruses in Secondary Wastewater Effluent. *Water Res.* **1995**, *29*(1), 95-100.

Wu, P.G.; Xie, R.C.; Imlay, J.A.; Shang, J. Antibacterial Behavior of High-Efficiency Visible-Light Photocatalytic Fiber. Unpublished work, 2006.

Wu, P.G.; Xie, R.C.; Shang, J. High-Efficiency Visible-Light Photocatalytic Fiber for Inactivation of *Bacillus Subtilis* Spores. Submitted to Acta Biomaterialica, 2006.

Wu, P.G.; Xie, R.C.; Imlay, J.A.; Shang, J. Visible-Light Photocatalytic Fibers for Inactivation of *Pseudomonas Aeruginosa*; *Proc. 30th International Conference on Advanced Ceramics and Composites*, 2006.

Yasui, N.; Kamiko, N. Comparison of Oxidation and Inactivation Caused by Photocatalyst to *E. Coli* and Coliphage; *Proc. IUVA 2005*, 2005.

Yu, B.; Wu, H.; Zhang, W. Application of Nano-Photocatalysts in Environmental Protection. *Shiyou Huagong/Petrochemical Technology* **2005**, *34*(5), 491-495.

Zhang, P.; Scrudato, R.J.; Germano, G. Solarcatalytic Inactivation of *Escherichia Coli* in Aqueous Solutions Using TiO_2 as Catalyst. *Chemosphere* **1994**, *28*(3), 607-611.

Combined Disinfection

Ahn, K.H.; Park, K.Y.; Maeng, S.K.; Song, K.G.; Kim, K.P.; Lee, S.H.; Kweon, J.H. Color Removal and Disinfection with UV/H_2O_2 System for Wastewater Reclamation and Reuse. *Water Sci. Technol.: Water Supply* **2005**, *5*(1), 51-57.

Ballester, N.A.; Malley, J.P. Sequential Disinfection of Adenovirus Type 2 with UV-Chlorine-Chloramine. *J. Am. Water Works Assoc.* **2004**, *96*(10).

Benjamin, L.; Landry, L.; Wobma, P. Integrating Ozone and UV Disinfection Processes at the Greater Vancouver Water District's Coquitlam Source; *Proc. IUVA Third International Congress on Ultraviolet Technologies*, International Ultraviolet Association, 2005.

Blume, T.; Neis, U. Improving Chlorine Disinfection of Wastewater by Ultrasound Application. *Water Sci. Technol.* **2005**, *52*(10-11), 139-144.

Blume, T.; Martinez, I.; Neis, U. Wastewater Disinfection Using Ultrasound and UV Light. *TU Hamburh-Harburg Reports on Sanitary Engineering* **2002**, *35*, 117-128.

Blume, T.; Neis, U. Improved Wastewater Disinfection by Ultrasonic Pre-treatment. *Ultrason. Sonochem.* **2004**, *11*(5), 333-336.

Caretti, C.; Lubello, C. Wastewater Disinfection with PAA and UV Combined Treatment: A Pilot Plant Study. *Water Res.* 2003, *37*(10), 2365-2371.

Carnimeo, D.; Di Marino, R.; Donadio, F.; Liberti, L.; Ranieri, E.; Pitzurra, M.; Savino, A. Comparison between H_2O_2/UV and ClO_2 Disinfection of Drinking Water. *Water Supply* **1995**, *13*(2), 159-169.

Diaz, M.E.; Law, S.E. Ultraviolet Photon Enhanced Ozonation for Microbiological Safety in Poultry Processing Water; *Proc. 1997 ASAE Annual International Meeting. Part 3 (of 3)*, ASAE, 1997.

DiBenedetto, P.J. Multi-Barrier Packaged: System Treats Potable and Industrial Water. *Water and Wastewater International* **2006**, *21*(2), 31-32.

Durance, C.; Hofmann, R.; Andrews, R.C.; Brown, M. Application of Ultraviolet Light for Inactivation of Adenovirus. *Proc. Disinfection 2005,* Water Environment Federation, 2005.

Ferron, S.; Brackin, J.; Bao, M. HiPox^{TM} Ozone-Peroxide Advanced Oxidation Water Treatment System for the Disinfection of Wastewater at a Los Angeles County Sanitation Joint Water Pollution Control Plant; *Proc. WEFTEC 2005*, Water Environment Federation, 2005.

Folch, M.; Huertas, E.; Tapias, J.C.; Salgot, M.; Brissaud, F. Wastewater Reclamation through a Physical-Chemical Pilot and Two Disinfection Systems (Ozone and Chlorine Dioxide) Combination. *Water Sci. Technol.: Water Supply* **2003**, *3*(3), 161-165.

Freese, S.D.; Nozaic, D.J.; Bailey, I.; Trollip, D.L. Alternative Disinfectants for Wastewater Effluents: Viable or Prohibitively Expensive? *Water SA* **2002**, *28*(SPEC ISS), 23-32.

Gori, R.; Lubello, C.; Ferrini, F.; Nicese, F. Reclaimed Municipal Wastewater as Source of Water and Nutrients for Plant Nurseries. *Water Sci. Technol.* **2004**, *50*(2), 69-75.

He, S.; Zhu, C.; Shan, M.; Feng, R. Recent Advances in Wastewater Treatment with Ultrasonic-Ozone Method. *Technical Acoustics* **2005**, *24*(3), 173-177.

Hua, I.; Thompson, J.E. Inactivation of *Escherichia Coli* by Sonication at Discrete Ultrasonic Frequencies. *Water Res.* **2000**, *34*(15), 3888-3893.

Huitric, S.J.; Tang, C.C. Sequential Chlorination to Minimize DBP; *79th California Water Environment Association Annual Conference*, 2007.

Khan, S.; Wintgens, T.; Sherman, P.; Zarlcky, J.; Schafer, A. A Performance Comparison of Individual and Combined Treatment Modules for Water Recycling. *Environ. Prog.* **2005**, *24*(4), 383-391.

Kinshella, P.; Fathali, B.; Doller, J.; Newhouse, W.; Salveson, A.; Mahar, E.; Stygar, S.; Heath, M. A Multiple Barrier: The Value and Performance of Multiple Disinfectants in Series for the City of Phoenix; *Proc. Disinfection 2007*, Water Environment Federation, 2007.

Koivunen, J.; Heinonen-Tanski, H. Inactivation of Enteric Microorganisms with Chemical Disinfectants, UV Irradiation and Combined chemical/UV Treatments. *Water Res.* **2005**, *39*(8), 1519-1526.

Kuo, J.; Smith, S.O. Disinfection. *Water Environ. Res.* **1996**, *68*(4), 503-510.

Landis, H.E.; Neemann, J.J.; Hulsey, B. Considerations for Combining Ozone and UV Disinfection; *Proc. Disinfection 2007*, Water Environment Federation: Pittsburg, PA, 2007.

Li, X.; Yang, Y.; Bai, Y.; Liu, R.; Li, G.; He, W.; Han, H. Application of Synergistic Disinfection of Potassium Permanganate and Chlorine in Wastewater Reuse Process. *Beijing Gongye Daxue Xuebao / Journal of Beijing University of Technology* **2005**, *31*(3), 276-280.

Linden, K.G.; Oliver, J.D.; Sobsey, M.D.; Shin, G.A. Fate and Persistence of Pathogens Subjected to Ultraviolet Light and Chlorine Disinfection. Report No. 98-HHE-2 (Water Environment Research Foundation Project), 2004.

Lubello, C.; Gori, R.; Nicese, F.P.; Ferrini, F. Municipal-Treated Wastewater Reuse for Plant Nurseries Irrigation. *Water Res.* **2004**, *38*(12), 2939-2947.

Lubello, C.; Caretti, C.; Gori, R. Comparison between PAA/UV and H_2O_2/UV Disinfection for Wastewater Reuse. *Water Sci. Technol.: Water Supply* **2002**, *2*(1), 205-212.

Mahar, E.; Salveson, A.; Pozos, N.; Ferron, S.; Borg, C. Peroxide and Ozone: A New Choice for Water Reclamation and Potable Reuse; *Proc. 2004 19th Annual WateReuse Symposium: Bringing Life to the Desert*, WateReuse Foundation, 2004.

Mahvi, A.H.; Dehghani, M.H.; Vaezi, F. Ultrasonic Technology Effectiveness in Total Coliforms Disinfection of Water. *J. Appl. Sci.* **2005**, *5*(5), 856-858.

Mason, T.J.; Duckhouse, H.; Joyce, E.; Lorimer, J.P. Uses of Ultrasound in the Biological Decontamination of Water; *Proc. World Congress on Ultrasonics*, 2003.

Mason, T.J.; Joyce, E.; Phull, S.S.; Lorimer, J.P. Potential Uses of Ultrasound in the Biological Decontamination of Water. *Ultrason. Sonochem.* **2003**, *10*(6), 319-323.

Mitch, W.A.; Sedlak, D.L. Formation of N-Nitrosodimethylamine (NDMA) from Dimethylamine during Chlorination. *Environ. Sci. Technol.* **2002**, *36*(4), 588-595.

Murphy, H.M.; Rand, J.L.; Payne, S.J.; Gagnon, G.A. Synergistic Effects of UV Irradiation in Combination with Chlorine and Chlorine Dioxide: Bench- and Lab-Scale Evaluation; *Proc. Disinfection 2007*, Water Environment Federation, 2007.

Ried, A.; Mielcke, J. Ozone and UV–a Tool for Multi-Barrier Concepts in Water Treatment. *Water Sci. Technol. Water Supply* **2006**, *6*(4), 17-25.

Ried, A.; Mielcke, J.; Kampmann, M.; Ternes, T.A.; Teiser, B. Ozone and UV Processes for Additional Wastewater Treatment to Remove Pharmaceuticals and EDCs; *Proc. IWA Leading Edge Technologies Conf.*, IWA Publishing, 2004.

Rittmann, D.; Tarquin, A. "Impact of Mixing Chlorine with Chlorine Dioxide on TTHMS Formation in Drinking Water"; *World Water and Environmental Resources Congress 2003, Jun 23-26 2003*, American Society of Civil Engineers: Philadelphia, PA, United States, 2003.

Rosenfeldt, E.; Linden, K.G. UV and UV/H_2O_2 Degradation of Endocrine Disrupting Chemicals in Water; *Proc. IUVA Second International Congress on Ultraviolet Technologies*, International Ultraviolet Association, 2003.

Royce, A.R.; Williamson, C.; Brown, N.; Patel, M. Performance Testing of a UV Disinfection and Oxidation System to Achieve Regulatory Approval for Wastewater Reinjection; *Proc. WEFTEC 2005*, Water Environment Federation, 2005.

Schreiber, I.M.; Mitch, W.A. Influence of the Order of Reagent Addition on NDMA Formation during Chloramination. *Environ. Sci. Technol.* **2005**, *39*(10), 3811-3818.

Snyder, S.A.; Wert, E.C.; Lei, H.; Westerhoff, P.; Yoon, Y. Removal of EDCs and Pharmaceuticals in Drinking and Reuse Treatment Processes. Report No. 2758 (AWWARF), 2007.

Tekippe, T.R.; Stutz-McDonald, S.; Matheson, R.J. Sodium Hypochlorite and Medium Pressure UV in Series: A Vallejo Case Study; *Proc. WEFTEC 1999*, Water Environment Federation, 1999.

U.S. EPA. Alternative Disinfectants and Oxidants Guidance Manual. Report No. EPA 815/R-99/014, 1999.

REFERENCES

Alig, C. Corporate Literature from Ferrate Treatment Technologies, LLC. Orlando, FL, 2007.

Amirhor, P. Overview of Wastewater Effluent Disinfection by Ultraviolet Irradiation. *J. New England Water Environment Association* **1999**, *33*(1), 27-38.

Asano, T.; Burton, F.L.; Leverenz, H.L.; Tshuchihashi, R.; Tchobanoglous, G. Water Reuse: Issues, Technologies, and Applications; McGraw-Hill: New York, 2007.

Awad, J. Ultraviolet Disinfection for Water Reuse. Planning, Design, and Operation of Effluent Disinfection Systems. *Proc. WEF Specialty Conference - Effluent Disinfection Systems*, Water Environment Federation, 1993.

Baiden, L.H.; Hunter, G.L.; Shurtz, K.B.; Wells, W.J. An Evaluation of UV and NaOCl for Wastewater Disinfection at the Pelham WWTP; *Proc. WEFTEC 2002*, Water Environment Federation, 2002.

Baiden, L.H.; Hunter, G.L.; Shurtz, K.B.; Wells, W.J. An Evaluation of UV and NaOCl for Wastewater Disinfection at the Pelham WWTP; *Proc. WEFTEC 2002*, 2002.

Baldry, M.G.C.; Cavadore, A.; French, M.S.; Massa, G.; Rodrigues, L.M.; Schirch, P.F.T.; Threadgold, T.L. Effluent Disinfection in Warm Climates with Peracetic Acid. *Water Sci. Technol.* **1995**, *31*(5-6), 161-164.

Ballester, N.A.; Malley, J.P. Sequential Disinfection of Adenovirus Type 2 with UV-Chlorine-Chloramine. *J. Am. Water Works Assoc.* **2004**, *96*(10).

Baron, J. Repair of Wastewater Microorganisms After Ultraviolet Disinfection Under Seminatural Conditions. *Water Environ. Res.* **1997**, *69*(5), 992-998.

Baron, J.; Bourbigot, M. Repair of *Escherichia Coli* and Enterococci in Sea Water After Ultraviolet Disinfection Quantification Using Diffusion Chambers. *Water Res.* **1996**, *30*(11), 2817-2821.

Basfar, A.A.; Abdel Rehim, F. Disinfection of Wastewater from a Riyadh Wastewater Treatment Plant with Ionizing Radiation. *Radiat. Phys. Chem.* **2002**, *65*(4-5), 527-532.

Bedner, M.; MacCrehan, W.A. Transformation of Acetaminophen by Chlorination Produces the Toxicants 1,4-Benzoquinone and N-Acetyl-p-Benzoquinone Imine. *Environ. Sci. Technol.* **2006**, *40*(2), 516-522.

Benjamin, L.; Landry, L.; Wobma, P. Integrating Ozone and UV Disinfection Processes at the Greater Vancouver Water District's Coquitlam Source; *Proc. IUVA Third International Congress on Ultraviolet Technologies*, International Ultraviolet Association, 2005.

Bess, J.W.; Casson, L.W.; Navin, T.J.; Lubie, D.B. The Effect of On-site Generated Ultra Pure Hypochlorite on Water and Wastewater Disinfection; *Proc. WEFTEC 2003*, Water Environment Federation, 2003.

Bidawid, S.; Farber, J.M.; Sattar, S.A.; Hayward, S. Heat Inactivation of Hepatitis A Virus in Dairy Foods. *J. Food Prot.* **2000**, *63*(4), 522-528.

Blatchley III, E.R.; Hunt, B.A.; Duggirala, R.; Thompson, J.E.; Zhao, J.; Halaby, T.; Cowger, R.L.; Straub, C.M.; Alleman, J.E. Effects of Disinfectants on Wastewater Effluent Toxicity. *Water Res.* **1997**, *31*(7), 1581-1588.

Blatchley, III, E.R.; Dumoutier, N.; Halaby, T.N.; Levi, Y.; Laine, J.M. Bacterial Responses to Ultraviolet Irradiation. *Water Sci. Technol.* **2001**, *43*(10), 179-186.

Blatchley, III, E.R.; Gong, W.L.; Rose, J.B.; Huffman, D.E.; Otaki, M.; Lisle, J.T. Effects of Wastewater Disinfection on Human Health. Report No. 99-HHE-1, 2005.

Blatchley, III, E.R.; Wood, W.L.; Schuerch, P. UV Pilot Testing: Intensity Distributions and Hydrodynamics. *J. Environ. Eng.* **1995**, *121*(3), 258-262.

Blume, T.; Martinez, I.; Neis, U. Wastewater Disinfection Using Ultrasound and UV Light. *TU Hamburh-Harburg Reports on Sanitary Engineering* **2002**, *35*, 117-128.

Blume, T.; Neis, U. Improved Wastewater Disinfection by Ultrasonic Pre-Treatment. *Ultrason. Sonochem.* **2004**, *11*(5), 333-336.

Blume, T.; Neis, U. Improving Chlorine Disinfection of Wastewater by Ultrasound Application. *Water Sci. Technol.* **2005**, *52*(10-11), 139-144.

Bolton, J.R.; Linden, K.G. Standardization of Methods for Fluence (UV Dose) Determination in Bench-Scale UV Experiments. *J. Environ. Eng.* **2003**, *129*(3), 209-215.

Boner, M.; Kim, J.Y.; Muller, R.J. Is Bromine Disinfection a Viable Wet Weather Solution; *Proc. Disinfection 2002*, 2002.

Booth, R.A.; Lester, J.N. The Potential Formation of Halogenated By-products during Peracetic Acid Treatment of Final Sewage Effluent. *Water Res.* **1995**, *29*(7), 1793-1801.

Borroum, Y.; Emerick, R.W. Bioassay Comparison of Similar Pilot- and Full-Scale UV Disinfection Systems; *Proc. WEFTEC 2005*, Water Environment Federation, 2005.

Boyce, D.S.; Sproul, O.J.; Buck, C.E. The Effect of Bentonite Clay on Ozone Disinfection of Bacteria and Viruses in Water. *Water Res.* 1981, *15*(6), 759-767.

Boyd, G.R.; Zhang, S.; Grimm, D.A. Naproxen Removal from Water by Chlorination and Biofilm Processes. *Water Res.* **2005**, *39*(4), 668-676.

Brunet, R.; Sakamoto, G.; Wright, H.; Cairns, W. UV Disinfection of Wastewater: Key Factors for Effective System Design; *2000 Annual Conference - Canadian Society for Civil Engineering, Jun 7-10 2000*, Canadian Society for Civil Engineering, Montreal, H3H 2R9, Canada: London, Ont., Canada, 2000.

Bukhari, Z.; LeChevallier, M. Assessing UV Reactor Performance for Treatment of Finished Water. *Water Sci. Technol.* **2003**, *47*(3), 179-184.

Bush, A. We Found the Solution, and We Worked Out the Bugs, Start-Up of a Major POTW Ultraviolet Disinfection System; *Proc. Disinfection 2005*, Water Environment Federation, 2005.

Butts, E. Chlorination: Is It Really That Important? *Water Well J.* **2004**, *58*(6), 30-33.

California Code of Regulations, Title 22 Chapter 3 Water Recycling Criteria; *60301.320,* 2000.

Calmer, J.; Neethling, J.B.; Murbach, D. Nitrification Effects of Chlorination and Disinfection, *Proc. Disinfection 1998* Water Environment Federation, 1998

Camacho, P.; Paul, E.; Salhi, M.; Ginestet, P. Does Reduction of Sludge Processes/Ozonation, Thermal Treatment have an Impact on Wastewater Treatment Efficacy? *Proc. Technology 2005 Second Joint Specialty Conference for Sustainable Management of Water Quality Systems for the 21st Century: Working to Protect Public Health*, 2005.

Caretti, C.; Lubello, C. Wastewater Disinfection with PAA and UV Combined Treatment: A Pilot Plant Study. *Water Res.* **2003**, *37*(10), 2365-2371.

Carlton, G. Personal Communication. 2006.

Chauret, C.; Liere, J.; Gillem, A.; Smith, C.; Hofmann, R.; Durance, C.; Andrews, R.C. Impact of Particulate Matter on the Inactivation by Chlorine and Monochloramine of *Escherichia Coli* O157:H7 and MS2 Bacteriophages; *Proc. 2004 AWWA Water Quality Technology Conference*, American Water Works Association, 2004.

Chen, G.H.; Saby, S.; Djafer, M.; Mo, H.K. New Approaches to Minimize Excess Sludge in Activated Sludge Systems; *Sludge Management Entering the Third Millenium*, 2001.

Chlorine Institute. Personal Protective Equipment for Chlorine and Sodium Hydroxide. Report No. 65, 1995.

Chlorine Institute. Sodium Hypochlorite Manual. Report No. 96, 2000.

Choi, J.; Valentine, R.L. Formation of N-Nitrosodimethylamine (NDMA) from Reaction of Monochloramine: A New Disinfection By-product. *Water Res.* **2002**, *36*(4), 817-824.

Connell, G. F. The Chlorination/Dechlorination Handbook; Water Environment Federation: Alexandria, VA, 2002.

Cooper, W.J.; Mezyk, S.P.; Cole, S.K.; Fox, R.V.; Mincher, B.J.; O'Shea, K.E.; Gardinali, P.R. Free Radical Chemistry and Treatment of Halonitromethanes: Possible Reaction By-products in Water Intended for Reuse; *Proc. Annual WateReuse Symposium*, WateReuse Foundation, 2006.

Cotton, C.A.; Owen, D.M.; Cline, G.C.; Brodeur, T.P. UV Disinfection Costs for Inactivating *Cryptosporidium*. *J. Am. Water Works Assoc.* **2001**, *93*(6), 82-94.

Crebelli, R.; Conti, L.; Monarca, S.; Feretti, D.; Zerbini, I.; Zani, C.; Veschetti, E.; Cutilli, D.; Ottaviani, M. Genotoxicity of the Disinfection By-products Resulting from Peracetic Acid- or Hypochlorite-Disinfected Sewage Wastewater. *Water Res.* **2005**, *39*(6), 1105-1113.

Cutler, D.; Chalmers, R.B.; Ng, H. Implementation of Emerging Technology UV Irradiation of NDMA for Recycled Water Projects; *Proc. 2002 Annual Conference*, American Water Works Association, 2002.

Darby, J.; Heath, M.; Jacangelo, J.; Loge, F.Swaim, P. Comparison of UV Irradiation to Chlorination: Guidance for Achieving Optimal UV Performance. Report No. WERF 91-WWD-1, 1995.

de St. Remy, E. Electron Injection Technology for Environmental, Industrial and Municipal Applications. 1995.

Dehghani, M.H. Effectiveness of Ultrasound on the Destruction of *E. Coli*. *Am. J. Environ. Sci.* 2005*, 1*(3), 187-189.

Dillert, R.; Siemon, U.; Bahnemann, D. Photocatalytic Disinfection of Municipal Wastewater. *Chem. Eng. Technol.* **1998**, *21*(4), 356-358.

Dodd, M.C.; Huang, C.H. Transformation of the Antibacterial Agent Sulfamethoxazole in Reactions with Chlorine: Kinetics, Mechanisms, and Pathways. *Environ. Sci. Technol.* **2004**, *38*(21), 5607-5615.

Drago, J.; Thompson, C. Historical Ozone use at Municipal Wastewater Treatment Plants in the United States; *Proc. Pan-American Group Annual Conference*, International Ozone Association, 2005.

Drury, D.D.; Snyder, S.A.; Wert, E.C. Using Ozone Disinfection for EDC Removal; *Proc. WEFTEC 2006*, Water Environment Federation, 2006.

Durance, C.; Hofmann, R.; Andrews, R.C.; Brown, M. Application of Ultraviolet Light for Inactivation of Adenovirus. *Proc. Disinfection2005,* Water Environment Federation, 2005.

Electric Power Research Institute. A Need for Disinfection Alternatives. Report No. TB-114560, 1999.

Elefritz, R.A. Jr. Comparison of Alternate Methods of Disinfection for Reclaim Effluents. *Florida Water Resource Journal* **2000**, 19-25.

Emerick, R.W.; Loge, F.J.; Thompson, D.; Darby, J.L. Factors Influencing Ultraviolet Disinfection Performance. Part II: Association of Coliform Bacteria with Wastewater Particles. *Water Environ. Res.* **1999**, *71*(6), 1178-1187.

Energy Information Administration of the U.S. Department of Energy. Electric Power Annual - Average Retail Price of Electricity to Ultimate Customers by End-use Sector. *http://www.eia.doe.gov/cneaf/electricity/epa/epat7p4.html* (accessed March, 2007).

Environment Canada. Review of Municipal Effluent Chlorination/Dechlorination Principles, Technologies and Practices. Report No. En84-36/2006E, 2006.

Esvelt, L.A.; Kaufman, W.J.; Selleck, R.E. Toxicity Removal from Municipal Wastewater. Report No. SERL 71-7 (Berkeley Sanitary Engineering Research Laboratory), 1971.

Fagan, J.; Waite, T.D. Biofouling Control with Ferrate (VI). *Environ. Sci. Technol.* **1983**, *17*(2), 123-125.

Farooq, S.; Kurucz, C.N.; Waite, T.D.; Cooper, W.J. Disinfection of Wastewaters: High-Energy Electron vs. Gamma Irradiation. *Water Res.* **1993**, *27*(7), 1177-1184.

Ferron, S.; Brackin, J.; Bao, M. HiPox™ Ozone-Peroxide Advanced Oxidation Water Treatment System for the Disinfection of Wastewater at a Los Angeles County Sanitation Joint Water Pollution Control Plant; *Proc. WEFTEC 2005*, Water Environment Federation, 2005.

Florida Administrative Code, *62-600.400,* 1996.

Fuchu, Y.; Watanabe, S.; Yasuhara, Y.; Okada, S. BCDMH Disinfection for CSO; *Proc. Disinfection 2005*, Water Environment Federation, 2005.

Gaines, W.A.; Kim, B.R.; Drews, A.R.; Bailey, C.; Loch, T.; Frenette, S. Hydrodynamic Cavitation Pilot Study for Controlling Cooling Water Quality; *Proc. WEFTEC 2006*, Water Environment Federation, 2006.

Gehr, R.; Cochrane, D.; French, M. Peracetic Acid (PAA) as a Disinfectant for Municipal Wastewaters: Encouraging Performance Results from Physicochemical as Well as Biological Effluents; *Disinfection 2002*, Water Environment Federation 2002.

Gehr, R.; Wagner, M.; Veerasubramanian, P.; Payment, P. Disinfection Efficiency of Peracetic Acid, UV and Ozone After Enhanced Primary Treatment of Municipal Wastewater. *Water Res.* **2003**, *37*(19), 4573-4586.

Gehr, R.; Wright, H. UV Disinfection of Wastewater Coagulated with Ferric Chloride: Recalcitrance and Fouling Problems. *Water Sci. Technol.* **1998**, *38*(3), 15-23.

Gillette, R., Marks, K. City of Roseville Alternative Disinfection Study; *Proc. WEFTEC 2003*, Water Environment Federation, 2003.

Gillette, R.; Marks, K.; Pawluk, D. City of Roseville Alternative Disinfection Study; *Proc. WEFTEC 2003*, Water Environment Federation, 2003.

Gogniat, G.; Thyssen, M.; Denis, M.; Pulgarin, C.; Dukan, S. The Bactericidal Effect of TiO_2 Photocatalysis Involves Adsorption onto Catalyst and the Loss of Membrane Integrity. *FEMS Microbiol. Lett.* 2006, *258*(1), 18-24.

Goodman, G.V.; Mills, J.A. Let There Be Light! *Water Environ. Technol.* **2002**, *14*(1), 42-46.

Gordon, G.L.; Adam, L.; Bubnis, B. Minimizing Chlorate Ion Formation in Drinking Water When Hypochlorite Ion Is the Chlorinating Agent; *Proc. 1995 AWWA Water Quality Technology Conference*, American Water Works Association, 1995.

Gori, R.; Lubello, C.; Ferrini, F.; Nicese, F. Reclaimed Municipal Wastewater as Source of Water and Nutrients for Plant Nurseries. *Water Sci. Technol.* **2004**, *50*(2), 69-75.

Great Lakes - Upper Mississippi River Board of State and Provincial Public Health and Environmental Managers. Ten State Standards: Recommended Standards for Wastewater Facilities. 2004.

Hamby, S.A Comparison of Operating Strategies - Chlorine vs. UV; *Proc. WEFTEC 2006*, Water Environment Federation, 2006.

Hancock, G.G.; Davis, E.M. Regrowth Potential of Coliforms After UV Disinfection of Municipal Wastewater. *J. Environmental Science and Health, Part A: Toxic/Hazardous Substances and Environmental Engineering* **1999**, *34*(9), 1737-1743.

Hanzon, B.; Hartfelder, J.; O'Connell, S.; Murray, D. Comparing UV and Chlorine for Wastewater Disinfection; *Proc. Disinfection 2005*, Water Environment Federation, 2005.

Hanzon, B.D.; Vigilia, R. Two Experts Offer Practical Guidance in Designing and Operating Ultraviolet Disinfection Systems. *Water Environ. Technol.* **1999**, *11*, 35-42.

Harp, D.L. Specific Determination of Inorganic Monochloramine in Chlorinated Wastewaters. *Water Environ. Res.* **2000**, *72*(6), 706-713.

Havelaar, A.H.; Nieuwstad, T.J.; Meulemans, C.C.E.; Olphen, M.F-Specific RNA Bacteriophages as Model Viruses in UV Disinfection of Wastewater. *Water Sci. Technol.* **1991**, *24*(2), 347-352.

Herrera Melian, J.A.; Dona Rodriguez, J.M.; Viera Suarez, A.; Tello Rendon, E.; Valdes do Campo, C.; Arana, J.; Perez Pena, J. Photocatalytic Disinfection of Urban Waste Waters. *Chemosphere* **2000**, *41*(3), 323-327.

Ho, C.H.; Pitt, P.; Mamais, D.; Chiu, C.; Jolis, D. Evaluation of UV Disinfection Systems for Large-Scale Secondary Effluent. *Water Environ. Res.* **1998**, *70*(6), 1142-1150.

Hoigne, J. Chemistry of Aqueous Ozone and Transformation of Pollutants by Ozonation and Advanced Oxidation Processes. *The Handbook of Environmental Chemistry: Quality and Treatment of Drinking Water II* 1998.

Hua, I.; Thompson, J.E. Inactivation of *Escherichia Coli* by Sonication at Discrete Ultrasonic Frequencies. *Water Res.* **2000**, *34*(15), 3888-3893.

Huber, M.M.; Gobel, A.; Joss, A.; Hermann, N.; Loffler, D.; McArdell, C.S.; Ried, A.; Siegrist, H.; Ternes, T.A.; von Gunten, U. Oxidation of Pharmaceuticals during Ozonation of Municipal Wastewater Effluents: A Pilot Study. *Environ. Sci. Technol.* **2005**, *39*(11), 4290-4299.

Huitric, S.J.; Kuo, J.; Creel, M.; Tang, C.C.; Snyder, D.; Horvath, R.W.; Stahl, J.F. Reclaimed Water Disinfection Alternatives to Avoid NDMA and THM Formation; *Proc. WEFTEC 2006*, Water Environment Federation, 2006.

Huitric, S.J.; Tang, C.C. Sequential Chlorination to Minimize DBP; *79th California Water Environment Association Annual Conference*, 2007.

Hunter, G. The History of UV Disinfection in the Last 20 Years; *Proc. UV – A Technical Symposium*, National Water Research Institute, 2000.

Hunter, G. The History of UV Disinfection in the Last 20 Years; *Proc. UV – A Technical Symposium*, National Water Research Institute, 2000.

Hunter, G. The Light at the End of the Tunnel – How to Successfully Operate a UV Disinfection System; *Proc. WEFTEC 2006*, Water Environment Federation, 2006.

Hunter, G.; Hulsy, B.; Mackey, H.; Neeman, J. Progress of Wastewater UV Systems - A Survey of Large Systems; *Proc. WEFTEC 2005,* Water Environment Federation, 2005.

Hunter, G.; Long, J. Ultraviolet Blues. *Water Environ. Technol.* **2003**, *15*(11), 51-53.

Hunter, G.L.; O'Brien, W.J.; Hulsey, R.A.; Carns, K.E.; Ehrhard, R. Emerging Disinfection Technologies. *Water Environ. Technol.* **1998**, *10*(6), 40-44.

International Consultants and Malcolm-Pirnie. Technologies and Costs for Control of Disinfection By-products. Report No. PB99-111486, 1998.

Iranpour, R.; Garnas, G.; Moghaddam, O.; Taebi, A. Hydraulic Effects on Ultraviolet Disinfection: Modification of Reactor Design. *Water Environ. Res.* **1999**, *71*(1), 114-118.

Jalali, Y.; Huitric, S.J.; Kuo, J.; Tang, C.C.; Garcia, A.; Thompson, S.; Horvath, R.W.; Stahl, J.F. A Large-Scale UV Pilot-Plant Study: Tertiary Effluent Disinfection and Effect on NDMA and Cyanide; *Proc. WEFTEC 2005*, Water Environment Federation, 2005.

Janex, M.L.; Xu, P.; Savoye, P.; Laine, J.M.; Lazarova, V. Ozonation as a Wastewater Disinfection Process to Meet Reuse Standards; *Proc. 14Th World Congress*, International Ozone Association: Stamford, CT, 1999.

Jessen, A.; Randall, A.; Reinhard, D.; Daly, L. Effectiveness and Kinetics of Ferrate as a Disinfectant for Ballast Water; *Proc. WEFTEC 2006*, Water Environment Federation, 2006.

Jeyanayagam, S.; Cotton, C. Practical Considerations in the Use of UV Light for Drinking Water Disinfection; *Proc. CASE/ASCE Joint Conference: An International Perspective on Environmental Engineering*, 2002.

Jiang, J.; Lloyd, B. Progress in the Development and Use of Ferrate(VI) Salt as an Oxidant and Coagulant for Water and Wastewater Treatment. *Water Res.* **2002**, *36*(6), 1397-1408.

Johns, F.; Lichtwardt, M.; Grundemann, P.; Gallegos, D. Selection and Installation of a UV Disinfection System as a Retrofit to an Existing Wastewater Treatment Plant; *Proc. WEFTEC 2002*, Water Environment Federation, 2002.

Jolis, D.; Lam, C.; Pitt, P. Particle Effects on Ultraviolet Disinfection of Coliform Bacteria in Recycled Water. *Water Environ. Res.* **2001**, *73*(2), 233-236.

Jolis, D.; Lam, C.; Pitt, P.; Hirano, R. Particle Size Effects on UV Light Disinfection of Filtered Reclaimed Wastewater; *Proc. 1996 WateReuse Conference*, WateReuse Foundation, 1996.

Joyce, E.; Phull, S.S.; Lorimer, J.P.; Mason, T.J. The Development and Evaluation of Ultrasound for the Treatment of Bacterial Suspensions. A Study of Frequency, Power and Sonication Time on Cultured *Bacillus* Species. *Ultrason. Sonochem.* **2003**, *10*(6), 315-318.

Kalisvaart, B.F. Re-use of Wastewater: Preventing the Recovery of Pathogens by Using Medium-Pressure UV Lamp Technology. *Water Sci. Technol.* **2004**, *50*(6), 337-344.

Kashimada, K.; Kamiko, N.; Yamamoto, K.; Ohgaki, S. Assessment of Photoreactivation Following Ultraviolet Light Disinfection. *Water Sci. Technol.* **1996**, *33*(10-11), 261-269.

Kavanaugh, M.; Dzombak, D.A.; Theis, T.L.; Young, T.C.; Luthy, R.G. Cyanide Formation and Fate in Complex Effluents and Its Relation to Water Quality Criteria. Report No. 98-HHE-5, 2003.

Kawamura, S. Integrated Design and Operation of Water Treatment Plant Facilities (2nd Ed.); John Wiley & Sons: New York, 2000.

Kazama, F. Inactivation of Coliphage Qβ by Potassium Ferrate. *FEMS Microbiol. Lett.* 1994, *118*(3), 345-349.

Kazama, F. Viral Inactivation by Potassium Ferrate. *Water Sci. Technol.* **1995**, *31*(5-6), 165-168.

Khoury, J.; Pang, M.; Young, C.; Pandit, A.; Carr, S.; Fischer, D. Cyanide Generation during Preservation of Chlorinated Wastewater Effluent Samples for Total Cyanide Analysis; *Proc. WEFTEC 2005*, Water Environment Federation, 2005.

Kim, Y.H.; Hensley, R. Effective Control of Chlorination and Dechlorination at Wastewater Treatment Plants Using Redox Potential. *Water Environ. Res.* **1997**, *69*(5), 1008-1014.

Kinshella, P.; Fathali, B.; Doller, J.; Newhouse, W.; Salveson, A.; Mahar, E.; Stygar, S.; Heath, M. A Multiple Barrier: The Value and Performance of Multiple Disinfectants in Series for the City of Phoenix; *Proc. Disinfection 2007*, Water Environment Federation, 2007.

Knapp, T. Personal Communication. Sanitation Districts of Los Angeles County. 2006.

Koivunen, J.; Heinonen-Tanski, H. Inactivation of Enteric Microorganisms with Chemical Disinfectants, UV Irradiation and Combined Chemical/UV Treatments. *Water Res.* **2005b**, *39*(8), 1519-1526.

Koivunen, J.; Heinonen-Tanski, H. Inactivation of Enteric Microorganisms with Chemical Disinfectants, UV Irradiation and Combined chemical/UV Treatments. *Water Res.* **2005**, *39*(8), 1519-1526.

Koivunen, J.; Heinonen-Tanski, H. Peracetic Acid (PAA) Disinfection of Primary, Secondary and Tertiary Treated Municipal Wastewaters. *Water Res.* **2005a**, *39*(18), 4445-4453.

Kopchynski, D.M.; Palmer, T.M.; Ross, M.C. Comparisons of On-line ORP and Chlorine Residual Monitoring/Control Systems for Wastewater Treatment Plant Final Effluent Chlorination; *Proc. WEFTEC 2001*, Water Environment Federation, 2001.

Koukouraki, E.; Diamadopoulos, E. Modeling the Formation of THM during Chlorination of Treated Municipal Wastewater; *IWA Regional Symposium on Water Recycling in the Mediterranean Region*, 2002.

Koukouraki, E.; Diamadopoulos, E. THM Formation during Chlorination of Treated Municipal Wastewater; *Proc. World Water Congress of the International Water Association*, 2001.

Krasner, S.W.; Weinberg, H.S.; Richardson, S.D.; Pastor, S.J.; Chinn, R.; Sclimenti, M.J.; Onstad, G.D.; Thurston, A.D., Jr. Occurrence of a New Generation of Disinfection Byproduct. *Environ. Sci. Technol.* **2006**, *40*(23), 7175-85.

Kuo, J.; Chen, C.; Nellor, M. Standardized Collimated Beam Testing Protocol for Water/Wastewater Ultraviolet Disinfection. *J. Environ. Eng.* **2003**, *129*(8), 774-779.

Kuo, J.; Stahl, J.F.; Burton, D.; El Jack, Z.; Horvath, R.W.; Tang, C.C. Chloramination of N/DN Effluent - Meeting the Ammonia, Coliform and THM Limits; *Proc. WEFTEC 2003*, Water Environment Federation, 2003.

Kurucz, C.N.; Waite, T.D.; Cooper, W.J.; Nickelson, M.J., Eds.; In *High Energy Electron-Beam Irradiation of Water, Wastewater, and Sludge;* Lewins, J., Becker, M., Eds.; Advances in Nuclear Science and Technology; Springer: New York, 1991.

Landis, H.E.; Neemann, J.J.; Hulsey, B. Considerations for Combining Ozone and UV Disinfection; *Proc. Disinfection 2007*, Water Environment Federation: Pittsburg, PA, 2007.

Larson, L.E. Water and Wastewater Technology Demonstration Projects. Report No. P600-00-010, 1999.

Lawryshyn, Y.A.; Cairns, B. UV Disinfection of Water: The Need for UV Reactor Validation. *Water Sci. Technol.: Water Supply* **2003**, *3*(4), 293-300.

Lazarova, V.; Janex, M. L.; Fiksdal, L.; Oberg, C.; Barcina, I.; Pommepuy, M. Advanced Wastewater Disinfection Technologies: Short and Long Term Efficiency. *Water Sci. Technol.* **1998**, *38*(12), 109-117.

Lee, S.; Nishida, K.; Otaki, M.; Ohgaki, S. Photocatalytic Inactivation of Phage Qβ by Immobilized Titanium Dioxide Mediated Photocatalyst. *Water Sci. Technol.* **1997**, *35*(11), 101-106.

Liberti, L.; Notarnicola, M. Advanced Treatment and Disinfection for Municipal Wastewater Reuse in Agriculture. *Water Sci. Technol.* **1999**, *40*(4-5), 235-245.

Liberti, L.; Notarnicola, M.; Lopez, A. Advanced Treatment for Municipal Wastewater Reuse in Agriculture. III - Ozone Disinfection. *Ozone: Sci. Eng.* **2000**, *22*(2), 151-166.

Liberti, L.; Notarnicola, M.; Petruzzelli, D. Advanced Treatment for Municipal Wastewater Reuse in Agriculture. UV Disinfection: Parasite Removal and By-product Formation. *Desalination* **2003**, *152*(1-3), 315-324.

Lin, L.; Johnston, C.T.; Blatchley, III, E.R. Inorganic Fouling at Quartz:Water Interfaces in Ultraviolet Photoreactors - I. Chemical Characterization. *Water Res.* **1999a**, *33*(15), 3321-3329.

Lin, L.; Johnston, C.T.; Blatchley, III, E.R. Inorganic Fouling at Quartz:Water Interfaces in Ultraviolet Photoreactors - II. Temporal and Spatial Distributions. *Water Res.* **1999b**, *33*(15), 3330-3338.

Linden, K.G. UV Dose Verification Using Chemical Actinometry and Biodosimetry Methods; *Proc. UV 2000: A Technical Symposium*, 2000.

Linden, K.G.; Oliver, J.D.; Sobsey, M.D.; Shin, G.A. Fate and Persistence of Pathogens Subjected to Ultraviolet Light and Chlorine Disinfection. Report No. 98-HHE-2 (Water Environment Research Foundation Project), 2004.

Linden, K.G.; Shemer, H.; Reckhow, D.A.; Makdissy, G. Ultraviolet Light Induced Disinfection Byproducts: Realities and Challenges; *Proc. Disinfection 2007*, 2007.

Loge, F.J.; Emerick, R.W.; Heath, M.; Jacangelo, J.; Tchobanoglous, G.; Darby, J.L. Ultraviolet Disinfection of Secondary Wastewater Effluents: Prediction of Performance and Design. *Water Environ. Res.* **1996**, *68*(5), 900-916.

Loge, F.J.; Emerick, R.W.; Thompson, D.E.; Nelson, D.C.; Darby, J.L. Factors Influencing Ultraviolet Disinfection Performance Part I: Light Penetration to Wastewater Particles. *Water Environ. Res.* **1999**, *71*(3), 377-381.

Lubello, C.; Gori, R.; Nicese, F.P.; Ferrini, F. Municipal-Treated Wastewater Reuse for Plant Nurseries Irrigation. *Water Res.* **2004**, *38*(12), 2939-2947.

MacCrehan, W.A.; Bedner, M.; Helz, G.R. Making Chlorine Greener: Performance of Alternative Dechlorination Agents in Wastewater. *Chemosphere* **2005**, *60*(3), 381-388.

MacCrehan, W.A.; Jensen, J.S.; Helz, G.R. Detection of Sewage Organic Chlorination Products That Are Resistant to Dechlorination with Sulfite. *Environ. Sci. Technol.* **1998**, *32*(22), 3640-3645.

Mackey, E.D.; Wright, H.; Hargy, T.; Fonseca, A.C.; Cabaj, A. Reducing UV Design Safety Factors - Optimization of UV Reactor Validation; *Proc. 2004 Water Quality Technology Conference*, 2004.

Madge, B.A.; Jensen, J.N. Disinfection of Wastewater Using a 20-kHz Ultrasound Unit. *Water Environ. Res.* **2002**, *74*(2), 159-169.

Madigan, M.T.; Martinko, J.M.; Parker, J. Biology of Microorganisms; Prentice Hall: Upper Saddle River, N. J., 1997.

Mahar, E.; Eaton, C.E.; Salveson, A. Velocity Profiling Versus Checkpoint Bioassay for Performance Testing of UV Disinfection Systems; *Proc. WEFTEC 2006*, Water Environment Federation, 2006.

Mahvi, A. H.; Dehghani, M. H.; Vaezi, F. Ultrasonic Technology Effectiveness in Total Coliforms Disinfection of Water. *J. Appl. Sci* **2005**, *5*(5), 856-858.

Malley, J.P.; Shaw, J.P.; Ropp, J.R. Evaluation of By-products Produced by Treatment of Groundwaters with Ultraviolet Irradiation. Report No. 90685 (AWWARF), 1995.

Mann, M.A.; Cramer, J.A. Disinfecting with Ultraviolet Radiation. *Water Environ. Technol.* **1992**, *4*(12), 40-45.

Marshall, T.H. On the Pulse. *Water Environ. Technol.* **2002**, *14*(6), 51-54.

Martin, N.; Gehr, R. Reduction of Photoreactivation with the Combined UV/Peracetic Acid Process or by Delayed Exposure to Visible Light, *Water Environ. Res.* **2007**, 79 (9), 991-999.

Marx, S. Sodium Hypochlorite Safer than Chlorine, but Must Be Approached with Respect. *J. New England Water Environment Association* **1996**, *30*(1), 49-52.

Mason, T.J.; Duckhouse, H.; Joyce, E.; Lorimer, J.P. Uses of Ultrasound in the Biological Decontamination of Water; *Proc. World Congress on Ultrasonics*, 2003a.

Mason, T.J.; Joyce, E.; Phull, S.S.; Lorimer, J.P. Potential Uses of Ultrasound in the Biological Decontamination of Water. *Ultrason. Sonochem.* **2003b**, *10*(6), 319-323.

Massachusetts Department of Environmental Protection. The Occurrence and Sources of Perchlorate in Massachusetts. Updated April 2006.

Melin, G., Ed.; In *Ultraviolet Disinfection Guidelines for Drinking Water and Water Reuse;* National Water Research Institute, in collaboration with the AWWA Research Foundation: Fountain Valley, CA, 2003.

Mitch, W.A.; Sedlak, D.L. Characterization and Fate of N-Nitrosodimethylamine Precursors in Municipal Wastewater Treatment Plants. *Environ. Sci. Technol.* **2004**, *38*(5), 1445-1454.

Mitch, W.A.; Sedlak, D.L. Formation of N-Nitrosodimethylamine (NDMA) from Dimethylamine during Chlorination. *Environ. Sci. Technol.* **2002**, *36*(4), 588-595.

Mitch, W.A.; Sharp, J.O.; Trussell, R.R.; Valentine, R.L.; Alvarez-Cohen, L.; Sedlak, D.L.N-Nitrosodimethylamine (NDMA) as a Drinking Water Contaminant: A Review. *Environ. Eng. Sci.* **2003**, *20*(5), 389-404.

Miyata, T.; Kondoh, M.; Minemura, T.; Arai, H.; Hosono, M.; Nakao, A.; Seike, Y.; Tokunaga, O.; Machi, S. High Energy Electron Disinfection of Sewage Wastewater in Flow Systems. *Radiat. Phys. Chem.* **1990**, *35*(1), 440-444.

Moce-Llivina, L.; Muniesa, M.; Pimenta-Vale, H.; Lucena, F.; Jofre, J. Survival of Bacterial Indicator Species and Bacteriophages After Thermal Treatment of Sludge and Sewage. *Appl. Environ. Microbiol.* **2003**, *69*(3), 1452-1456.

Moffa, P.E.; Davis, D.P.; Somerlot, C.; Sharek, D.; Gresser, B.; Smith, T. Alternative Disinfection Technology Demonstrates Advantages for Wet Weather Applications - a Pilot Study of Powdered Bromine Technology; *Proc. WEFTEC 2006*, Water Environment Federation, 2006.

Moffa, P.E.; Somerlot, C.A Comparison Between High Rate Chlorine and Bromine Disinfection; *Proc. Disinfection 2007*, Water Environment Federation, 2007.

Mofidi, A.A.; Baribeau, H.; Green, J.F.; De Leon, R.; Wolfe, R.L. Disinfection Using High-Intensity Pulsed-Ultraviolet Irradiation; *Proc. AWWA 1998 Water Quality Technology Conference*, American Water Works Association, 1998.

Monarca, S.; Feretti, D.; Collivignarelli, C.; Guzzella, L.; Zerbini, I.; Bertanza, G.; Pedrazzani, R. Influence of Different Disinfectants on Mutagenicity and Toxicity of Urban Wastewater. *Water Res.* **2000**, *34*(17), 4261-4269.

Moreland, V.; Ono, P. Primary Effluent Disinfection – Necessary Procedures in UV System Design for Low Water Quality Applications (to Pilot or Not to Pilot?); *Proc. WEFTEC 2001*, Water Environment Federation, 2001.

Morowitz, H.J. Absorption Effects in Volume Irradiation of Microorganisms. *Science* **1950**, *111*(2879), 229-230.

Murmann, R.K.; Robinson, P.R. Experiments Utilizing FeO_4^{2-} for Purifying Water. *Water Res.* **1974**, *8*(8), 543-547.

Murphy, H.M.; Rand, J.L.; Payne, S.J.; Gagnon, G.A. Synergistic Effects of UV Irradiation in Combination with Chlorine and Chlorine Dioxide: Bench- and Lab-Scale Evaluation; *Proc. Disinfection 2007*, Water Environment Federation, 2007.

Na, C.; Olson, T.M. Mechanism and Kinetics of Cyanogen Chloride Formation from the Chlorination of Glycine. *Environ. Sci. Technol.* **2006**, *40*(5), 1469-1477.

Naber, J. Switching to Sodium Hypochlorite Disinfection. *Water Environ. Technol.* **2003**, *15*(9), 100-102.

National Research Council. Prudent Practices in the Laboratory: Handling and Disposal of Chemicals; National Academy Press: Washington, D.C., 2000.

Neis, U.; Blume, T. Ultrasonic Disinfection of Wastewater Effluents for High-Quality Reuse. *Water Sci. Technol.: Water Supply* **2003**, *3*(4), 261-267.

Nelson, K.L. Ultraviolet Light Disinfection of Wastewater Stabilization Pond Effluents. *Water Sci. Technol.* **2000**, *42*(10-11), 165-170.

Norman, J.E.; Patel, M.V.; Leslie, G.L. UV Disinfectability of Membrane Treated Wastewater; *Proc. 1999 Membrane Technology Conference*, American Water Works Association, 1999.

NSF International. Protocol for Equipment Verification Testing for Inactivation of Microbiological Contaminants. Report No. 03/9204/EPADWCTR, 2003.

Nurizzo, C.; Bonomo, L.; Malpei, F. Some Economic Considerations on Wastewater Reclamation for Irrigation, with Reference to the Italian Situation. *Water Sci. Technol.* 2001, *43*(10), 75-81.

Oguma, K.; Katayama, H.; Ohgaki, S. Photoreactivation of *Legionella Pneumophila* After Inactivation by Low- or Medium-Pressure Ultraviolet Lamp. *Water Res.* **2004**, *38*(11), 2757-2763.

Oliver, M. Validation of Cleaning Methods for UV Disinfection Systems; *Proc. WEFTEC Disinfection 2002*, 2002.

Pandit, A.; Young, C.; Pang, M.; Khoury, J.; Carr, S.; Fischer, D.; Stahl, J. Factors Affecting Cyanide Generation in Chlorinated Wastewater Effluent Matrix; *Proc. WEFTEC 2006*, Water Environment Federation, 2006.

Paraskeva, P. The Treatment of a Secondary Municipal Effluent by Ozone. Ph.D. Thesis, Imperial College, University of London, 1998.

Paraskeva, P.; Graham, N.J. Ozonation of Municipal Wastewater Effluents. *Water Environ. Res.* **2002**, *74*(6), 569-581.

Parker, J.A.; Darby, J.L. Particle-Associated Coliform in Secondary Effluents: Shielding from Ultraviolet Light Disinfection. *Water Environ. Res.* **1995**, *67*(7), 1065-1075.

Peng, J.; Qiu, Y.; Gehr, R. Characterization of Permanent Fouling on the Surfaces of UV Lamps Used for Wastewater Disinfection. *Water Environ. Res.* **2005**, *77*(4), 309-322.

Petri, B.M.; Lawryshyn, Y.A. Validating UV Reactors for Low Dose Wastewater Applications; *Proc. Disinfection 2005*, Water Environment Federation, 2005.

Qualls, R.G.; Johnson, J.D. Bioassay and Dose Measurement in UV Disinfection. *Appl. Environ. Microbiol.* **1983**, *45*(3), 872-877.

Reimers, R.S.; Reinhard, D.R.; Sharma, V.K.; Austin, G.C. The Application of the Green Oxidant Ferrate for Wastewater Disinfection and Reuse to Be Utilized for Wetland Restoration, Irrigation, and Groundwater Recharge; *Proc. Disinfection 2007*, Water Environment Federation, 2007.

Rein, D.A.; Jamesson, G.M.; Monteith, R.A. Toxicity Effects of Alternate Disinfection Processes; *Proc. WEFTEC 1992*, Water Environment Federation, 1992.

Reisch, M., Assessing Rita, *Chemical and Engineering News* **2005**, 83(40), 9.

Ried, A.; Mielcke, J.; Kampmann, M.; Ternes, T.A.; Teiser, B. Ozone and UV Processes for Additional Wastewater Treatment to Remove Pharmaceuticals and EDCs; *Proc. IWA Leading Edge Technologies Conf.*, IWA Publishing, 2004.

Rincon, A.; Pulgarin, C. Bactericidal Action of Illuminated TiO_2 on Pure *Escherichia Coli* and Natural Bacterial Consortia: Post-Irradiation Events in the Dark and Assessment of the Effective Disinfection Time. *Applied Catalysis B: Environmental* 2004, *49*(2), 99-112.

Rincon, A.; Pulgarin, C. Use of Coaxial Photocatalytic Reactor (CAPHORE) in the TiO_2 Photo-Assisted Treatment of Mixed *E. Coli* and *Bacillus* Sp. and Bacterial Community Present in Wastewater. *Catalysis Today* **2005**, *101*(3-4 SPEC ISS), 331-344.

Rincón, A.G.; Pulgarin, C.; Adler, N.; Peringer, P. Interaction Between *E. coli* Inactivation and DBP-precursors – Dihydroxybenzene Isomers – in the Photocatalytic Process of Drinking-Water Disinfection with TiO_2. *J. Photochem. Photobiol. A.* **2001**, *139*(2), 233-241.

Roberts, P.V.; Aieta, E.M.; Berg, J.D.; Chow, B.M. Chlorine Dioxide for Wastewater Disinfection: A Feasibility Evaluation. Report No. EPA-600/2-81-092, 1981.

Robson, C.M.; Rice, R.G. Wastewater Ozonation in the USA. - History and Current Status - 1989. *Ozone: Sci. Eng.* **1991**, *13*(1), 23-40.

Rosenfeldt, E.; Linden, K.G. UV and UV/H_2O_2 Degradation of Endocrine Disrupting Chemicals in Water; *Proc. IUVA Second International Congress on Ultraviolet Technologies*, International Ultraviolet Association, 2003.

Rossi, S.; Antonelli, M.; Nurizzo, C. PAA Disinfection: A Feasible Alternative to Wastewater Chlorination; *Disinfection 2005*, Water Environment Federation, 2005.

Royce, A.R.; Williamson, C.; Brown, N.; Patel, M. Performance Testing of a UV Disinfection and Oxidation System to Achieve Regulatory Approval for Wastewater Reinjection; *Proc. WEFTEC 2005*, Water Environment Federation, 2005.

Ryan Pasteurization & Power. RP&P Wastewater Pasteurization System Validation Report, Submitted to the California Department of Health Services. 2006.

Ryan, G. Personal Communication. Ryan Pasteurization and Power. 2006.

Ryder, R.A.; Faller, J.A. Disinfection Challenges with Partially Nitrifying Fixed Film Reactor Effluent, *Proc. Planning, Design, & Operations of Effluent Disinfection Systems*, Water Environment Federation, 1993.

Ryu, H.; Alum, A.; Abbaszadegan, M. Microbial Characterization and Population Changes in Nonpotable Reclaimed Water Distribution Systems. *Environ. Sci. Technol.* **2005**, *39*(22), 8600-8605.

Safe Drinking Water Committee, Ed.; In *Safe Drinking Water and Health;* National Academy Press: Washington, D.C., 1980.

Sakaji, R.H. Reclaimed Water: The California Experience; *Proc. UV – A Technical Symposium*, National Water Research Institute, 2000.

Salveson, A.; Reina, J.; Ryan, G. Pasteurization – A Sustainable Approach to Wastewater Disinfection; *Proc. WateReuse Foundation 10th Annual Water Reuse Research Conference*, WateReuse Foundation, 2006.

Samel, L.S.; Norton, J.M.; McLelland, J. Implementation of Best-Value UV Systems through Competitive Life Cycle Cost Bidding Procedures; *Proc. WEFTEC 2003*, Water Environment Federation, 2003.

Sanchez-Ruiz, C.; Martinez-Royano, S.; Tejero-Monzon, I. Evaluation of the Efficiency and Impact of Raw Wastewater Disinfection with Peracetic Acid Prior to Ocean Discharge. *Water Sci. Technol.* **1995**, *32*(7), 159-166.

Sanitation Districts of Los Angeles County. Pomona Virus Study - Final Report. 1977.

Santoro, D.; Gehr, R.; Bartrand, T. A.; Liberti, L.; Notarnicola, M.; Dell'Erba, A.; Falsanisi, D.; Haas, C. N. Wastewater Disinfection by PAA: Relating Residual Measurements and Inactivation; *WEF Disinfection Specialty Conference*, 2005.

Scheible, O.K.; Lawryshyn, Y.A. UV Reactor Validation: Matching the Microbe with the Target Dose; *Proc. WEFTEC 2004*, Water Environment Federation, 2004.

Scheible, O.K.; Mignone, E.J. Final Draft (Version 3.4) of the Generic Verification Protocol for Secondary Effluent and Water Reuse Disinfection Applications. 2002.

Scherba, G.; Weigel, R.M.; O'Brien, W.D. Quantitative Assessment of the Germicidal Efficacy of Ultrasonic Energy. *Appl. Environ. Microbiol.* **1991**, *57*(7), 2079-2084.

Schink, T.; Waite, T.D. Inactivation of f2 Virus with Ferrate(VI). *Water Res.* **1980**, *14*(12).

Schreiber, I.M.; Mitch, W.A. Influence of the Order of Reagent Addition on NDMA Formation during Chloramination. *Environ. Sci. Technol.* **2005**, *39*(10), 3811-3818.

Scruggs, C.; Hunter, G.; Wallis-Lage, C. Could EDC Control Change Disinfection Practices at WWTPs? *Proc. Disinfection 2005,* Water Environment Federation, 2005.

Sharma, V.K. Use of Iron(VI) and Iron(V) in Water and Wastewater Treatment. *Water Sci. Technol.* **2004**, *49*(4), 69-74.

Sharma, V.K.; Kazama, F.; Jiangyong, H.; Ray, A.K. Ferrates (Iron(VI) and Iron(V)): Environmentally Friendly Oxidants and Disinfectants. *J. Water Health* **2005**, *3*(1), 45-58.

Shaw, J.P.; Malley Jr., J.P.; Willoughby, S.A. Effects of UV Irradiation on Organic Matter. *J. Am. Water Works Assoc.* **2000**, *92*(4), 157-167.

Singer, P.C. Disinfection Byproducts in Drinking Water: Additional Science and Policy Considerations in the Pursuit of Public Health Protection. *The 2006 Clarke Lecture*, National Water Research Institute, 2006.

Snyder, S.; Wert, E.; Zegers, R.; Drury, D. Oxidation of EDCs and Pharmaceuticals in Reuse Water; *Proc. Disinfection 2007*, Water Environment Federation, 2007.

Snyder, S.A.; Wert, E.C.; Lei, H.; Westerhoff, P.; Yoon, Y. Removal of EDCs and Pharmaceuticals in Drinking and Reuse Treatment Processes. Report No. 2758 (AWWARF), 2007.

Soroushian, F.; Abramson, C.; Ferris, M.; Mohammed, A. Pilot-Scale Studies of High-Intensity UV Disinfection By-products; *Proc. WEFTEC 1996*, Water Environment Federation, 1996.

Soroushian, F.; Norman, J.; Patel, M.; Leslie, G.; Tchobanoglous, G. Comparison of UV Disinfection Technologies for Meeting California Wastewater Reclamation Criteria; *Proc. 2000 Water Reuse Conference*, 2000.

Soroushian, F.; Shen, Y.; Patel, M.; Wehner, M. Evaluation and Pilot Testing of Advanced Treatment Processes for NDMA Removal and Reformation Prevention; *Proc. AWWA Annual Conference*, American Water Works Association, 2001.

Spivey, N. Personal Communication. 2006.

State of North Carolina. Ultraviolet Disinfection Systems – Reliability Guidelines. 2001.

Stedman, L.A Unanimous Vote of Confidence? *Water 21* **2004**(Feb), 31-32.

Steinberg, L.J.; Albrecht, J.M.; Basolo, V. The Diffusion of Environmental Technology: A Case Study of Ultraviolet Disinfection for Wastewater; *Proc. WEFTEC 2000*, Water Environment Federation, 2000.

Swift, J.; Emerick, R.; Scheible, K.; Soroushian, F.; Putnam, L.B.; Sakaji, R. Treat, Disinfect, Reuse. *Water Environ. Technol.* **2002**, *14*(11), 21-25.

Swift, J.L.; Wilson, J.P.; Johnson, M. Assessment of Operation and Maintenance Costs for Ultraviolet Disinfection Systems; *Proc. WEFTEC 2000*, Water Environment Federation, 2000.

Swift, J.L.; Wilson, J.P.; Johnson, M. The Impact of UV-Absorbing Wastewater from a Printed Circuit Board Manufacturing Facility on the Performance of a Municipal UV Disinfection System; *Proc. WEFTEC 2001*, Water Environment Federation, 2001.

Swinwood, J.F.; Waite, T.D.; Kruger, P.; Rao, S.M. Radiation Technologies for Waste Treatment: A Global Perspective. *IAEA Bull.* **1994**, *36*(1).

Tang, C.C.; Kuo, J.; Huitric, S.J.; Jalali, Y.; Horvath, R.W.; Stahl, J.F. UV Systems for Reclaimed Water Disinfection – from Equipment Validation to Operation; *Proc. WEFTEC.06 79Th Annual Technical Exhibition and Conference*, Water Environment Federation, 2006.

Tanner, B.D.; Kuwahara, S.; Gerba, C.P.; Reynolds, K.A. Evaluation of Electrochemically Generated Ozone for the Disinfection of Water and Wastewater. *Water Sci. Technol.* **2004**, *50*(1), 19-25.

Tchobanoglous, G.; Burton, F.L.; Stensel, D.H. Wastewater Engineering Treatment and Reuse; McGraw Hill: Boston, 2003.

Tchobanoglous, G.; Loge, F.; Darby, J.; Devries, M. UV Design: Comparison of Probabilistic and Deterministic Design Approaches. *Water Sci. Technol.* **1996**, *33*(10-11), 251-260.

Tekippe, T.R.; Stutz-McDonald, S.; Matheson, R.J. Sodium Hypochlorite and Medium Pressure UV in Series: A Vallejo Case Study; *Proc. WEFTEC 1999*, Water Environment Federation, 1999.

Ternes, T.A.; Stuber, J.; Herrmann, N.; McDowell, D.; Ried, A.; Kampmann, M.; Teiser, B. Ozonation: A Tool for Removal of Pharmaceuticals, Contrast Media and Musk Fragrances from Wastewater? *Water Res.* **2003**, *37*(8), 1976-1982.

Thomas, L.; Hajda, P.; Zehner, S. To UV or Not to UV, Is That the Question? Or Is It: When Is Clear Water Not Transparent? *Proc. WEFTEC 2002*, Water Environment Federation, 2002.

Thompson, C.; Drago, J. Ozone use Trends in the Water Treatment Industry in the United States; *Proc. Pan-American Group Annual Conference*, International Ozone Association, 2005.

Thompson, C.; Leong, LY.C. 30 Years of Experience Using Ozone in the United States Disinfecting Wastewater; *Proc. Disinfection 2007*, Water Environment Federation, 2007.

Thompson, S.S.; Jackson, J.L.; Suva-Castillo, M.; Yanko, W.A.; El Jack, Z.; Kuo, J.; Chen, C.L.; Williams, F. P.; Schnurr, D. P. Detection of Infectious Human Adenoviruses in Tertiary-Treated and Ultraviolet-Disinfected Wastewater. *Water Environ. Res.* **2003**, *75*(2), 163-170.

Tikkanen, M.; Schroeter, J.H.; Leong, L.Y.C.; Ganesh, R.; Guidance Manual for Disposal of Chlorinated Water, AWWA Research Foundation, 2001.

Tobin, A.; Burror, J.; Yahya, M.; Wistrom, A.; Nedic, S. Alternatives for Disinfection Effluent Generated at Orange County Sanitation District Wastewater Treatment Plants; *Proc. WEFTEC 2005*, Water Environment Federation, 2005.

Tone, K.; Tamburini, R.; Ramey, W.W. So You're Thinking about an UV Disinfection System: Here Are Some Considerations for You; *Proc. 1998 WEFTEC*, Water Environment Federation, 1998.

Topnik, J.M.; Buchanan, I.D.; Smith, D.W. The Effect of Iron in Wastewater on UV Lamp Sleeve Fouling; *Proc. WEFTEC 2000 Conference*, Water Environment Federation, 2000.

Tosa, K.; Hirata, T. Photoreactivation of Enterohemorrhagic *Escherichia Coli* Following UV Disinfection. *Water Res.* **1999**, *33*(2), 361-366.

Tree, J.A.; Adams, M.R.; Lees, D.N. Chlorination of Indicator Bacteria and Viruses in Primary Sewage Effluent. *Appl. Environ. Microbiol.* **2003**, *69*(4), 2038-2043.

Tree, J.A.; Adams, M.R.; Lees, D.N. Virus Inactivation during Disinfection of Wastewater by Chlorination and UV Irradiation and the Efficacy of F^+ Bacteriophage as a 'Viral Indicator'. *Water Sci. Technol.* 1997, *35*(11-12), 227-232.

U.S. EPA Interim Enhanced Surface Water Treatment. *Federal Register* **1998**, *63*(241), 69477-69521.

U.S. EPA Long-Term 2 Enhanced Surface Water Treatment Rule. *Federal Register* **2006**, *71*(3), 653-702.

U.S. EPA. Alternative Disinfectants and Oxidants Guidance Manual. Report No. EPA 815/R-99/014, 1999.

U.S. EPA. Alternative Disinfectants and Oxidants Guidance Manual. Report No. EPA 815/R-99/014, 1999a.

U.S. EPA. Combined Sewer Overflow Technology Fact Sheet: Alternative Disinfection Methods. Report No. EPA 832-F-99-033, 1999.

U.S. EPA. Design Manual: Municipal Wastewater Disinfection. Report No. EPA 625/1-86/021, 1986.

U.S. EPA. Electron Beam Treatment for the Removal of Benzene and Toluene from Aqueous Streams and Sludge. Report No. EPA/540/F-93/502, 1993.

U.S. EPA. Electron Bream Treatment for Removal of Trichloroethylene and Tetrachloroethylene from Streams and Sludge. Report No. EPA/540/M5-92/009, 1992.

U.S. EPA. Part 503 Implementation Guidance. Report No. 833-R-95-001, 1995.

U.S. EPA. RED Facts: Peroxy Compounds. Report No. EPA-738-F-93-026, 1993.

U.S. EPA. Ultraviolet Disinfection Guidance Manual for the Final Long Term 2 Enhanced Surface Water Treatment Rule. Report No. EPA 815-R-06-007, 2006.

U.S. EPA. User's Manual for UVDIS, Version 3.1 UV Disinfection Process Design Manual. 1992.

U.S. EPA. Wastewater Technology Fact Sheet – Ozone Disinfection. Report No. EPA 832/F-99-063, 1999b.

U.S. EPA. Wastewater Technology Fact Sheet – Ultraviolet Disinfection. Report No. EPA 832-F-99-064, 1999.

U.S. Department of Homeland Security, Chlorine Gas Decision Tool for Water and Wastewater Utilities; 2006.

Veschetti, E.; Cutilli, D.; Bonadonna, L.; Briancesco, R.; Martini, C.; Cecchini, G.; Anastasi, P.; Ottaviani, M. Pilot-Plant Comparative Study of Peracetic Acid and Sodium Hypochlorite Wastewater Disinfection. *Water Res.* 2003, *37*(1), 78-94.

Wagner, M.; Brumelis, D.; Gehr, R. Disinfection of Wastewater by Hydrogen Peroxide or Peracetic Acid: Development of Procedures for Measurement of Residual Disinfectant and Application to a Physicochemically Treated Municipal Effluent. *Water Environ. Res.* **2002**, *74*(1), 33-50.

Waite, T.D. Feasibility of Wastewater Treatment with Ferrate. *J. Environmental Engineering Division* **1979**, *105*(6).

Walsh, D.S.; Sproul, O.J.; Buck, C.E. Ozone Inactivation of Floc Associated Viruses and Bacteria. *J. Environ. Eng. Division, ASCE* **1980**, *106*(4), 711-726.

Watts, R.J.; Kong, S.; Orr, M.P.; Miller, G.C.; Henry, B.E. Photocatalytic Inactivation of Coliform Bacteria and Viruses in Secondary Wastewater Effluent. *Water Res.* **1995**, *29*(1), 95-100.

Weavers, L.; Pee, G.Y.; Frim, A.; Yang, L.; Rathman, J.F. Ultrasonic Destruction of Surfactants: Application to Industrial Wastewaters; *Proc. WEFTEC 2002*, Water Environment Federation, 2002.

Weber, E.; Scheible, K. Special Section: Security – Pulsed-UV Unit May Inactivate Biological Agents. *J. Am. Water Works Assoc.* **2003**, 34-46.

WEF. Wastewater Disinfection - MOP FD-10. 1996.

WEF. Wastewater Disinfection – Training Manual. 2006.

WEF. Water Environment Federation Disinfection Process Survey Summary. *Water Environ. Technol.* **2004**, *16*(7), 47-49.

WEF. Water Environment Federation Disinfection Process Survey Summary. *Water Environ. Technol.* **2004**, *16*(7), 47-49.

Wert, E.; Snyder, S.; Zegers, R.; Drury, D. Formation of Ozonation DBPs in Tertiary Wastewater; *Proc. Disinfection 2007*, Water Environment Federation: Pittsburgh, Pennsylvania, 2007.

Whitby, E. The History of UV through Patents; *Proc. Disinfection 2002*, Water Environment Federation, 2002.

White, G.C. Handbook of Chlorination and Alternative Disinfection; Wiley & Sons: New York, 1999.

Williamson, C. UV-Oxidation Disinfects Water and Treats Chemicals Simultaneously. *Water and Wastewater International* **2002**, *17*(6), 33.

Wolfe, R.L. Ultraviolet Disinfection of Potable Water. *Environ. Sci. Technol.* 1990, *24*(6), 768-773.

Wood, P.; Hunter, G.; Kobylinski, E. To PLC or Not to PLC UV Systems – Alternatives for UV System Control; *Proc. Disinfection 2005*, Water Environment Federation, 2005.

Wright, H.B. Comparison and Validation of UV Dose Calculations for Low- and Medium-Pressure Mercury Arc Lamps. *Water Environ. Res.* **2000**, *72*(4), 439-443.

Wu, P.G.; Xie, R.C.; Imlay, J.A.; Shang, J. Antibacterial Behavior of High-Efficiency Visible-Light Photocatalytic Fiber. Unpublished work, 2006a.

Wu, P.G.; Xie, R.C.; Imlay, J.A.; Shang, J. Visible-Light Photocatalytic Fibers for Inactivation of *Pseudomonas Aeruginosa*; *Proc. 30th International Conference on Advanced Ceramics and Composites*, 2006c.

Wu, P.G.; Xie, R.C.; Shang, J. High-Efficiency Visible-Light Photocatalytic Fiber for Inactivation of *Bacillus Subtilis* Spores. Submitted to Acta Biomaterialica, 2006b.

Yamamoto, T.; Mikami, K.W.S. Feasibility Study of Bromine Disinfection to Combined Sewer Overflows by Full-Scale Experiment; *Proc. WEFTEC 2003*, Water Environment Federation, 2003.

Yang, X.; Shang, C.; Huang, J. DBP Formation in Breakpoint Chlorination of Wastewater. *Water Res.* 2005, *39*(19), 4755-4767.

Yasui, N.; Kamiko, N. Comparison of Oxidation and Inactivation Caused by Photocatalyst to *E. Coli* and Coliphage; *Proc. IUVA 2005*, 2005.

York, D.W.; Potts, E.A. UV Disinfection: Florida's Perspective; *Proc. UV – A Technical Symposium*, National Water Research Institute, 2000.

Zenz, D.R.; Law, K.; Lanyon, R.; Sobanski, J.; Kunetz, T.; Bouchard, A.B.; Hass, C.; Schwab, K.; Marinas, B. Study of Effluent Disinfection for Urban Rivers in Chicago; *Proc. Disinfection 2007*, Water Environment Federation, 2007.

WASTEWATER UTILITY

Alabama
Montgomery Water Works & Sanitary Sewer Board

Alaska
Anchorage Water & Wastewater Utility

Arizona
Glendale, City of, Utilities Department
Mesa, City of
Peoria, City of
Phoenix Water Services Dept.
Pima County Wastewater Management
Safford, City of

Arkansas
Little Rock Wastewater Utility

California
Central Contra Costa Sanitary District
Corona, City of
Crestline Sanitation District
Delta Diablo Sanitation District
Dublin San Ramon Services District
East Bay Dischargers Authority
East Bay Municipal Utility District
Eastern Municipal Water District
El Dorado Irrigation District
Fairfield-Suisun Sewer District
Fresno Department of Public Utilities
Inland Empire Utilities Agency
Irvine Ranch Water District
Las Virgenes Municipal Water District
Livermore, City of
Los Angeles, City of
Los Angeles County, Sanitation Districts of
Napa Sanitation District
Orange County Sanitation District
Palo Alto, City of
Riverside, City of
Sacramento Regional County Sanitation District
San Diego Metropolitan Wastewater Department, City of
San Francisco, City & County of
San Jose, City of
Santa Barbara, City of
Santa Cruz, City of
Santa Rosa, City of
South Bayside System Authority
South Coast Water District
South Orange County Wastewater Authority
Stege Sanitary District
Sunnyvale, City of
Union Sanitary District
West Valley Sanitation District

Colorado
Aurora, City of
Boulder, City of
Greeley, City of
Littleton/Englewood Water Pollution Control Plant
Metro Wastewater Reclamation District, Denver

Connecticut
Greater New Haven WPCA
Stamford, City of

District of Columbia
District of Columbia Water & Sewer Authority

Florida
Broward, County of
Fort Lauderdale, City of
Jacksonville Electric Authority (JEA)
Miami-Dade Water & Sewer Authority
Orange County Utilities Department
Reedy Creek Improvement District
Seminole County Environmental Services
St. Petersburg, City of
Tallahassee, City of
Tampa, City of
Toho Water Authority
West Palm Beach, City of

Georgia
Atlanta Department of Watershed Management
Augusta, City of
Clayton County Water Authority
Cobb County Water System
Columbus Water Works
Fulton County
Gwinnett County Department of Public Utilities
Savannah, City of

Hawaii
Honolulu, City & County of

Idaho
Boise, City of

Illinois
Greater Peoria Sanitary District
Kankakee River Metropolitan Agency
Metropolitan Water Reclamation District of Greater Chicago
Wheaton Sanitary District

Iowa
Ames, City of
Cedar Rapids Wastewater Facility
Des Moines, City of
Iowa City

Kansas
Johnson County Wastewater
Unified Government of Wyandotte County/ Kansas City, City of

Kentucky
Louisville & Jefferson County Metropolitan Sewer District
Sanitation District No. 1

Louisiana
Sewerage & Water Board of New Orleans

Maine
Bangor, City of
Portland Water District

Maryland
Anne Arundel County Bureau of Utility Operations
Howard County Bureau of Utilities
Washington Suburban Sanitary Commission

Massachusetts
Boston Water & Sewer Commission
Massachusetts Water Resources Authority (MWRA)
Upper Blackstone Water Pollution Abatement District

Michigan
Ann Arbor, City of
Detroit, City of
Holland Board of Public Works
Saginaw, City of
Wayne County Department of Environment
Wyoming, City of

Minnesota
Rochester, City of
Western Lake Superior Sanitary District

Missouri
Independence, City of
Kansas City Missouri Water Services Department
Little Blue Valley Sewer District
Metropolitan St. Louis Sewer District

Nebraska
Lincoln Public Works and Utilities Department

Nevada
Henderson, City of
Las Vegas, City of
Reno, City of

New Jersey
Bergen County Utilities Authority
Ocean County Utilities Authority
Passaic Valley Sewerage Commissioners

New York
New York City Department of Environmental Protection

North Carolina
Charlotte/Mecklenburg Utilities
Durham, City of
Metropolitan Sewerage District of Buncombe County
Orange Water & Sewer Authority
University of North Carolina, Chapel Hill

Ohio
Akron, City of
Butler County Department of Environmental Services
Columbus, City of
Metropolitan Sewer District of Greater Cincinnati
Northeast Ohio Regional Sewer District
Summit, County of

Oklahoma
Oklahoma City Water & Wastewater Utility Department
Tulsa, City of

Oregon
Albany, City of
Clean Water Services
Eugene, City of
Gresham, City of
Portland, City of Bureau of Environmental Services
Water Environment Services

Pennsylvania
Hemlock Municipal Sewer Cooperative (HMSC)
Philadelphia, City of
University Area Joint Authority

South Carolina
Charleston Water System
Mount Pleasant Waterworks & Sewer Commission
Spartanburg Water

Tennessee
Cleveland Utilities
Knoxville Utilities Board
Murfreesboro Water & Sewer Department
Nashville Metro Water Services

Texas
Austin, City of
Dallas Water Utilities
Denton, City of
El Paso Water Utilities
Fort Worth, City of
Houston, City of
San Antonio Water System
Trinity River Authority

Utah
Salt Lake City Corporation

Virginia
Alexandria Sanitation Authority
Arlington, County of
Fairfax County
Hampton Roads Sanitation District
Hanover, County of
Henrico, County of
Hopewell Regional Wastewater Treatment Facility
Loudoun Water
Lynchburg Regional WWTP
Prince William County Service Authority
Richmond, City of
Rivanna Water & Sewer Authority

Washington
Everett, City of
King County Department of Natural Resources
Seattle Public Utilities
Sunnyside, Port of
Yakima, City of

Wisconsin
Green Bay Metro Sewerage District
Kenosha Water Utility
Madison Metropolitan Sewerage District
Milwaukee Metropolitan Sewerage District
Racine, City of
Sheboygan Regional Wastewater Treatment
Wausau Water Works

Australia
ACTEW (Ecowise)
South Australian Water Corporation
South East Water Limited
Sydney Water Corporation
Water Corporation of Western Australia

Canada
Edmonton, City of/Edmonton Waste Management Centre of Excellence
Lethbridge, City of
Regina, City of, Saskatchewan
Toronto, City of, Ontario
Winnipeg, City of, Manitoba

New Zealand
Watercare Services Limited

STORMWATER UTILITY

California
Fresno Metropolitan Flood Control District
Los Angeles, City of, Department of Public Works
Monterey, City of
San Francisco, City & County of
Santa Rosa, City of
Sunnyvale, City of

Colorado
Aurora, City of
Boulder, City of

Florida
Orlando, City of

Georgia
Griffin, City of

Iowa
Cedar Rapids Wastewater Facility
Des Moines, City of

Kansas
Overland Park, City of

Kentucky
Louisville & Jefferson County Metropolitan Sewer District

Maine
Portland Water District

North Carolina
Charlotte, City of, Stormwater Services

Pennsylvania
Philadelphia, City of

Tennessee
Chattanooga Stormwater Management

Texas
Harris County Flood Control District, Texas

Washington
Bellevue Utilities Department
Seattle Public Utilities

STATE

Arkansas Department of Environmental Quality
Connecticut Department of Environmental Protection
Kansas Department of Health & Environment
Kentucky Department of Environmental Protection
New England Interstate Water Pollution Control Commission (NEIWPCC)
Ohio River Valley Sanitation Commission
Urban Drainage & Flood Control District, CO

CORPORATE

ADS Environmental Services
Advanced Data Mining International
Alan Plummer & Associates
Alpine Technology Inc.
Aqua-Aerobic Systems Inc.
Aquateam–Norwegian Water Technology Centre A/S
ARCADIS
Associated Engineering
Black & Veatch
Blue Water Technologies, Inc.
Boyle Engineering Corporation
Brown & Caldwell
Burgess & Niple, Ltd.
Burns & McDonnell
CABE Associates Inc.
The Cadmus Group
Camp Dresser & McKee Inc.
Carollo Engineers Inc.
Carpenter Environmental Associates Inc.
Carter & Burgess, Inc.
CET Engineering Services
CH2M HILL
CRA Infrastructure & Engineering
CONTECH Stormwater Solutions
D&B/Guarino Engineers, LLC
Damon S. Williams Associates, LLC
Earth Tech Inc.
Ecovation
EMA Inc.
Environmental Operating Solutions, Inc.
Environ/The ADVENT Group, Inc.
Fay, Spofford, & Thorndike Inc.
Freese & Nichols, Inc.
ftn Associates Inc.
Gannett Fleming Inc.
Garden & Associates, Ltd.
Geosyntec Consultants
GHD
Golder Associates Ltd.
Greeley and Hansen LLC
Hazen & Sawyer, P.C.
HDR Engineering Inc.
HNTB Corporation
Hydromantis Inc.
HydroQual Inc.
Infilco Degremont Inc.
Jacques Whitford NAWE, Inc.
Jason Consultants LLC Inc.
Jordan, Jones, & Goulding Inc.
KCI Technologies Inc.
Kelly & Weaver, P.C.
Kennedy/Jenks Consultants
KMK Consultants
Komline Sanderson Engineering Corporation
Larry Walker Associates
Limno-Tech Inc.
The Low Impact Development Center Inc.
Malcolm Pirnie Inc.
Material Matters
McKim & Creed
Metcalf & Eddy Inc.
Monteco Corporation
MPR Engineering Corporation, Inc.
MWH
O'Brien & Gere Engineers Inc.
Odor & Corrosion Technology Consultants Inc.
Original Engineering Consultants, Ltd.
Paradigm Environmental Technologies, Inc.
Parametrix Inc.
Parsons
Post, Buckley, Schuh & Jernigan
Praxair, Inc.
Ring Industrial Group
RMC Water & Environment
Ross & Associates Ltd.
SAIC
Siemens Water Technologies
The Soap & Detergent Association
Smith & Loveless, Inc.
Stantec Consulting Inc.
Stearns & Wheler, LLC
Stone Environmental Inc.
Stratus Consulting Inc.
Synagro Technologies Inc.
Tetra Tech Inc.
Trojan Technologies Inc.
Trussell Technologies, Inc.
URS Corporation
Westin Engineering Inc.
Weston Solutions Inc.
Woodard & Curran
Zoeller Pump Company

INDUSTRY

American Electric Power
American Water
Anglian Water Services, Ltd.
ChevronTexaco Energy Research & Technology Company
The Coca-Cola Company
Dow Chemical Company
DuPont Company
Eastman Chemical Company
Eli Lilly & Company
Merck & Company Inc.
Procter & Gamble Company
Suez Environnment
United Utilities North West (UUNW)
United Water Services LLC

Note: List as of 4/10/08

IWA Publishing's authorised EU representative for General Product Safety Regulations is Diane D'Arras, 15 rue Duret, 75116 Paris, France, e-mail: safety@iwap.co.uk.

Printed and bound by CPI Group (UK) Ltd, Croydon, CR0 4YY
12/05/2026
02108865-0008